Routledge Revivals

Historical Geography:
Progress and Prospect

Historical geography has been a major area of activity in recent years. Much of the recent work and research findings have been extremely valuable to historians and archaeologists and as background to the study of contemporary geography. This reissue, first published in 1987, presents an overview of contemporary developments in all the major branches of the discipline. As such it provides a valuable introduction to the subject, a review of the latest state of the art and a pointer to future research directions.

Historical Geography: Progress and Prospect

Edited by
Michael Pacione

Routledge
Taylor & Francis Group

First published in 1987
by Croom Helm Ltd.

This edition first published in 2011 by Routledge
4 Park Square, Milton Park, Abingdon, Oxon OX14 4RN

Simultaneously published in the USA and Canada
by Routledge
605 Third Avenue, New York, NY 10017

Routledge is an imprint of the Taylor & Francis Group, an informa business

Chapter One © 1987 R. A. Butlin
Remaining Material © Michael Pacione

Publisher's Note
The publisher has gone to great lengths to ensure the quality of this reprint but
points out that some imperfections in the original copies may be apparent.

Disclaimer
The publisher has made every effort to trace copyright holders and welcomes
correspondence from those they have been unable to contact.

A Library of Congress record exists under LC Control Number: 86024301

ISBN 13: 978-0-415-61533-4 (hbk)
ISBN 13: 978-0-415-61534-1 (pbk)

HISTORICAL
GEOGRAPHY:
Progress and Prospect

Edited by Michael Pacione

CROOM HELM
London • Sydney • Wolfeboro, New Hampshire

Chapter One © 1987 R.A. Butlin
Remaining Material © 1987 Michael Pacione
Croom Helm Ltd, Provident House, Burrell Row,
Beckenham, Kent, BR3 1AT
Croom Helm Australia 44-50 Waterloo Road,
North Ryde, 2311, New South Wales

British Library Cataloguing in Publication Data

Historical geography: progress and prospect.
  — (Croom Helm progress in geography series)
  1. Geography, Historical
  I. Pacione, Michael
  911 G141

  ISBN 0-7099-4046-7

Croom Helm, 27 South Main Street,
Wolfeboro, New Hampshire 03894-2069, USA.

Library of Congress Cataloging-in-Publication Data

Historical geography.

  (Croom Helm progress in geography series)
  Includes index.
  1. Geography, Historical. I. Pacione, Michael.
II. Series.
G141.H49 1987   911   86-24301
ISBN 0-7099-4046-7

Printed and bound in Great Britain by Mackays of Chatham Ltd, Kent

CONTENTS

CONTENTS

TABLES AND FIGURES

Tables

Figures

to

Christine, Michael John and Emma Victoria

PREFACE

Historical studies have featured prominently in geography
since the emergence of the subject as an academic discipline.
The fact that historical geography studies both space
(chorography) and time (chronology) leads some to contend
that it is as much a mode of analysis as a discrete branch of
knowledge. Certainly, the field can appear almost boundless
since, in contrast to other sub-branches of geography, it is
defined not by concentration on a particular area or theme
but by its focus on the fourth dimension – that of time. It
has been argued that all geography is historical geography
since a full understanding of the present requires an
appreciation of the past. In practice, however, the subject
matter of historical geography has been defined by practitioners in a slightly less eclectic manner, typically
summarised as concern with (a) geographies of the past, (b)
changing landscapes, (c) the past in the present, and (d)
geographical history. Methodologically these may be distributed into static or synchronic reconstruction of past periods
and diachronic or process studies of geographical change. In
addition to the centrality of time, historical geography has
also traditionally been characterised as an academic as
opposed to applied discipline, by empirical-based analysis,
dependency upon data sources often compiled for non-geo-
graphic purposes, and conservativism towards theoretical and
methodological developments affecting other branches of human
geography. Over the last fifteen years, however, historical
geography has incorporated humanist and structuralist theoretical perspectives and has embraced computer-based information systems and statistical methods. The balance of attention has also swung in favour of process-oriented analyses as
researchers seek to understand the principles underlying
geographical patterns. In addition the value of lessons from
the past for an applied historical geography has been
suggested. In terms of content, the traditionally strong
interest in agrarian and rural themes and in the nineteenth
century has been complemented by investigations of other

PREFACE

themes, (such as vernacular architecture, ethnic segregation and landscape perception), and periods. Finally, for some, the healthy eclecticism of the subject provides a potential foundation for a reconvergence of human and physical geography in regional-based historical studies.

This collection of original essays is designed to encapsulate the major themes and recent developments in a number of areas of central importance in historical geography. The volume is a response to the need for a text which reviews the progress and current state of the subject and which provides a reference point for future developments in historical geography.

Michael Pacione
University of Strathclyde
Glasgow

INTRODUCTION

The centrality of historical geography in relation to the discipline as a whole has waxed and waned over the last half century. With the introduction of humanist and structuralist philosophies, historical geography has regained some of the ground lost during the period of ascendency of quantification in geography. In Chapter 1 Robin Butlin surveys recent developments in the theory and methodology of historical geography. A pre-eminent trend has been the move away from an empirically-oriented positivist analysis of the historic past to more theoretically-informed investigations, and consideration of a wider range of possible interpretations of historic events. While the value of international study is emphasised it is acknowledged that most developments in theory and methodology occur at the intra-national scale; and an indication of the volume and orientation of work in several different cultural-political realms is provided. The 'classical' phase of historical geography was characterised by the study of landscapes and settlement patterns, the mapping of historical data sources, and reconstruction of past regional geographies. While such work was largely empiricist-positivist in nature it elevated historical geography to a prominent position in the discipline, and furthered the development of methodology. This tradition continues to make a valuable contribution in the search for new data sources, the re-evaluation of existing sources, and development of new analytical techniques. Though criticised as being data-led, such studies have stimulated debate on matters central to the practice of historical investigation relating to written records (e.g. the Domesday Book, probate inventories, Tithe files, population records), cartographic evidence, and field work (ranging from excavation to remote sensing). Further development of methodologies for the evaluation and utilisation of historical data sources is required. Care is required, however, to avoid the excesses of the quantitative revolution which introduced a variety of mathematical modelling (e.g. general systems theory, catastrophe theory) and statistical analytical techniques (e.g. nearest neighbour analysis, regression). Problems encountered included the fact that many modelling techniques related to static patterns rather than dynamic circumstances, and doubts over the suitability of historical data sets for input to rigorous statistical techniques. A general criticism is that much of the work implied objectivity, value-freedom and the passivity of space, adopted a linear approach to change, and failed to make explicit the underlying theory. Rejection of naturalism led to calls for a hermeneutic approach. This attempts to reconstitute historical events in terms of the meanings attributed to them by the agents involved. Contemporary historical geography has embraced humanist (e.g. idealism, and phenomenology) and structuralist (e.g. Marxism, structuration) perspectives and has reaffirmed the goal of regional synthesis. The conception of space has also altered from being viewed as a physical entity to a position where it exists in a dialectic relationship with society. The characteristics of Marxist and humanist theory are discussed. It is concluded that, in both cases, while the potential contribution to explanation in historical geography is great, progress requires the completion of more 'worked examples'. Several themes deserving of further attention are identified, including the role of women in the past, the historical geography of power, and detailed study of regions and communities. It is concluded that the foundations laid by the humanist-structuralist approaches affords both a major challenge and opportunity for historical geographers to develop, via a critical realist perspective, a broader framework for study which considers the life experiences of a wide spectrum of the population.

Despite restrictions on access and certain confidentiality constraints historical geographers in Britain are presented with a vast and ever-increasing volume of data, while in the USA and European countries, such as Sweden, an even wider range of material falls within the public domain. In Chapter 2 John Hamshere discusses a number of key issues relating to the use of these disparate sources. Traditionally, historical geographers have been more concerned with the manipulation of data rather than with analysis of the source itself but the need to understand the language (e.g. foreign script or medieval shorthand), methods (e.g. structured survey, personal observation or recording of oral evidence), form of transmission (e.g. diary or parliamentary report), and purposes and motives of the compiler is crucial to its proper use. While such background information may be available in guides to record sources, its derivation often requires study of the general historical, economic and social structure of the period. In addition, for the geographer the spatial units employed are of paramount importance but these are often never explicitly stated and may have to be inferred from supplementary sources. In short, the researcher must formulate a mode of analysis which meets the goals of his investigation and is compatible with the defined capabilities of the source material. Consideration is then given to the way in which the recent methodological debate in historical geography has been reflected in changes in the type of source material employed. This is illustrated by considering research undertaken over the period 1950-1985. Between 1950 and 1970 work was characterised by a declining recourse to relict features with the reconstruction of past landscapes primarily based on map evidence (e.g. settlement morphology, place names and field boundaries). Medieval sources providing quantitative (e.g. Domesday Book, Hundred Rolls, and taxation records) as opposed to qualitative (e.g. manorial court records or chronicles) data were also used extensively; with the preference for objective information also evidenced in the use made of later public records such as parish registers and crop returns.

The nature of these data lent itself to cartographic portrayal and the mapping of data extracted from historical sources was a major element in most studies. Since 1970 computer-assisted methods of data handling have been introduced, and late nineteenth century census materials have become available. Such objective documentary sources have been complemented by increased use of qualitative sources (e.g. personal correspondence, literature and art) stimulated in part by North American work on the process and perception of colonisation and by the humanistic perspective in human geography. The general trend in research over the period 1950-85, of an increased utilisation of nineteenth and twentieth century sources and a relative decline in the use of medieval and earlier materials, resulted in a shift of emphasis from studies of rural society and landscape towards analysis of urban and industrial society. The active search for appropriate theoretical frameworks for this new focus was in contrast to the reduction in importance attached to map-based research. It is suggested that the problems associated with the construction of historical maps and atlases is an area worthy of renewed investigation. Particular difficulties such as the absence of fixed spatial units and the lack of standardised units of measurement (e.g. cartloads!) underline the fundamental influence source materials exert on analytical results. The introduction of computer techniques in historical geography has proceeded slowly and has largely been confined to the application of standard statistical packages to relatively small data sets. The full potential of computer technology to store, manipulate and analyse large data sets (such as the Domesday Book) has yet to be realised. It is concluded that historical geography's inherent conservativism with respect to data sources and methodology may be challenged by external pressures (technological change, restricted research funding, and fewer young researchers)

INTRODUCTION

favouring more multi-disciplinary research. The historical geographer must develop a methodological framework which demonstrates the particular value of a geographical perspective.

The Dark Ages are characterised by fragmentary written records, a consequent dependence upon relict features and, in academic terms, scant attention within historical geography in which the medieval and subsequent periods dominate research. In Chapter 3 Graeme Whittington considers the problems and potential for geographical analysis of the Dark Ages. He first traces seminal European contributions to Dark Age studies. Particular attention is given to innovatory Swedish work which devised methods such as morphometric analysis of farms and fields, and the metrological technique to address the question of continuity in field and settlement farms. This school of thought was generally responsible for bringing together appropriate techniques from different disciplines (e.g. pollen analysis, phosphate analysis, aerial photography, radio carbon dating) and applying them to the study of the Dark Ages in Scandinavia. The main focus of Dark Age studies has been on agrarian systems and settlement patterns in a rural setting. Far fewer studies of early urban development have been undertaken and this imbalance is worthy of attention. The seeds planted by the early geographical investigations of the Dark Ages have grown only sporadically, with the productivity of the Scandinavian school in marked contrast to the relative paucity of work in Britain. It is suggested that the future for Dark Age historical geography may be bleak given the apparent absence of young researchers interested in the period. In order to locate the contribution of historical geographers within the wider context of Dark Age studies attention is focused on work undertaken in other disciplines, with specific reference to Dark Age Britain. The break down of traditional disciplinary barriers, which has effectively removed geography's exclusive claim to the spatial dimension, is identified as a major reason for the decline in geographical involvement in the period. This is illustrated by considering developments in the study of place name evidence. Whereas early work was based on philology more recent studies have incorporated consideration of geographical factors. The consequent re-interpretation of knowledge on Dark Age settlement processes, including the origin and distribution of Anglo-Saxons in England, has been achieved by place name scholars embracing techniques which historical geographers have traditionally regarded as characteristic of their discipline. The limited attention devoted to urbanisation in the Dark Ages, (including the continuity - catastrophe debate over the fate of Roman urban creations), is primarily the result of data deficiencies which confront researchers. Although much of the available evidence on the Dark Ages comes from archaeological excavation it is contended that this dependence on primary data is not sufficient to account for the lack of geographical involvement in Dark Age analyses. Several additional reasons are suggested to explain why the historical geographer's expertise in medieval studies has not been extended into the earlier period. These include the move by archaeologists away from investigations of single sites and single periods towards a 'total landscape' (geographical) perspective. Concurrently within geography the contemporary emphasis on relevance in research has acted to reduce the supply of future historical geographers, with the Dark Ages, already a minority specialisation, suffering disproportionately. Despite such difficulties, however, historical geographers have an important role to play in Dark Age studies. The complexity of reconstructing the life styles of the period means that the task is outwith the capability of a single discipline. A multi-disciplinary collaborative approach is required in which the historical geographer can make a valuable contribution to the study of the 'total landscape' by bringing his own expertise to complement the excavation skills of the archaeologist. What is required is for a mutually-beneficial symbiosis to be effected between geography and archaeology.

In Chapter 4 Ian Whyte first identifies the temporal and geographical limits of the term medieval and then examines a range of themes relating to the landscape, rural economy, population and rural society, transport and communications, and towns. Geographers have made a significant contribution to the study of rural settlement and field systems, and the reconstruction of past landscapes remains a prominent theme. Recent work has, for example, identified the role of pre-historic and Roman populations in creating the modern land-scape, forced a revision of the view that nucleated villages and open field systems were introduced in early Saxon times, and provided further evidence for considerable change as opposed to stability in the medieval landscape. Despite the difficulties posed by lack of documentary evidence further efforts to illuminate the processes underlying observed patterns are essential. Study of medieval field systems has involved both case studies which have added to the store of knowledge on regional variations and, more recently, contri-butions to the theoretical debate over their origin and nature. Themes addressed have included the relationship between piecemeal colonisation, subdivided fields and communal farming; the influence of inheritance practices and the land market on field fragmentation; the possibility that the open field system represented a risk avoidance strategy; and study of the links between lordship and the organisation of field systems. The difficulty of establishing a causal relationship between events is exemplified with reference to a study of field systems at the regional scale. Despite the wealth of case studies there remains an absence of concensus on the relative strength of possible explanatory factors. The influence of climate on socio-economic change has also been examined, and particular attention has been focused on marginal environments in an attempt to reduce the problem of isolating cause and effect. Studies of population and rural society are hampered by data deficiencies prior to the more general availability of parish registers in the late sixteenth century. Despite this, however, the relationship between population change and the medieval economy is a major research theme. Another important research focus is the growing international interest in community studies. Detailed reconstructions of local communities have been undertaken in an effort to piece together family and household structures, kinship and neighbouring patterns, crime levels, birth control practices and the incidence of debt and credit. In general recent population research has suggested higher levels of mobility in medieval society than had been anticipated. Issues which demand further investigation include the study of power and conflict within feudal society and the relative importance of urban and rural influences in the transition from feudalism to capitalism. In the field of transport and communications much more work has been devoted to international trade than to regional and local activity and transport routes. This clearly reflects the limited availability of extant evidence and suggests that the reconstruction of medieval road networks remains an enduring challenge. Work on medieval towns has tended to focus on individual places rather than developing a comparative perspective. While significant advances have been made in recent decades as a result of archaeological excavation, the changing importance of urban centres throughout the period remains open to debate. At the regional level work continues on the development of urban hierarchies, but comparatively little attention has been given to the study of urban social patterns, and this offers considerable scope for further research. In general the emphasis in medieval historical geography has been descriptive rather than analytical, with comparatively little attention to conceptual issues, (such as the transition from feudalism to capitalism.) Theoretical and methodological debates have had, as yet, a relatively slight impact on medieval research. It is concluded that while data-based investigations will remain of central importance in medieval studies recognition of the limitations of positivism - in particular that research and data sources are not value free - will facilitate the incorporation of alternative theoretical perspectives of considerable potential utility in the attempt to explain observed patterns.

In Chapter 5 John Walton examines recent contributions to the historical geography of agriculture and rural society for the period 1500-1900. An international review of research issues contrasts the situation in Britain with that in North America where work in the cultural tradition is of particular importance (as, for example, in the study of folk housing, settler-environment interaction, and the definition of cultural regions). In general, it is suggested that despite the appeal of the world-system approach there remains an enduring preference for localised empirical studies, albeit increasingly informed by theory. The dynamics of change in the British countryside represent a major research area, within which considerable attention has been focused on the decay of feudalism and the advent of new capitalistic systems of production - as a result of changes in techniques, products, market demand, distribution of ownership, and relations between landowner, tenant and labourer. Prior to consideration of this theme, however, several important research topics not encompassed by this organising structure are examined. The theme of landscape change, which has occupied historical geographers for over fifty years, remains of central importance, although the traditional holistic approach has been augmented by a categoric approach which focuses on specific landscape features. Among those of particular interest to the study of agriculture and rural society are field systems (increasingly interpreted in the context of local economies and societies); vernacular architecture (in which an initial concern with description and classification has been complemented by interpretive studies, for example, of the relationship between new building and agricultural living standards); and historical ecology (e.g. reconstruction of past land uses, the effect of climatic change on agriculture, and the relationship between local agricultural and social conditions and the timing, nature and location of industrial growth). Research into each of these themes is discussed before attention turns to the question of change in post-medieval British agriculture. Four particular issues are examined (a) the agricultural revolution, (b) the rise of agrarian capitalism, (c) enclosure, and (d) innovation diffusion. The validity of the term "agricultural revolution" is questioned and evidence for a process of evolution as opposed to revolution discussed. It is concluded that reification of the term revolution assigns insufficient importance to the conservative forces inherent in the agricultural sector. Research on the advent of agrarian capitalism has generally focused on the themes of changes in tenure and landownership, the decline of communal forms of organisation, the fate of the peasantry, and the relationship between different social groups and classes. Recent work in each of these fields is surveyed. The causal factors underlying enclosure have also been the subject of lengthy debate. Evidence for a conspiracy theory behind parliamentary enclosures (with vested-interest MPs forcing landless peasants into wage labour) tends to confirm the important role of power elites, and supports the view that parliamentary enclosure was a main factor in the destruction of the traditional economy. Attention has also been directed towards the effects of pre-parliamentary enclosures which in terms of land area affected exceeded the later acts. Innovation diffusion is a key element in the transformation of rural society and each factor in the process of agricultural change displays its own chronology and geography. Unfortunately for the historical geographer data limitations mean that it is much easier to reconstruct past chronologies than past spatial patterns of diffusion. The problems and prospects for research into the geography of innovation diffusion are discussed. Despite the diversity of research traditions and themes which characterise historical geography a common thread in recent work on agricultural and rural society has been an emphasis on the dynamics of change. It is suggested, however that immobility and inactivity are more general characteristics of rural life and that it might now be appropriate to afford more consideration to those conservative aspects of agrarian society which survive the pressures for change.

In Chapter 6 Colin Pooley examines a number of major issues in the historical geography of industrial change. Four main research areas are identified - (1) the study of specific industries (including the organisation and operation of firms, working conditions, and locational decisions), (2) the regional structure of industrial change (e.g. the effect of industrialisation on traditional economies, and the emergence of problem areas), (3) mechanisms of industrial change (the role of transport and communications, labour migration, innovation diffusion, and inter-regional capital flows), and (4) the impact of industrial change on the contemporary social formation (urbanisation, class relations, employment trends). Recent research in each field is then considered. Study of the historical geography of specific industries provides valuable case study information as well as contributing to our understanding of both the spatial structure of nineteenth century British industry and the regional impact of industrial change. The utility of map-based research in this context is acknowledged. The influence of scale on research results and the utility of conducting analyses at the local level and of examining a wide range of industries (including, for example, the service sector of the Victorian economy) is emphasised. The effect of industrialisation on regional distinctiveness represents a second important research area. Analysis of the regional geography of England during the industrial revolution identifies the central role of transport and underlines the continual importance of regional economies during the development of a national economic system. It is suggested that the regional perspective may be extended to examination of the growth of problem areas in inter-war Britain, but, to date, such recent historical events remain an under-researched theme. Inter-regional transfers of capital, information, labour and products are key elements in the mechanism of industrial change, and each represents a poten-

tially fruitful area of geographical study. Thus the significance of transport for the movement of people and goods has
been examined. Labour migration has also been well
researched, although more detailed work is required on
specific relationships (e.g. the effect of migration on
individual and group employment opportunities). This will
require longitudinal analysis of micro-level data and
integration of material from a wide range of sources. Less
attention has been devoted to the diffusion of information and
ideas, and their effect on urbanisation and industrialisation.
Circulation of capital is a fundamental component of the
capitalist system and while some analysis of institutional
structures (e.g. banking) and of capital formation in specific
areas and industries has been undertaken greater consideration needs to be given to spatial flows of capital and to the
development of other factors of production (such as power,
technology and entrepreneurship). The fourth research
question, concerning the impact of industrial change, can be
investigated at several levels, ranging from the effect of new
technology on individuals to the impact of industrialisation on
urban development. While some work has been carried out at
the micro-level, (using for example structuration theory and
the concept of time-geography), geographers have been more
concerned with issues at the macro-scale (e.g. segregation,
social structure and health in cities). The essential reciprocal
relationship between industrialisation and social change has
only rarely been addressed explicitly however. Reasons for
this include the complexity of the linkages between economy
and society, the latency inherent in some causal forces, and
an over-dependence on census data and consequent neglect of
less quantifiable earlier sources. It is concluded that
geographers have made only a limited contribution to the
understanding of the processes underlying industrial change.
Three priority areas for future research are identified. First,
there is a need for further empirical investigation of the
process and impact of industrial change, working at a variety
of scales, encompassing a range of industrial sectors, and
covering a sufficient time period to allow identification of
gradual as well as "revolutionary" changes. Secondly, it is
suggested that the development of appropriate theory must
proceed in conjunction with empirical research, and finally
more explicit study of the relationship between economic
change and social phenomena is required.
    The form and functioning of the city in industrial society
represents a major research focus in historical geography,
with particular attention being devoted to the Victorian city.
Despite statistical analyses based on large census-based data
sets, empirical investigation of the relationship between
economy, society and urban spatial structure is an under-
researched area. In Chapter 7 Richard Dennis first reviews
work on the quantitative analysis of nineteenth century cities,

then addresses questions relating to community and segregation, class and ethnicity, and housing and mobility, before finally drawing a number of comparisons between the situation in Britain and that in North America. The major sources of data on the social geography of nineteenth and early twentieth century cities are identified, a principal source being census enumerators' books which provide information on family and household structure, social status and birthplace. The lack of data on housing conditions, tenure or recent migration, however, confounds efforts to relate social space to physical space. Other important sources include rate books (which do provide information on housing quality and tenure), directories (which indicate home and workplace locations of businessmen), and registers (e.g. marriage registers which can shed light on inter-generational social mobility). Such quantitative records are complemented by qualitative sources in literature, art, diaries, autobiographies and oral histories which seek to uncover the thought processes underlying decision-making and human behaviour. The potential and problems of these various sources are reviewed. (See also Chapter 2). Segregation by class, culture or ethnicity is a major feature of urban life and historical geographers have studied such patterns using location quotients and indices of dissimilarity and isolation. The significance of size, type of city, and scale of analysis for such analyses is illustrated. More recently, attention has been focused on the economic and social processes underlying revealed patterns, and on the implications of segregation in terms of, for example, differential access to urban resources. As yet few geographers have examined the views of urban socio-spatial structure held by contemporary observers. With reference to the investigation of social class, a major difficulty stems from the mobility of simple classifications to detect complex intra-class differences. It is suggested that in order to approach the question of how changing social structure (e.g. the emergence of a middle class in the nineteenth century) is related to spatial structure it is necessary to investigate human behaviour (e.g. voting behaviour, intermarriage, club membership) within particular class areas of the city. Attention then turns to the issues of housing and mobility. Stimulated initially by the concept of "housing class", considerable research effort has been directed towards documenting the production of the built environment and the role played by various private (e.g. building clubs, speculative builders, landlords) and public (e.g. local councils) interests. The management of dwellings and issues such as landlord-tenant relations and the effects of rent legislation are, however, less well researched. The changing tenure structure in British cities is discussed. The importance of this research is indicated by the fact that the geography of housing exerts a causal influence on patterns of residential differentiation, and that "housing class" is related to both patterns of social interaction and residential mobility. It is concluded that integrated study of issues such as segregation, community and mobility is required in order to approach a full understanding of the processes underlying the observed pattern of people and housing in industrial societies.

Historical demography seeks to advance our understanding of past societies through the reconstruction of the demographic behaviour of individuals and communities. In Chapter 8 Philip Ogden draws upon an international range of research to illustrate developments in this field. He first considers the relationship between history, demography and geography and traces the emergence of a distinctive sub-discipline of historical demography to developments in the post war era. Particular importance is attached to methodological advances (such as the introduction of longitudinal analysis made possible by the technique of family reconstruction, and the complementary technique of back projection) and to changes in research perspective, (such as a recognition of the utility of an inter-disciplinary range of both quantitative and qualitative data sources, and a belief in the central importance of demographic as opposed to economic factors in social change). Attention then focuses on four major themes in contemporary historical geography. The first research stream emanates from Malthus' work on the relationship between population and resources in traditional societies and in particular the debate over the relative importance of positive (e.g. mortality) and preventive (nuptuality and fertility) checks to population growth. Consideration is given to recent work on the link between demography and economy, but the lack of concensus underlines the need for further investigation of marriage behaviour as well as study of mortality trends and the impact of catastrophic events. A second focus of research activity is the demographic transition which seeks to represent a process in which high uncontrolled levels of mortality and fertility in traditional societies (e.g. nineteenth century England or the Third World at present) give way to controlled levels in industrial societies as a consequence of modernisation. While the evidence for such a change is irrefutable the veracity of transition theory itself has been questioned (e.g. its over-generalised representation of pre-transition societies). The fact that recent evidence from Europe demonstrates no simple correlation between economic development and population transition suggests a greater importance for cultural factors (e.g. diffusion of contraceptive practice) - a finding with clear implications for contemporary policy initiatives in the Third World. The study of families and households represents a third major research area. Three approaches may be identified, (a) the 'demographic', concerned with marriage, childbearing and household size (b) the 'sentimental', with an interest in the emotional meaning of marriage and the family, and (c) the 'economic', concentrating on the economic behaviour of household members. The results of recent work in all three spheres are discussed. Research in the demographic tradition, for example, has dispelled the myth of the extended family as the typical pre-industrial form of social organisation, with the nuclear family coming to prominence after industrialisation. It is suggested that further investigation is required into the significant geographical variations on this dimension throughout Europe. The fourth main research focus refers to the theme of migration and mobility. Major contributions have been made on migration in traditional rural society, on the role of migration in urban growth, and on international movements including slavery. Recent empirical evidence, for example, has debunked the notion of an immobile pre-industrial village population, with considerable movement (both permanent and seasonal) recorded within rural areas as well as to the growing towns. Areas of further research include the relationship between migration and the wider demographic system (e.g. the distortion of local age-sex structures and the resulting effect on nuptuality, fertility and mortality), and investigation of forced migration and its demographic impact on sourse and host regions. A wider geographical remit is also required with most current research being confined to N W Europe. The question of the appropriate geographic scale of analysis is also considered and it is suggested that further studies at the regional scale are required to complement national- and parish- level investigations. More generally, for geographers working in the field of historical demography the greatest challenge remains the identification and interpretation of spatial variations and the development of an historical population geography.

The study of urban morphology involves consideration of the town plan, building form and the pattern of land and building use. In Chapter 9 Jeremy Whitehand identifies three main schools of urban morphological inquiry in Germany, Britain and North America, and reviews the origins, development and principal characteristics of each. In Germany an initial focus on the study of urban form to the neglect of function was counterbalanced from the inter-war period on by process-oriented investigation, which developed into the post-war emphasis on in-depth analyses of individual cities. The German morphogenetic tradition exerted an influence on urban historical geography in Britain, as witnessed by the studies of Whitby, Alnwick and Newcastle and the introduction of concepts such as the burgage cycle and fringe belt. This was complemented by an indigenous concern with descriptive generalisation of urban morphology which later developed into quantitative morphographic analyses. Work in North America is characterised by two themes, the first emanating from cultural geography (concerned, for example, with architectural styles) and the second with foundations in urban socio-economic analyses and demonstrated by the classical ecological models of land use. Each of these three major strands has contributed to current research in urban morphology. Three major elements of contemporary work - town plan analysis, the cyclical approach, and agents of change - are examined in detail. Most town plan analyses employ Conzenian concepts developed in the 1960s, since when theory development has proceeded slowly. Much of the work in this tradition is undertaken in Europe when topographical, archaeological and documentary evidence are often combined in the attempt to reveal the details of town plan development. Detailed work on burgages in Britain has suggested the merits of plot measurement and reconstruction as a means of identifying an outline pattern of development, but the burgage cycle has received comparatively little attention. Overall an integrated body of concepts has yet to be attained. The primary link between town plan analysis and the study of the cyclical character of land use and building form was provided by the fringe-belt concept. Subsequent attempts have been made to relate the changing form of urban areas to economic fluctuations in general and building cycles in particular. The examination of constructs such as rent theory and its relationship to urban growth cycles led to conceptualisations of the links between building cycles, land values, fringe belt formation and residential density. Consideration has also been given to the influence of innovation diffusion on the urban landscape. Some attempt has been made to develop a general model of the fringe belt process independent of a particular historical context and consistent with current theories of urban economic structures. Such formulations however await empirical verification. Generally, the individuals and organisations responsible for urban development have received more attention from historians than geographers but this imbalance is being redressed by the decision-making approach to town plan analyses. To date the main emphasis has been on the relationship between the various agents of change and building form, with the majority of studies dealing with central urban rather than suburban areas. Particular attention has been focused on speculative investors and developers whose activities have direct implications for the timing and form of building development. Explicating the development of commercial cores over the last half century represents a major research challenge. Finally consideration is given to the suggestion that the morphogenetic tradition can provide the conceptual basis from which historical analysis can be linked to the future management of the urban landscape. This perspective views the urban scene as an historical phenomenon and contends that it is necessary to understand the past better to plan the future. While there are obvious links between the identification of urban landscape units and the definition of conservation areas the wider practical applications of this proposal have yet to be elaborated. Another major task for urban morphologists is to articulate the nature of the links between the three traditions of town plan analysis, the cyclical perspective, and study of the agents of change.

In the final chapter Brian Roberts identifies three major approaches to the study of rural settlement. The first, empirical reconstruction of settlement conditions, involves four areas of inquiry (a) the physical character of places (e.g. layout, building material), (b) functional relationships (e.g. marketing and transport systems, social characteristics), (c) settlement classification, and (d) the evolution of a settlement through time. The second approach concentrates on the dynamic theme of process and seeks to understand general processes of geographic change, to relate these forces to specific situations, and to develop relevant conceptual frameworks. The search for theory represents the third of the major approaches and includes consideration of, for example, core-periphery, diffusion and hierarchical models. Clearly all three approaches may be applied within any particular geographical setting. The emphasis attached to each within different regional schools of inquiry is partly dependent upon the nature of data available. Despite the ubiquity of rural settlement, it is suggested that the major challenge for the historical geographer is to construct models to account for the complexity of settlement in the old settled regions of the world. This is illustrated with reference to a transect across part of Yorkshire in which evidence of occupation dates back to the mesolithic period. In areas where evidence is limited the advantage of a long time perspective is noted and the relevance of methods of archaeological inference suggested. A distinction is drawn between large and small-scale studies (using a threshold of 1 sq km), and the addition of the three-fold (empirical, process, theory) classification to this size dimension provides an organising framework for rural settlement studies. Attention is then given to small-scale (intra-settlement) studies of place, form and function; and the important links between historical geography and social anthropology (e.g. living space, inheritance roles), architecture (building styles, barn types) and archaeology (evolution of the peasant house) are emphasised. Examples of work in the morphologic tradition demonstrate both the breadth of activity and the need for additional research to further understanding of the regional diversity of past rural settlement. There is a clear functional link between rural settlement and the surrounding land resource. Recent studies demonstrating the close ties between settlement plans, field systems and social structures also suggest the possibility of

undertaking detailed analyses of the historical village through
some incorporation of techniques from related disciplines. The
distribution of property rights, and in particular the balance
between private and collective rights, is a key element underlying settlement forms, and the way in which land ownership
affects the social, economic and physical characteristics of
rural settlement is a research question of enduring importance. The examination of settlement patterns at the macro-
scale increases the possibility for generalisation and theory
construction and a number of settlement models are critically
examined. A major dilemma is the possibility that the search
for general laws of rural settlement may be confounded by
culturally-specific factors. The theme is illustrated with
reference to a regional study of desertion and land colonisation in Scandinavia which attempts to link rural settlement
change to a series of factors related to population (size,
migration), land (proportion arable, climate, technology),
political and economic structure (e.g. taxation legislation) and
legal and administrative arrangements (e.g. tenure, inheritance practices). It is concluded that the future health of
rural settlement studies in historical geography requires
progress in all three approaches - empirical, process
analysis, and theory construction. A number of problems,
however, present barriers to such an advance. Perhaps most
fundamental is the significant gap between empirical investigations and existing theory, with few studies successfully
merging the two. Related to this is the absence of sound
comparative work which would facilitate testing the appli-
cability of theories in a variety of situations. In a methodological context it is suggested that the decline in use of
maps is to be regretted since this traditional geographic tool
can be of considerable value for both retrogressive and
prospective studies. It is advocated that in addition to
developing theoretical and methodological links to cognate
disciplines, historical geographers should devote greater
attention to sharpening the analytical weapons in their own
armoury.

Chapter One

THEORY AND METHODOLOGY IN HISTORICAL GEOGRAPHY

R.A. Butlin

INTRODUCTION

The Nature of 'Progress' in Historical Geography
It is difficult to produce a consensus view of progress within
an academic sub-discipline. Even the most fundamental of
premisses regarding the concept of "progress" and the nature
of the sub-discipline or field of inquiry are likely to be
disputed, and the matter is further complicated by the social,
institutional, linguistic and cultural contexts of the reviewer.
What follows, therefore, is a highly personal, perhaps even
idiosyncratic, overview of what appear to have been some of
the more interesting developments in the theory and methodology of historical geography over a period of about a
decade, that is from the mid-1970s to the mid-1980s.
    The essence of the argument is that considerable
interest, excitement and stimulus has been generated within
and beyond historical geography in the last ten years or so
by a discernible movement from the interrogation of a
narrowly-conceived past or pasts by means of a limited range
of largely empiricist and positivist methodologies, reflecting a
narrow ideology which generally admitted only the powerful
and the prominent to its narratives, towards the questioning
and investigation of a number of possible historical geographies of a variety of pasts by means of a wider range of
theoretically informed perspectives.
    In addition to these largely internal developments in
theory and methodology in historical geography, the broader
role of historical geography within geography as a whole, and
human geography in particular, deserves attention as a
salient feature of change or progress. Of particular interest
is the way in which historical geography seems to have moved
from a central position in geography, to a peripheral position
during the quantitative revolution, and back to a more central
position in the humanistic phase. One might indeed go further
and say that much of the advance of human geography away
from positivism was spearheaded by some critically important

work by historical geographers, and that historical geography, at least in its more 'progressive' manifestations, is a leading cutting edge of human geography. One outcome of this changing role seems to have been the increased historical sensitisation of human geography, together with the considerable extension of dialogue and debate with other social sciences. In some senses the wheel might be thought to have gone full circle, with the work of historical geographers having facilitated a return to forms (albeit in modern style) of traditional human and cultural geography: what might be termed 'new human geography' and 'new cultural geography'.

Before moving to a detailed study of the evidence for 'progress' in theory and methodology in historical geography, some further preliminary consideration - of the meaning of progress - is necessary.

Dennis, in a review of theory and progress in historical geography (Dennis, 1984), referring to a " 'Whig view of history' - history as progress, and history with hindsight, written in the knowledge of what happened next", says that this 'view' is "applicable to most historical geographers", and contends that "at the very least, historical geographers should adopt a more circumspect view of the 'progress' they describe". The implicit and explicit difficulties that attach to a progressive view of historical geography are fairly obvious: change is thought always to be for the better, with failure or 'retreat' generally omitted. Kennedy, in a broad critical review of two books dealing with changes in academic geography over the last hundred and fifty years has expressed this graphically (Kennedy, 1982): "Perhaps the prime defect of these volumes is that the story they depict leans too heavily on successive invasions by Waves of Saints. In 150 years there must, surely, have been more Bad Things, even if a few invaders actually managed to go through the groves of academe with fire and the sword" (p.76). McCaskill similarly, in a review of the fourth volume of Clark's History of Australia (McCaskill, 1981), speaks of the type of historical writing "especially economic histories and historical geographies [that] tends to be about successes rather than failures, of developments rather than destruction and degradation", while commending Clark's reversal of this trend in his analysis, through a "despairing view of past events", of Australia in the period 1851-1888.

Further problems derive from the sheer volume of material in the field of historical geography, for it would be impossible, even in several full-length books, to avoid the wrenching of samples of research work from intellectual, institutional, and cultural/political contexts, in order to parade or to embed them in a stiff and inflexible progressivist historiographic base. Interesting new ideas and techniques emerge while the institutional ground is shifting beneath the scholars who produce them, and political control and fettering of the academic right of free enquiry severely restricts the dissemination of ideas and the free movement of scholars in and from totalitarian states of various kinds. The notion that progress is mainly to be achieved through technology and applied science in 'democratic' states leads to a gross under-valuation of the social sciences and humanities, so that the basic provisions are reduced and with them the opportunities for young scholars to gain a place in and livelihood from such fields as historical geography. These are some of the many hidden agendas of contemporary scholarship which are all too easily neglected in the 'progress in' type of article.

The notion of progress, therefore, poses problems from both ideological and practical perspectives. In the ideological context we must acknowledge the long tradition of the idea of progress, at least in Western thinking, and can perhaps agree with Arendt's statement that "Progress as an ideal of action cannot be precisely identical with progress as a fact or object of actual or possible knowledge", for "observed progress is mainly technical, whereas believed progress is mainly spiritual" (Arendt, 1954, cited by Pollard, 1968). It is easier, and maybe of greater importance, to acknowledge this 'spiritual' aspect of progress as a necessary condition of humanity, than it is to measure the technical aspects of progress in scholarship.

Progress, then, is a variegated and relative concept. In practical terms, progress, in the sense of the perceived intensification of active scholarship and the presentation of 'new' ideas in active discourse and debate, may be measured in a number of ways. One way is the simple quantitative assessment of 'activity'. Baker (1982) has suggested, for example, that in the course of the 1970's historical geography became "more institutionalised and increasingly international-ised, although still essentially operating within a domestic rather than a factory system of production. The last ten years or so have seen the growth - and in some instances the birth - of local, regional, national and international research groups in historical geography, promoting the subject intrinsically and contact among its practitioners extrinsically".

Although much work in historical geography is carried out at local and regional scales, national and international dialogue has been facilitated by the growth of the research groups described by Baker. These include the Historical Geography section of successive International Geographical Union congresses, specifically those at Moscow (1976), Tokyo (1980) and Paris (1984), the work and specific meetings of the IGU's Working Group in Historical Changes in Spatial Organisation, which existed and organised meetings from 1976 to 1984 (it had a fixed term, and therefore did not stop in 1984 because it was moribund), and a number of more localised international exchanges such as the successive CUKANZUS meetings (mainly, in spite of the acronym, between geographers from the UK and North America) - see, for example, Newcomb (1983) - and the periodic Franco-British and European Rural Landscape research seminars (Cleary, 1982; Simms, 1982). In this international context one would support strongly Baker's contention (Baker, 1983), that "Progress in historical geography and international co-operation among historical geographers are related objectives. Journeying towards them is sometimes frustratingly slow, along a route encumbered with foreign languages and alien concepts. But such explorations in historical geography are arguably vital". The increasing intensity of dialogue at international level has not led - nor should it - to a more homogenous view of historical geography, but has indicated a fascinating and stimulating (and at times bewildering) variety of traditions and perspectives. As far as theory and methodology are concerned, much of the interesting new work has been developed at intra-national levels. The Historical Geography Research Group of the Institute of British Geographers, for example, has been in existence, in effect, since the late 1960's, and is active in discussion and dissemination of research ideas and publications, notably through its Research Paper series. There has, in addition, been an increase in journals covering the subject: the best-known in English being the Journal of Historical Geography (started in 1975) and Historical Geography (formerly The Historical Geography Newsletter, started in 1971). Other publications include the recently-established Dutch journal Historisch Geographisch Tijdschrift (1983), the publications of the Arbeitskreis für Genetische Siedlungsforschung in Mitteleuropa, edited from the Seminar für Historische Geographie of the University of Bonn, the Czech Academy of Sciences publication Historicka Geografie, and the Japanese journal Historical Geography.

The general progress of work on theory and methodology has varied in many countries. On the whole, greatest attention to theory, and perhaps methodology, seems to have been paid in the United Kingdom, Sweden, the United States and Canada, at least in respect of what might be termed 'Western' historical geography. Within socialist countries methodologies and theories of a somewhat different kind influence the nature of historical geography advocated and practised, so that, for example, Jelecek (1980) has indicated that in Czechoslovakia, work on and reviews of historical atlases indicate and express "the connection of our historico-geographic research with the practice of developing an advanced socialist society in Czechoslovakia". Jelecek states that the prime orientation of the Czech publication Historicka Geografie is towards historical economic geography, with particular interest in atlases of historical maps in East European countries and the USSR. While scholars in capitalist countries are progressing theory via studies, inter alia, of

the transition from feudalism to capitalism, it seems that
scholars in socialist countries are focussing on the transition
from capitalism to socialism and beyond. The contexts of
evaluation of progress in theory and methodology in historical
geography obviously vary very widely: a notion which needs
strong and repeated emphasis. Even a superficial reading of
the contributions to the volume Progress in Historical
Geography, published in 1972 (Baker, 1972), notwithstanding
the understandable under-representation of overviews from
socialist, African and Asian countries, indicates the variety of
appraisals of progress, and more recent overviews show this
still to be the case. Studies of work in historical geography
in France, for example, indicate a continued attachment to
traditional sources and region-based analyses, with some
significant measure of antipathy to more general types of
theory. Hence Baker's reference to the "limited amount of
methodological debate within French historical geography",
related to "the relative absence of an explicitly reflective
critique" (Baker, 1984). Some recognition of the need for
further development in theory and methodology has been made
by Claval (1984), against a background of what he considers
to be "la puissante école de geographie historique
brittanique", and the necessity "de rompre avec les sentiers
battus en insistant d'avantage sur les determinations sociales
et culturelles des formes d'organisation" (Claval, 1981),
though his views would not be widely supported in France.
While the strong tradition of landschaftskunde and a stress on
environmental relations are to be found in modern studies of
historical geography in Germany, there are signs of an
increasing interest in theory and methodology (Fehn, 1982).
Progress in historical geography in Japan seems at present to
comprise an attempted accommodation of process studies with
the construction of cross-sections, based on detailed records
of land ownership (Senda, 1982) and there is obviously a
growing awareness in Japan of the methodological developments elsewhere. In the United States and Canada, progress
in methodology and theory are largely related to refinements
of the cultural historical traditions, established, inter alia, by
Sauer and Clark, though methodological and theoretical
debates, which will be discussed later, have developed around
humanistic and idealist approaches. A not dissimilar picture
may be presented of work in Australia and New Zealand
(Powell, 1981; Wynn, 1977).

It may well be argued that false images may be promoted
by attempts to portray 'national' trends as schools of
thought. In most countries progress in historical geography
occurs through the influences of individuals and groups.
Hence, for example, Simms (1982) has listed various 'schools'
of influence in Britain, and a number of appraisals of the
influence of individuals on traditions within the subject have
been published (e.g. Slater and Jarvis, 1982; Gourou, 1982;

Harris, 1976). Periodically, election of historical geographers to chairs in universities allows for exegeses of the subject by means of inaugural lectures (Jones, 1976a; Butlin, 1982a). It is essential, therefore, that criteria of 'progress' in theory and methodology in historical geography be related to the cultural, historical and even political contexts of the research work evaluated. Though this essay reflects what might be seen as a Western view and evaluation of significant changes in approach, a view from a socialist bloc country would almost certainly order the criteria 'success' and 'progress' in a very different way.

Theory and Methodology within the 'Classical' and 'Neo-Classical' Traditions of Historical Geography

Progress may be evaluated both within and between various traditions, methodologies and ideologies in historical geography. An initial difficulty and danger in this type of exercise is to apply general descriptive labels which appear to bind together for critical and evaluative purposes what in practice are usually extremely heterogeneous types of approaches and works. Terminology is a major problem particularly in a rapidly-changing subject, and the state of the maturing wine is not easily accommodated by the labels of the conceptual bottles into which it is poured - it is difficult to have an agreed 'appellation controlee' for historical geography.

The strongest tradition in historical geography, notwith-standing the work of contemporary conceptual pioneers, remains that of the 'classical' phase, characterised by Gregory (1981a), as a period in which historical geography occupied a central place in geography, with a strong commitment to historical study of landscape and to the mapping of historical data sources, with occasional ventures into the reconstructions of regional geographies of the past. While it is all too easy to apply the labels 'empiricist' and 'positivist' to the types of work representative of these traditions, such applications are rather misleading, and must be qualified. Work in this 'classical' phase - associated indelibly in England with the work of H.C. Darby and a small band of pioneers, has to be seen in the context of attempts to establish historical geography as a credible academic pursuit and at a time of anti-environmental determinist sentiments. The work produced was experimental, of meticulously high standards of scholarship, undoubtedly exciting, and highly successful in its achievements. Darby (1983), refers to the 'ferment' of the New Geography in Britain in the 1920s and 1930s, with which "came the rise of historical geography as a self-conscious discipline. We 'new geographers' realised that every past had once been a present. There was a high degree of unanimity among us. We had

21

something of the dogmatic fervour of new converts to a faith, heightened by the fact that the position of geography was not all that well established in our universities". Pragmatic attempts at new methodologies were made in An Historical Geography of England before 1800, using cross-sectional reconstructions of the geography of past periods, with sources of material and the problems discussed determining the methodologies, which were extremely varied. Darby, the editor of the book, describes how he "pondered over the possibility of writing a philosophical introduction, but came to the conclusion that the time was not yet ripe" (Darby, 1983). While the methodology of historical geography was advanced by that volume, theory was not: a somewhat puzzling feature in view of the amount of contemporary writing on the theory of history, but the temerity of those young pioneers must be recognised. The outcome, as Baker (1984) has indicated was that "Darby's search for a method in historical geography discovered a diversity of problems, of courses, of approaches and techniques. His principal finding was an experimental and pragmatic scholarship."

Within these terms of reference - of an experimental and pragmatic scholarship, in which methodology is determined by problem and source but within which explicit theory plays a small part - it is possible to indicate progress of a methodological kind in the last few years or so. This methodological progress has largely been evidenced by the use of new sources, the re-evaluation and refinements of known sources, and the employment of new analytical techniques. The extent to which the interpretation of sources of evidence remains a hall-mark of historical geography is clearly indicated in Prince's review of historical geography in 1980 (Prince, 1980), where he draws our attention to the search by historical geographers for new sources of evidence and their re-evaluation of known sources. Archive material has been particularly important, notwithstanding the daunting prospect that "additions to the Public Record Office at present fill approximately one mile of shelving per year" (Tosh, 1984), or of the size of the National Archives in Washington. The outstanding single-source based study has probably been Darby's work on the Domesday geography of England, the publication of which has been completed in the period under review. Hence Glasscock's observation (1978) that "The Domesday Geography has become a lynchpin of English historical geography... In the last thirty years it has become the archetypal example of work based on a single source for a single year. Domesday Book ... has lent itself to this kind of treatment and, in a sense, has produced its own methodology". Glasscock counters the criticism of this type of study as merely source-based empiricism by pointing to the value of formal presentation of scarce (medieval) sources as a basis for detailed analytical study, and followed this route himself in his edition of the 1334 Lay Subsidy (Glasscock, 1978). Domesday and related work has, in fact, led to discussion and debate on aspects of methodology of use of medieval sources (Stanley, 1980; Biddick, 1983).

The use of computers has considerably aided the development of sophisticated analyses of large data sets. Outstanding examples from Britain are: the various works on the 1801 crop returns; the Tithe files and awards of the nineteenth century; the data pertaining to parliamentary enclosures; probate inventories; Domesday Book itself; and major works on the analysis of census and other population records. While much of the analysis of this type is justifiably concerned with the accuracy of data and the methodology of its employment - often in relation to other sources - it is probably foolish to dismiss the approach as solely a type of source-bound empiricism, for broader issues are always raised which lead to discussion and debate. In the case of Overton's analyses of probate inventories, for example (Overton, 1977; 1985), the technical difficulties of analysis and their solution are necessary and intermediate steps (the preliminary steps being the formation of problems) to the solution of important problems of agricultural change seen in broader historical and intellectual contexts. The derivation of accurate data bases from important sources as means rather than ends in themselves does appear to have been the object of many works by historical geographers, including work by Turner on the 1801 crop return (Turner, 1981), by Turner and Chapman on the parliamentary enclosures (Turner, 1980; Chapman and Harris, 1982), by Kain and Prince on the Tithe surveys (Kain, 1979; Kain and Prince, 1985), by Lawton (1978) and others on the census returns, and the reconstruction of the population history of England by the Cambridge group (Wrigley and Schofield, 1982). Data source evaluations of similar kinds are to be found in and for many countries other than Britain, many in relation to census data and to the historical geography of agriculture and land-use, with particular reference to records of land ownership and occupance (many examples are given in: Norton, 1984; Wynn, 1977; Ehrenberg, 1975; Conzen, 1980; Ward, 1979).

Additional and much worked sources of information of particular significance are cartographic sources and the evidence of the land and landscape, as interpreted by the use of field work (including excavation) air photography/remote sensing, and documentary evidence. The history of cartography has become almost a separate sub-discipline (or, more likely, multi-disciplinary area of study) and is offering new vistas and methodologies in historical geography (sceptics might even see the future of historical geography in the United States being partly dependent on the health of cartographic history). Work such as that of Harley (1982a, 1982b), Andrews (1975) and De Vorsey (1980) has amply exemplified

THEORY & METHODOLOGY IN HISTORICAL GEOGRAPHY

the need for careful evaluation of sourse materials. The fieldwork/settlement evolution tradition, strong in most countries where historical geography has a significant presence, provides plentiful evidence of methodological advance within varying intellectual and cultural traditions. Hence the emphasis has very much been on spatial form or morphology, and the typology and classification of such forms. Work of this ilk in Britain includes that of Jones on Celtic settlement forms (Jones, 1976b, 1984, 1985), of Roberts on planned settlements (Roberts, 1977) and of Thorpe on settlement morphology (Thorpe, 1973). The broader traditions of landscape analysis within cultural historical geography, partly in the Sauer tradition, have been maintained and developed in North America and Australia (Williams, 1974, 1975; Ward, 1979; Bowen, 1978; Davidson, 1974; Harris and Warkentin, 1974; Meinig, 1978), and, from different origins and traditions in Central and Western Europe (Fehn, 1982).

What then, may be said by way of evaluation of progress in methodology and theory of the tiny sample of source- and landscape-related type of research in historical geography briefly outlined above? A proper evaluation of these types of research requires both a critical assessment of them within their own apparent terms of reference and also within the broader context of the changing intellectual fashions and history of geography in the last ten years or so.

Within their own self-admitted (or implicit) terms of reference, there are grounds for both dissatisfaction and satisfaction with progress. A potent series of criticisms has been made by Harley (1982a) namely: that "there is no properly developed tradition within historical geography of editing original documents as a research or teaching activity"; and that "some historical geography is coy in the critical description of evidence". These limitations are ascribed to our tendency towards excessive polarisation of theoretical and empirical approaches, and to the relatively late date of development of historical geography, whose short period of genesis was prematurely terminated - before full reign could be allowed to the scholarly evaluation of data sources - by the advent of scientism in human geography. Harley's view is that the development of new methodologies for the scholarly evaluation and employment of evidence in historical geo-graphy, with particular reference to the language and contexts of that evidence, is of vital importance. This echoes a sentiment expressed some years earlier by Clark (1975) who, in an evaluation of the significance of historical data, regretted that "a consequence of emphasising theory on the part of social or behavioural scientists has been the implication that we should de-emphasise facts and with that de-emphasis, turn away from much of the meticulous training in their discovery, handling and interpretation. This training, traditionally, has been associated with the best historical scholarship".

The danger hinted at is that of throwing away a carefully nurtured empirical baby with the epistemological bathwater. If there was a real danger of this, it probably occurred with the relatively few attempts by historical geographers to associate their research with the quantitative 'revolution'. A sober statement of the potential of a more scientific geography for historical geography was given in the introduction to the volume Geographical Interpretations of Historical Sources (Baker, Hamshere and Langton, 1970), thus: "Empirical analysis has always been overwhelmingly prevalent in English historical geography, and there is no reason to believe that this will not continue to be the case ... the logical conclusion of empirical analysis is the inductive derivation of theoretical principles. One of the major current objectives of human geography is the foundation and testing of formal hypotheses, the anticipated culmination of which is the establishment of dynamic theories of location. The contribution of historical geography to this end is potentially massive, especially if, as seems likely, the systems concept is to provide a framework for the analysis of dynamism". This statement, however, does beg the question of how this analysis is to be conducted. If it is conducted on an empiricist basis, the assumption is made that theories and observations are independent, and that conventional hypothesis - testing methodologies can be used. If this assumed independence is not accepted, then such methodologies cannot be used. The significance of systems theory for the explanation of dynamic processes of change was developed by Langton (1972), who stressed that it was essentially an empirical method (though it may be argued that this is true of systems analysis but not of systems theory).

General systems theory, however, has not been extensively used by historical geographers (see, however, Pred, 1966; Baker and Butlin, 1973; Langton, 1979), in spite of its attractiveness in relation to the complexity of the problems which they sought to analyse, which could not be solved by use of essentially closed-system models to which linear techniques were applied. At this point, however, it has to be said that the concept of historical geography as part of a spatial science, and mainly concerned with the analysis of spatial structure, did prove attractive to a number of practitioners, and continues to do so. A number of attempts, some more successful than others, have been made to apply a variety of mathematical modelling and analytical techniques to historical geography, including graph theory, nearest neighbour analysis, central place theory, the rank-size rule and location theory. Examples have been listed by Gregory (1981a), who also observed that "for some historical geographers, clearly, this was a denial of the classical conception of their sub-discipline, a reduction of its richness and complexity to a one-dimensional contemplation and point patterns". This is not, of course, to deny the utility of powerful statistical and analytical techniques and method- ologies as aids to understanding measurable aspects of the past, if used with considerable care. There are some basic difficulties, however, which relate to this problem, one of which is that the techniques used have not only been inappropriate and inadequate, but in some cases simply wrong in their underpinning mathematics. A debate on the use of cusp-catastrophe theory, (a non-linear and therefore poten- tially very useful mathematical concept, capable of dealing with very much larger numbers of variables than the essen- tially linear techniques so much favoured by geographers in the 1960's and early 1970's), as applied to a study of the changes in settlements in Southern Greece (Wagstaff, 1978, 1979a; Baker, 1979b) was of interest, allowing that dis- continuity in settlement might be perceived by the use of the model. A subsequent review (Alexander, 1979, 1981) indicated that the application of the mode of analysis was probably wrong anyway. More recent work on polis settlements in late prehistoric Greece (Rihll and Wilson, 1985) has indicated further possibilities of modelling settlement structures in the past, with particular reference to the assessment of settlement hierarchy on the limited archaeological and historical evidence of the existence of sites. A useful general review of the geographical analysis of data pertaining to ancient historical research in the classical Greek and Roman worlds prefaces Doorn's (1985) account of the Strouza Region Project in Central Greece.

Regression analysis remains a popular technique among historical geographers, particularly those concerned with the study of aspects of past population. A recent example is the examination of changes in patterns of apprenticeship migration to Edinburgh between 1675 and 1799, using Poisson regression analysis to assess the significance of the size and distance of towns from which migrants came to Edinburgh (Lovett, Whyte and Whyte, 1985).

Another difficulty which attends the use of mathematical models and some statistical techniques is that of submitting limited evidence to rigorous processing via procedures which are appropriate to data-rich contemporary societies but inappropriate to a data-poor past (see Harley, 1973). A major problem is that many of the modelling techniques used relate to static equilibrium positions and not to complex dynamic changes.

Notwithstanding the continuing popularity of historical geography of the classical and neo-classical varieties (the latter still perhaps a minority interest), the limitations of the implicit theory and the methodologies involved have been recognised. While it is difficult to make explicit connections between these types of approach and logical positivist and empiricist philosophies, nonetheless one has to acknowledge the tacit connection, and what that means in terms of the assumptions about the notions of space, evidence and process. Many of the conventional data-related and field-work related studies do assume a posture of attempted 'objectivity' and a neutrality and passivity of place and space, do take a linear/progressive approach to the past ('improvement' - which is in fact an ideologically-loaded concept - being a 'key word' here), and generally do not make explicit the role of theory and the type of past of which the recovery is attempted. In some respects this represents not only a failure to establish a reflexive critique which would result in alternative approaches being developed, but also a failure to follow up H.C. Darby's important statements of the problems of geographical explanation and the nature of the relations of geography and history (Darby, 1953, 1962).

Within the contexts of classical and neo-classical historical geography attempts have been made to incorporate supposedly 'behavioural' and 'phenomenological' approaches, but there has been considerable confusion in the use of these terms, for there is no necessary link between 'behavioural' approaches and a 'scientific', 'positivist' geography, and there is certainly no relationship between such a mode of geographical inquiry and 'phenomenology'. Phenomenological or behavioural methodologies are negations of positivism. Thus Gregory (1976, 1978b) and Billinge (1977) have argued that the 'behavioural' and 'phenomenological' stances which geographers have used have been total misnomers, in the sense that their essence of anti-positivism has been ignored or suppressed.

Attempts to modify a positivist approach in historical geography have been made, using theoretical realism, by Gregory in his book on the West Riding of Yorkshire (1982b). The essential features of this approach include: a concern with careful conceptions of structures and mechanisms (systems of social practice); an interpenetration of 'theoretical' and 'empirical' methods; and an emphasis on the significance of context and of time space structures - to the operation and outcome of processes. The use of realist philosophy (Urry, 1985) does hold considerable promise.

A NEW SOCIAL HISTORICAL GEOGRAPHY

Increasing reaction against an explicit or tacit attachment to the theory and methodology of the natural sciences, on the grounds that they offer models which are unhelpful or harmful to the social sciences and humanities, has been well under way since the early 1970's, though warning anti-positivist shots were fired very much earlier. The rejection of

inductive or hypothetico-deductive reasoning, related to the
assumption that theory and explanation springs virtually
unaided from supposedly neutral and unbiased 'facts' (and in
the case of geography related to a Kantian/Newtonian view of
space) was and is a widespread phenomenon in the social
sciences and the humanities. The alternative, as Skinner
(1985) has expressed it, has been a return to 'Grand
Theory': "From many different directions the cry has gone up
for the development of a hermeneutic approach to the human
sciences, an approach that will do justice to the claim that
the explanation of human action must always include - and
perhaps even take the form of - an attempt to recover and
interpret the meanings of social actions from the point of view
of the agents performing them". The actual form of response
has varied widely, one characteristic within human and historical geography being a turning to the philosophical and
theoretical writings of scholars in other disciplines, notably
social history, sociology, social psychology and politics.
Much-favoured sources of inspiration include the unorthodox
Marxisms of Habermas and the Frankfurt School, the Marxist
theory of the French philosopher Louis Althusser, the histories or 'archaeologies' of the human sciences by Michel
Foucault, the works of the French 'Annales' school of history,
the challenging writings of E.P. Thompson and Raymond
Williams, and, in more recent time, the rise of the feminist
critique and the women's movement. A complex and stimulating
set of debates continues, "and amidst all this turmoil the
empiricist and positivist citadels of English-speaking social
philosophy have been threatened and undermined by successive waves of hermeneuticists, structuralists, post-empiricists,
deconstructionists and other invading hordes" (Skinner,
1985).

As far as human and historical geography are concerned,
the indication of a variety of ways forward has occurred,
inevitably, because of the abilities and insights of a small
group of pioneers, whose work has been of crucial importance, though still perhaps not as widely debated as it should
be. Gregory's Ideology, Science and Human Geography,
published in 1978, was an important critique of the positivist
positions within geography, an advocation and demonstration
of the need for committed and reflexive structural explanation
in geography. An earlier statement by Harris (1971) had
similarly argued, specifically from the viewpoint of historical
geography, that the main objective of historical geography
was regional synthesis and not spatial science. It is important, for contextual purposes, to remind ourselves of
Johnston's judgement (Johnston, 1979), that "from the early
1970's, work by cultural and historical geographers has been
presented as an attack on the positivism of spatial science":
historical (and cultural) geography was back in the centre,

having in some senses been out in the positivist cold for a while.

The new directions in historical geography have varied, and are still not well explored and mapped. They are, however, primarily to do with Marxist, humanist, idealist, structuralist/structurationist perspectives, with related preferences for orientation to regional geography, together with a new form of environmentalism. An important associated innovation is the changing view of space and spatial structure, that is away from the scientistic geographical view of space as a fixed entity which 'organises' and influences societies and economies towards a conception of space as a flexible, socially transferable and organisationally manipulable entity, produced in the image of complex social and material production systems, the most potent of which is capitalism.

The many new directions in historical geography have been described in a group of texts (Dodgshon and Butlin, 1978) and review papers, including: those by Baker (1977, 1978, 1979) and Dennis (1984) in the journal Progress in Human Geography; Prince (1980); the collections of essays edited by Baker and Billinge (1982) and Baker and Gregory (1984); and Butlin's short book and review essay (Butlin, 1982b, 1982c); but an attempt will be made to provide a digest of the principal concepts in which historical geographers are becoming increasingly interested.

The greatest interest has probably been attracted by a wide heterogeneous range of concepts associated with Marxist and humanist perspectives, which do have much in common, however defined or regardless of which of the very many Marxist perspectives is involved, except perhaps for Althusser's structural Marxism, which was a critique of humanism. In essence, both focus on the reality of the human condition, and those processes and forces with which human beings are reflexively engaged, consciously or unconsciously, in the reproduction of their cultural and material environments, from which they have been alienated. Marxist theory involves complex attempts to understand society in a comprehensive way, placing particular emphasis on a materialist conception of history (historical materialism), that is one in which features and structures of a society are seen as the outcome of processes of historical development, closely linked to the prevailing socio-economic structure or modes of production. A basic entity is human labour, and much of the understanding of the historical experience of human labour is to be found in its alienation from that which it produces or the surplus value of that which it produces, that is by its appropriation to limited and largely privileged classes, notably and specifically under capitalism. Central to the Marxist theories is the distinctive mode of production labelled capitalism, with which is associated deprivation, commodification, proletarianisation and immiseration of the majority of

THEORY & METHODOLOGY IN HISTORICAL GEOGRAPHY

the populace, agricultural and industrial, which come under
its sway, the result, as Marx said, of capitalism creating "a
world after its own image". While there has been much
debate, especially among Marxist historians, on the nature,
causes and chronology of the transition from the feudal to the
capitalist mode of production, what perhaps is more important
to geographical studies of the past is the fact that "the
creation of wage labour in an antagonistic class relation to
capital fundamentally altered the way in which society
produced and sought to satisfy its material needs, fundamentally altered the social relations with nature, and therefore fundamentally altered the geography of a given area"
(Harvey, 1981). Marxist theory, in its many and developing
forms, though conceived in the mid-nineteenth century, does
obviously offer an attractive and powerful analytic theory and
method, with its particular emphasis on society, alienation,
and freedom in a complex mesh of theory and historical
materials. The theory offers a 'text' which is open to a
variety of readings, and which will accommodate concepts of
social structure and human agency.
  There is no lack of advocacy of the adoption of Marxist
approaches in human and historical geography: the difficulty
is that there are very few examples of its application in
recent literature. Exceptions are Dunford and Perrons' The
Arena of Capital (1983), and Harvey's The Limits to Capital
(1983), works which Kearns (1984) suggests "propose a
marriage of human geography and capitalism". The Arena of
Capital offers a brave, if limited, attempt, to read the
historical geography of Britain through a particularly
unyielding Marxist framework; Limits to Capital is a mainly
abstract work, though clear focus on substantive issues is
provided by Harvey's The Urbanisation of Capital (1985).
What is necessary, according to Kearns (1984), is "a need to
re-open the dialogue between evidence and concepts in order
to develop the theory of historical materialism at a scale and
in directions more appropriate to the central concerns of
human geography", though it could be argued that this is
exactly what such geographers as Massey, Thrift, Sayer and
others have been demonstrating since the 1970's. Strong
criticism of dialectical materialism of a different kind has been
made by Wagstaff (1979b), who in effect accuses human (and
presumably historical) geographers of selling their birthright:
"the adoption of Marxism by geographers is a confession of
their own failure to generate ideas and methodologies. Not
only is it shameful, it is also the beginning of a road that
can only lead to the victory of 'a barren scholasticism' in our
subject". Few would agree with this curious statement, for
one can be a Marxist and a geographer just as easily as being
a positivist and a geographer. Granted that Marxism has
intellectually a radical consequence for the nature of human
geography, but its concern with praxis is hardly a 'scholasticism' (especially in comparison with catastrophe theory).
30

THEORY & METHODOLOGY IN HISTORICAL GEOGRAPHY

The application of Marxist theory will undoubtedly be one of the most important developments in historical geography. It should be said at this stage that there have been some useful by-products - of the Marxist concept of the central importance of capitalism and of the creation of tension and class-struggle as a condition and consequence - for the historical geographer. Wallerstein's (1974, 1980) lengthy exposition of the growth of a capitalist world-economy from the sixteenth century has provided the basis for a spatial perspective of the possible historical geography of capitalism, and for an extensive debate on both concept and chronology (see Dodgshon, 1977), and there has been an increased interest among historical geographers in the geography of protest and riot, an example of which is Charlesworth's Atlas of Rural Protest in Britain, 1548-1900 (1982), which demonstrates that ordinary people sometimes intervened in their own history, in attempts to free themselves from oppression and deprivation.

The humanistic perspective has a broad base, and has close (though not inevitable or inextricable) links with Marxism. It is a critical and radical philosophy in the sense of rejecting the tenets of positivism and spatial science in geography, though it has been allied to philosophies such as phenomenology, existentialism and perhaps idealism, which were in some respects either constraining, and restrictive or too liberal in relation to the question of freedom of human agency from structural or institutional influences. There are deep and varied roots for the various forms of humanism, including, according to Ley and Samuels (1978): "Jewish, Christian, Muslim, Confucian, Hellenistic, scientific, Marxist, existential" roots, with modern Western humanism owing much to the humanism of the Italian Renaissance of the fourteenth century, but differing from it because of greater emphasis on secular and transcendental meaning. For historical geographers, Harris offered advice on the application of the habits of 'the historical mind', to the practice of humanistic geography (Harris, 1979), this 'habit of mind' being an implicit rather than explicit methodology, exemplified by the work of Sauer and Braudel, and amounting almost to a total cultural/historical geography, steeped in the land and experiential evidence of place in the past. This and other recommendations for historically focussed studies of the role, awareness, value and meaning of human life and experiences in its cultural, political, social, contexts can be readily accepted in principle. Exemplification of this type of approach is provided, inter alia by Guelke's attempt to demonstrate an idealist approach to understanding the past.

Guelke's consistent attempts to introduce an idealist alternative in human and historical geography (Guelke, 1974, 1975, 1982) have not met with much support. Relying on R.G. Collingwood's Idea of History, published in 1946, Guelke seeks to focus attention on the thoughts and actions of individuals in history, and from them to create and understand the theories behind the action. Difficulties arise from the flawed logic of this argument (see Cosgrove, 1984a) and the unconvincing attempt to exemplify this approach in his study of the Dutch settlement of the Cape Colony in the period 1652-1780.

A more promising line of inquiry has been opened by Gregory's (e.g. 1981b, 1982a) advocacy of a structurationist approach in human and historical geography. Linked closely to the ideas of the Cambridge sociologist Giddens, the notion of structuration ties together the voluntarism characteristic of humanistic geography and the notion of constraining/facilitating structures or institutions pertinent to human experiences, such as systems of social practices and structures of social relations. Individuals and groups of 'actors' remain important, but are studied in a context of interaction with structures of social relations and systems of social practice both, consciously and unconsciously, thereby effecting transformation and themselves experiencing changes in view, action and reaction. This structurationist perspective examines the wide questions of structure and agency and the dialectic between action and structure in studies of situated human experiences, which involve the admittance of "the existence of unacknowledged conditions and unintended consequences of action in this way demands a move towards structural explanation distanced from conventional empiricism" (Gregory, 1982a). It is important to note how this set of ideas emerges out of Marxism, and how it is symptomatic of the now very close reciprocal connections between social theory and human geography, for questions of time and space are central to structuration theory. Exploration and development of this complex structure continues, but so far there are few examples which demonstrate ways of working with this system of concepts. One exception, perhaps, is Gregory's Regional Transformation and Industrial Revolution (1982b). A key element, demonstrated in this work, is the significant revival of a theoretically-informed narrative, but a persistent problem is that while the use of narrative to capture place is not generally difficult, its ability to convey spatial structure is more problematic, and this is central to structuration theory.

We may acknowledge the recognition within historical geography of Marxist/humanist perspectives, albeit of differing hues and degrees of conviction. The best examples so far are those which relate to the study of landscape ideas and symbolism and various forms of community study.

The Ideology of Symbolism and Landscape
There has of late been a stimulating growth of work on landscapes, rooted in the Sauerian tradition of historical

cultural geography but more closely related to theory of change in the idea of landscape and the representation of landscape in literature and art. The landscape as ideology, as symbol and as a moral statement are elements of this new cultural/historical geography of landscape. The historical study of landscape as evidence has a long tradition in human and historical geography in Europe, and the work of economic historians like W.G. Hoskins has been of the utmost importance, as has that of J.B. Jackson in reading the meanings of past and present landscapes. It is with the meaning and messages of landscape that much recent work has been concerned. In a major work Social Formation and Symbolic Landscape (1984b), Cosgrove has analysed the history of the meaning of landscape in a materialist/humanist perspective, and proposed a theory of symbolic landscape, which is illustrated by studies from Italy, England and North America. The essence of his argument is "that landscape is a social and cultural product, a way of seeing projected on to land and having its own techniques and compositional forms; a restrictive way of seeing that diminishes alternative modes of experiencing our relations with nature. This way of seeing has a specific history in the West ... in the context of the long processes of transition in western social formations from feudal use values in land to capitalist production of land as a commodity for the increase of exchange value. Within that broad history there have been, in different social formations, specific social and moral issues addressed through landscape images". The origins of the landscape idea as a way of seeing has its origin in renaissance humanism (Cosgrove, 1984b). The intellectual contexts of this work on landscape owe much to the pioneering work of such scholars as Lowenthal, Prince, Yi-Fu Tuan and Relph, and the work of scholars of landscape literature such as John Barrell and Raymond Williams (particularly his book The Country and the City, which portrays the representation of town and country in literature). Interesting recent developments include Daniels' studies (Daniels, 1982a, 1982b) of the 'morality' of landscapes designed by the English landscape gardener Humphrey Repton.

The Historical Geography of Communities

Another interesting, though highly heterogeneous, set of pieces of work related closely to Marxist/humanist perspectives is concerned with the historical geography of groups and communities. The term 'people's history' has been used by social historians to describe this revived emphasis on the historical experiences of the majority of the population (as opposed to the history of the elite and the privileged), though the problematic is very much wider than this (Samuel, 1981; Wolf, 1982). Work by J.T. Lemon, especially his The

Best Poor Man's Country: A Geographical Study of Early South-Eastern Pennsylvania (1972) pointed the way to at least a humanistic approach to the historical geography of communities and regions. More positivist work on 'social areas' within nineteenth-century cities has stimulated a reaction in the form of a more structural/humanist approach. Billinge's (1982) paper advocates an alternative approach to the study of historical communities which pays more attention to human agency, and focuses on class, community and institutions and exemplifies this possibility in a later study of the Manchester Literary and Philosophical Society (Billinge, 1984). This approach is gathering momentum, particularly in urban and 'industrial' historical geography (see Chapter 7 of this book), and also in the study of rural communities, linked in the case of some Swedish-based work, to the concepts of time-geography. The Locknevi project, for example, an interdisciplinary research project looking at the rural society of Locknevi parish, in Småland, Sweden, has resulted in publications which interrogate archival and field evidence by use of various types of theory, including Marxist theory (Fogelvik, Gerger and Hoppe, 1981; Miller and Gerger, 1985). A combination of a theory of place as a historically contingent process, with structurationist theory and time-geography has been used by Pred (1985) in an interesting regional and community-related study of enclosures in Skåne, Sweden, in the eighteenth and nineteenth centuries.

It is clear that studies of historical geography 'from the bottom up', of complex urban and rural societies, have much to offer, and that further studies of such communities, of the rural and urban proletariat are needed. One particularly interesting and exciting development is the application of the idea of a 'people's historical geography' to work in Africa, and the "task which confronts the historical geographer is to understand how [for example] South Africa's brutal landscapes were forged in subordination and struggle, and how the geography of privilege and deprivation has saturated the experience of both overlords and underlings in that society" (Crush, 1986). If the Marxist/humanist/structurationist theories are to be employed, however, they need considerable refinement beyond what Thrift has called 'jumbo Marxism' (Thrift, 1983), and this can only be achieved by a reflexive engagement between theory and problem. The same can be said of time-geography and diffusion theory, which have re-engaged the attention of historical geographers (Pred, 1985; Gregory, 1985). One consequence will be the widening of the range of scales at which research in historical geography takes place, though the rediscovery of the region (Langton, 1979, 1984; Gregory, 1982b) seems to be the strongest possibility.

THEORY & METHODOLOGY IN HISTORICAL GEOGRAPHY

Environmental Perspectives in Historical Geography

The last topic for which space is available for review is that
of the evidence of the revival of an environmentalist
perspective in historical geography, free from the deterministic connotation with which it was once associated. It is
hardly surprising that the bulk of the work has come from
studies of pioneer or frontier societies, where the perception
of and responses to new and unknown environments (and
their inhabitants) led to a wide range of patterns of
occupance. Powell has outlined (Powell, 1977), in a fascinating and important work, and within a humanistic framework, the role of images and image-makers in the settlement
process in the New World, and in a review of work in
Australia has also drawn attention to what he describes as
'the painful and engrossing theme of race relationships'
(Powell, 1981) as a major related theme. The frontier and new
environmental experiences are a constant theme in work on
the United States, Canada and Australia, and well symbolised
in the readings edited by Ward - Geographic Perspectives on
America's Past (1979), Harris and Warkentin's Canada before
Confederation (1974) and the impressive essays in the
festschrift for Andrew Clark (Gibson, 1978). An important
study of the strategy and ecology of the American fur trade
of the Trans-Missouri West has been undertaken by Wishart
(1979), and studies by Christopher (1984) and Harris and
Guelke (1977) have charted man-environment relations in
Africa. Overlap with work on landscape idea and myth has
been interesting, vide Kay and Brown's study (1985) of
Mormon beliefs about land and natural resources, and Olwig's
study of the Danish heathland, which provide a link with the
applied historical geography of Newcomb's Planning the Past
(1979).

An ecological approach in historical geography is
evidenced also in recent work on health, including Dobson's
study of malaria in England (Dobson, 1980) and Kearns' study
of cholera in London (Kearns, 1985). Links with physical
geography and economic history may also be found in a range
of work on changing physical environments and the historical
geography of hazards, especially climatic hazards.

PROSPECT

As far as specific themes are concerned there are a number
of aspects of the past which require attention. One of the
most important is the need for historical geographers to think
(and write) more carefully and specifically about the kind of
past which they seek to create. This has tended to be an
implicit theme, notwithstanding the massive literature on
history and historical writing, and which now should be made
more explicit. Lowenthal's massive work The Past is a Foreign

35

Country (1985) provides some interesting ideas and insights. A related theme is the possible use of narrative as a means and form of explanation in historical geography, fine examples of which are Spate's work on the Pacific (Spate, 1979, 1983) and Mead's Historical Geography of Scandinavia (1981). Daniels (1985) has indicated the appeal of narrative to humanistic geographers, in that "narratives conserve a more seamless sense of the fluency of relations between people and between people and place than do systems or structural modes of temporal explanation" (p.153). The writing of historical geography certainly needs more careful thought in the light of the difficulty of reading and understanding much modern theoretical work: in the craft of writing about the past, much can be learned from reading the prose of Darby, Pounds, Spate, Raymond Williams, E.P. Thompson and the history- and place-sensitive modern novelists such as John Fowles and Graham Swift.

A second theme, deserving of much wider attention, is that of the role of women in the past, which has hardly been mentioned in the literature of historical geography, notwith- standing a massive literature on women's studies and women in history (Hufton, 1983; Scott, 1983). Within this context, advances are to be made not simply from one general feminist perspective, but from many, in relation to differing models of the role of women in past societies, to different scales of study and to differing modes of production and geographical and social situations. The scandalous under-representation of women in academic institutions will not be rectified overnight, but the increase in the number of women undertaking post- graduate research in geography is leading to an increased interest in the feminist perspectives, which ought, in turn, to be reflected more strongly in historical geography. Rose's study (Rose, 1984) of a geography of working-class women in East London in the late nineteenth century is a signpost to a way forward.

Other interesting themes for the future undoubtedly include the historical geography of popular culture, (includ- ing music and art), more intensive studies of the historical geography of regions, communities, and the historical geo- graphy of power.

Over the next decade, we will undoubtedly see the continuation and acceleration of a more problematic approach to the past, one in which the theme of human agency will retain a central position, indicating a surer and more certain acknowledgement, for example, that the landscapes of the past were not empty and devoid of people (as is implicit in some classical studies of historic landscapes), but were made by them as part of their making of history. Closely allied will be the increasing acknowledgement that space was not and is not either template or product, a consequence being a greater concern with spatial structure, that is in the way in which space is implicated in the operation and outcome of processes.

In the sense that historical geography has emerged from the quantitative 'revolution' and a longer-lasting period of positivism and empiricism, in the vanguard of a new post-positivist and theoretically-informed human geography, a crucial consequence has been the increased sensitivity of the whole discipline of geography to historical factors and contextual explanations. As these trends will undoubtedly continue and accelerate, an interesting question is whether historical geography will survive in its present form as a distinctive sub-discipline, for there seems to be a distinct movement, within human geography, away from resolutely historical processes defined temporally towards a more critical human geography for whom contextual and historical considerations are an essential but not a sole concern. The division between historical and contemporary processes and explanations may no longer be a valid or legitimate strategy. This may be especially important for a time of continuing economic depression, for our increasing sensitivity to the uneven and socially unjust allocations and deprivations of western capitalism in crisis, will surely focus our attention on a past in which the use-rights (or lack of them) of individuals and communities become a more central theme. A less comfortable, more critical and sensitive approach to the past is particularly important to offset the deceptive influence of the 'past as therapeutic nostalgia' trend which always accompanies times of economic crisis, notably in Britain. The warning signs have been powerfully expressed by Cannadine (1985), in his analysis of the history of public reaction to periods of economic depression in Britain, which 'suggests a recognisable and distinctive public mood, which has twice come and gone, and which is now firmly entrenched in Britain once again: withdrawn, nostalgic and escapist, preferring conservation to development, the country to the town, and the past to the present. Not surprisingly, the version of the past that catches and crystallises these sentiments is itself as conservative as the prevailing political climate ... The idea of a "national" heritage which is somehow "threatened" and must be "saved" is sometimes little more than a means of preserving an essentially elite culture by claiming - often quite implausibly - that it is really everybody's. The claim is usually accompanied by a highly value-laden version of the past, not so much history as myth, in which there is no room (and no need) for dissent or a different point of view'. One major challenge and duty for an historically sensitive human geography or a continuing historical geography will be to demolish such comfortable and cosmetic myths and replace them with more cogent, credible, realist alternative views, centred on the lived experiences of a wider spectrum of the populace.

Acknowledgement
I am particularly grateful to Mike Heffernan, Denis Cosgrove and Derek Gregory for their helpful criticisms of the first draft of this chapter.

REFERENCES

Alexander, D. (1979) 'Catastrophic Misconception?', Area 11, 3, 228-9

Alexander, D. (1981) 'Letting the 'Castastasis' Out of the Bag', Area 13, 1, 22-4

Andrews, J.H. (1975) A Paper Landscape: the Ordnance Survey in Nineteenth-Century Ireland, Oxford University Press, Oxford

Arendt, H. (1954; English edition 1961) Between Past and Future, Faber, London

Baker, A.R.H. (1977) 'Historical Geography', Progress in Human Geography, 1, 465-74

Baker, A.R.H. (1978) 'Historical Geography: Understanding and Experiencing the Past', Progress in Human Geography, 2, 495-504

Baker, A.R.H. (1979a) 'Historical Geography: a New Beginning', Progress in Human Geography, 3, 560-70

Baker, A.R.H. (1979b) 'Settlement Pattern Evolution and Catastrophe Theory: a Comment', Transactions, Institute of British Geographers, N.S. 4, 3, 435-7

Baker, A.R.H. (1982) Book Review, Journal of Historical Geography 8, 1, 98-9

Baker, A.R.H. (1983) 'Conference Report: IGU Working Group on Historical Changes in Spatial Organisation: Warsaw Symposium', Journal of Historical Geography, 9, 307-8

Baker, A.R.H. (1984) 'Reflections on the Relations of Historical Geography and the Annales School of History' in A.R.H. Baker and D. Gregory (eds.), Explorations in Historical Geography, pp. 1-27

Baker, A.R.H. (ed.) (1972) Progress in Historical Geography, David and Charles, Newton Abbot

Baker, A.R.H., and Butlin, R.A. (1973) Studies of Field Systems in the British Isles, Cambridge University Press, London

Baker, A.R.H., and Billinge, M. (eds.) (1982) Period and Place. Research Methods in Historical Geography, Cambridge University Press, Cambridge

Baker, A.R.H., and Gregory, D. (eds.) (1984) Explorations in Historical Geography, Cambridge University Press, Cambridge

Baker, A.R.H., Hamshere, J.D., and Langton, J. (eds.) (1979) Geographical Interpretations of Historical Sources, David and Charles, Newton Abbot

Biddick, K. (1983) 'Aspects of Medieval Economy and Society:
    Seminar Convened by the Historical Geography Research
    Group at Exeter, July 1983', Journal of Historical Geo-
    graphy 4, 9, 399-401
Billinge, M. (1977) 'In Search of Negativism: Phenomenology
    and Historical Geography', Journal of Historical Geo-
    graphy 3, 55-68
Billinge, M. (1982) 'Reconstructing Societies in the Past: the
    Collective Biography of Local Communities' in A.R.H.
    Baker and M. Billinge (eds.), Period and Place, pp.
    19-32
Billinge, M. (1984) 'Hegemony, Class and Power in Late
    Georgian and Early Victorian England: Towards a Cul-
    tural Geography' in A.R.H. Baker and D. Gregory
    (eds.), Explorations in Historical Geography, pp. 28-67
Bowen, W.A. (1978) The Willamette Valley: Migration and
    Settlement on the Oregon Frontier, University of
    Washington Press, Seattle
Brown, E.H. (ed.) (1980) Geography Yesterday and
    Tomorrow, Oxford University Press, Oxford
Butlin, R.A. (1982a) A Sense of Place, Loughborough Univer-
    sity of Technology
Butlin, R.A. (1982b) 'Historical Geography in the 1970's' in
    A.R.H. Baker and M. Billinge (eds.), Period and Place,
    Research Methods in Historical Geography, pp. 10-16
Butlin, R.A. (1982c) The Transformation of Rural England c.
    1580-1800: a Study in Historical Geography, Oxford
    University Press, Oxford
Cannadine, D. (1985) 'Brideshead Re-Revisited', New York
    Review of Books 32 (20), 19 December, pp. 17-20
Chapman, J., and Harris, T.M. (1982) 'The Accuracy of
    Enclosure Estimates: Some Evidence from Northern
    England', Journal of Historical Geography 8, 3, 261-4
Charlesworth, A. (ed.) (1983) An Atlas of Rural Protest in
    Britain 1548-1900, Croom Helm, Beckenham
Christopher, A.J. (1984) Colonial Africa, Croom Helm,
    Beckenham
Clark, A.H. (1975) 'First Things First' in Ehrenberg (ed.),
    Pattern and Process, pp. 9-21
Claval, P. (1981) 'Géographie Historique', Annales de
    Géographie 90, 669-71
Claval, P. (1984) 'The Historical Dimension of French Geo-
    graphy', Journal of Historical Geography 10, 3, 229-45
Cleary, M. (1982) 'Franco-British Historical Social Geography,
    Journal of Historical Geography 8, 4, 411-12
Conzen, M.P. (1980) 'Historical Geography: North American
    Progress during the 1970's', Progress in Human Geo-
    graphy 4, 549-59
Cosgrove, D.E. (1984a) Review of Guelke's Historical Under-
    standing in Geography, Landscape History, pp. 100-1
Cosgrove, D.E. (1984b) Social Formation and Symbolic Land-

scape, Croom Helm, Beckenham

Cosgrove, D.E. (1985) 'Prospect, Perspective and the Evolution of the Landscape Idea', Transactions, Institute of British Geographers, N.S. 10, 1, 45-62

Crush, J. (1986) 'Towards a People's Historical Geography for South Africa', Journal of Historical Geography 12, 1, 2-4

Daniels, S. (1982a) 'Humphrey Repton and the Morality of Landscape' in J.R. Gould and J. Burgess (eds.), Valued Environments, George Allen and Unwin, London, pp. 124-44

Daniels, S. (1982b) 'The Political Landscape' in G. Carter, P. Goode and L. Kedron (eds.), Humphrey Repton, Landscape Gardener 1752-1818, Sainsbury Centre for Visual Arts, Norwich, pp. 110-31

Daniels, S. (1985) 'Arguments for a Humanistic Geography' in R.J. Johnston (ed.), The Future of Geography

Darby, H.C. (1953) 'On the Relations of Geography and History', Transactions, Institute of British Geographers, 19, 1-11

Darby, H.C. (1962) 'The Problem of Geographical Description', Transactions, Institute of British Geographers, 30, 1-14

Darby, H.C. (1983) 'Historical Geography in Britain, 1920-1980: Continuity and Change', Transactions, Institute of British Geographers, N.S., 8, 421-8

Davidson, W.V. (1974) Historical Geography of the Bay Islands, Honduras: Anglo-Hispanic Conflict in the Western Caribbean, Southern University Press, Birmingham, Alabama

Dennis, R. (1984) 'Historical Geography: Theory and Progress', Progress in Human Geography 8, 4, 536-43

De Vorsey, L. (1980) 'Dating the Emergence of a Savannah River Island: an Hypothesis in Forensic Historical Geography', Environmental Review 4, 2, 6-19

Dobson, M. (1980) '"Marsh Fever" - the Geography of Malaria in England', Journal of Historical Geography 6, 4, 357-89

Dodgshon, R.A. (1977) 'The Modern World-System. A Spatial Persepctive', Peasant Studies VI, 1, 8-19

Dodgshon, R.A., and Butlin, R.A. (eds.) (1978) An Historical Geography of England and Wales, Academic Press, London

Doorn, P.K. (1985) 'Geographical Analysis of Early Modern Data in Ancient Historical Research: the Example of the Strouza Region Project in Central Greece', Transactions, Institute of British Geographers, N.S. 10, 275-91

Ehrenberg, R.E. (1975) Pattern and Process. Research in Historical Geography, Howard University Press, Washington

Fehn, K. (1982) 'Die Historische Geographie in Deutschland nach 1945', Erdkunde 36, 2, 65-70

Fogelvik, S., Gerger, T., and Hoppe, G. (1981) Man, Land-

scape and Society. An Information System, Almqvist and
Wiksell, Stockholm
Gibson, J. (ed.) (1978) European Settlement and Development
in North America, University of Toronto Press, Toronto
Glasscock, R.E. (ed.) (1975) The Lay Subsidy of 1334,
British Academy Records of Social and Economic History,
New Series II, London
Glasscock, R.E. (1978) Review of H.C. Darby, Domesday
England (1977), Journal of Historical Geography 4, 4,
395-7
Gourou, P. (1982) 'Roger Dion, 1896-1981', Journal of His-
torical Geography 8, 2, 182-4
Gregory, D. (1976) 'Re-Thinking Historical Geography', Area
8, 295-8
Gregory, D. (1978a) Ideology, Science and Human Geography,
Hutchinson, London
Gregory, D. (1978b) 'The Discourse of the Past: Phenomen-
ology, Structuralism and Historical Geography', Journal
of Historical Geography, 4, 161-73
Gregory, D. (1981a) 'Historical Geography' in R.J. Johnston
(ed.), The Dictionary of Human Geography, Blackwell,
Oxford, pp. 146-50
Gregory, D. (1981b) 'Human Agency and Human Geography',
Transactions, Institute of British Geographers 6, 1, 1-18
Gregory, D. (1982a) 'Action and Structure in Historical
Geography' in A.R.H. Baker and M. Billinge (eds.),
Period and Place, pp. 244-50
Gregory, D. (1982b) Regional Transformation and Industrial
Revolution, Macmillan, London
Gregory, D. (1985) 'Suspended animation: the Stasis of
Diffusion Theory' in D. Gregory and J. Urry (eds.),
Social Relations and Spatial Structures, pp. 296-336
Gregory, D., and Urry, J. (eds.) (1985) Social Relations and
Spatial Structures, Macmillan, London
Guelke, L. (1974) 'An Idealist Alternative in Human Geo-
graphy', Annals, Association of American Geographers,
64, 193-202
Guelke, L. (1975) 'On Re-Thinking Historical Geography',
Area, 7, 135-8
Guelke, L. (1982) Historical Understanding in Geography: An
Idealist Approach, Cambridge University Press,
Cambridge
Harley, J.B. (1973) 'Change in Historical Geography: a
Qualitative Impression of Quantitative Methods', Area, 5,
69-74
Harley, J.B. (1982a) 'Historical Geography and its Evidence:
Reflections on Modelling Sources' in A.R.H. Baker and
M. Billinge (eds.), Period and Place, Research Methods
in Historical Geography, pp. 261-73
Harley, J.B. (1982b) 'The Ordnance Survey 1:528 Board of
Health Town Plans in Warwickshire 1848-1854' in Slater

and Jarvis (eds.), Field and Forest, (1979), pp. 347-84

Harris, R.C. (1971) 'Theory and Synthesis in Historical Geography', Canadian Geographer, 19, 157-72

Harris, R.C. (1976) 'Andrew Hill Clark, 1911-1975', Journal of Historical Geography 2, 1, 1-2

Harris, R.C. (1979) 'The Historical Mind and the Practice of Geography' in D. Ley and M.S. Samuels (eds.), Humanistic Geography. Prospects and Problems, Croom Helm, London

Harris, R.C., and Warkentin (1974) Canada before Confederation: a Study in Historical Geography, Oxford University Press, New York

Harris, R.C., and Guelke, L. (1977) 'Land and Society in Early Canada and South Africa', Journal of Historical Geography, 3, 135-53

Harvey, D. (1981) 'Marxist Geography' in R.J. Johnston (ed.), The Dictionary of Human Geography, Blackwell, Oxford, pp. 209-12

Hufton, O. (1983) 'Women in History, Early Modern Europe', Past and Present, 101, 125-41

Jelecek, L. (1980) 'Current Trends in the Development of Historical Geography in Czechoslovakia', Historicka Geografie (Prague), 19, 59-102

Johnston, R.J. (1979) Geography and Geographers, Anglo-American Geography since 1945, Arnold, London

Johnston, R.J. (ed.) (1985) The Future of Geography, Methuen, London

Jones, G.R.J. (1976a) 'Historical Geography and our Landed Heritage', University of Leeds

Jones, G.R.J. (1976b) 'Multiple Estates and Early Settlement' in P.H. Sawyer (ed.), Medieval Settlement: Continuity and Change, pp. 15-40

Jones, G.R.J. (1984) 'The Multiple Estate: a Model for Tracing the Inter-Relationships of Society, Economy and Habitat' in K. Biddick (ed.), Archaeological Approaches to Medieval Europe, pp. 9-41

Jones, G.R.J. (1985) 'Multiple Estates Perceived', Journal of Historical Geography 11, 4, 352-63

Kain, R.J.P. (1979) 'Compiling an Atlas of Agriculture in England and Wales from the Tithe Surveys', The Geographical Journal, 145, 225-35

Kain, R.J.P., and Prince, H.C. (1985) The Tithe Surveys of England and Wales, Cambridge University Press, Cambridge

Kay, J. and Brown, C.J. (1985) 'Mormon Beliefs about Land and Natural Resources, 1847-1877', Journal of Historical Geography 11, 3, 253-67

Kearns, D. (1984) 'Making Space for Marx' (Review Article), Journal of Historical Geography 10, 4, 411-17

Kearns, G. (1985) 'Urban Epidemics and Historical Geography: Cholera in London, 1848-9', Historical Geography Research Series, Geo Books, Norwich, 15

Kennedy, B. (1982) Review Article, Journal of Historical Geography 8, 1, 1984, 75-6

Langton, J. (1972) 'Potentialities and Problems of Adopting a Systems Approach to the Study of Change in Human Geography', Progress in Geography, 4, 127-79

Langton, J. (1979) Geographical Change and Industrial Revolution: Coal-Mining in South-West Lancashire, Cambridge University Press, Cambridge

Langton, J. (1984) 'The Industrial Revolution and the Regional Geography of England', Transactions, Institute of British Geographers, N.S. 9, 145-67

Lawton, R. (ed.) (1978) The Census and Social Structure, Cass, London

Ley, D., and Samuels, M.S. (1979) 'Introduction: Contexts of Modern Humanism in Geography' in D. Ley and M.S. Samuels (eds.), Humanistic Geography. Prospects and Problems, pp. 1-17

Lovett, A.A., Whyte, I.D., and Whyte, J.A. (1985) 'Poisson Regression Analyses and Migration Fields: the Example of the Apprenticeship Records of Edinburgh in the Seventeenth and Eighteenth Centuries', Transactions, Institute of British Geographers, N.S. 10, 317-32

McCaskill, M. (1981) Review Article, Journal of Historical Geography 7, 4, 435-8

Mead, W.R. (1981) A Historical Geography of Scandinavia, Academic Press, London

Meinig, D.W. (1978) 'The Continuous Shaping of America: a Prospectus for Geographers and Historians', American Historical Review, 83, 1186-217

Miller, G., and Gerger, T. (1985) Social Change in 19th Century Swedish Agrarian Society, Almqvist and Wiksell International, Stockholm

Newcomb, R.M. (1983) Journal of Historical Geography 9, 4, 396-401

Norton, W. (1984) Historical Analysis in Geography, Longman, London

Olwig, K. (1984) Nature's Ideological Landscape: a Literary and Geographical Perspective on the Development and Preservation on Denmark's Jutland Heath, George Allen and Unwin, London

Overton, M. (1977) 'Computer Analysis of an Inconsistent Data Source: the Case of Probate Inventories', Journal of Historical Geography 3, 4, 317-26

Overton, M. (1985) 'The Diffusion of Agricultural Innovation in Early Modern England: Turnips and Clover in Norfolk and Suffolk, 1580-1740', Transactions, Institute of British Geographers, N.S. 10, 205-21

Pollard, S. (1968) The Idea of Progress, History and Society, Penguin, London

Powell, J.M. (1977) Mirrors of the New World, Dawson,

Folkestone
Powell, J.M. (1981) 'Wide Angles and Convergences: Recent Historical-Geographical Interaction in Australia', Journal of Historical Geography 7, 4, 407-13
Pred, A. (1966) The Spatial Dynamics of US Urban-Industrial Growth, 1800-1914, Massachusetts Institute of Technology Press, Boston
Pred, A. (1985) 'The Social becomes the Spatial, the Spatial becomes the Social: Enclosures, Social Change and the becoming of Places in the Swedish Province of Skane' in D. Gregory and J. Urry (eds.), Social Relations and Spatial Structures, pp. 337-65
Prince, H.C. (1980) 'Historical Geography in 1980' in E.H. Brown (ed.), Geography, Yesterday and Tomorrow, pp. 229-50
Rihll, T.E., and Wilson, A.G. (1985) 'Settlement Structures in Ancient Greece: Model-Based Approaches to Analysis', Working Paper, School of Geography, University of Leeds, 424
Roberts, B.K. (1977) Rural Settlement in Britain, Dawson, Folkestone
Rose, G. (1984) '"For Twopence and Blood and Slavery": A Geography of Working-Class Women in East London, 1880-1890', B.A. dissertation, Department of Geography, University of Cambridge
Samuel, R. (ed.) (1981) People's History and Socialist Theory, Routledge, London
Scott, J.W. (1983) 'Women in History', Past and Present, 101, 141-157
Senda, M. (1982) 'Progress in Japanese Historical Geography', Journal of Historical Geography 8, 2, 170-181
Simms, A. (1982) 'Die Historische Geographie in Gross Britannien; a Personal View', Erdkunde, 71-78
Simms, A. (1982) 'European Perspectives on Rural Land-scape', Journal of Historical Geography 8, 4, 409-412
Skinner, Q. (1985) 'Introduction: The Return of Grand Theory', in Q. Skinner (ed.), The Return of Grand Theory in the Human Sciences, 3-20, Cambridge University Press, Cambridge
Slater, T.R. and Jarvis, P.J. (eds.) (1982) Field and Forest. An Historical Geography of Warwickshire and Worcestershire, Geo Books, Norwich
Smith N. (1984) Uneven Development, Blackwell, Oxford
Spate, O.H.K. (1979) The Pacific Since Magellan: I. The Spanish Lake, Croom Helm, London
Spate, O.H.K. (1983) The Pacific Since Magellan: II. Monopolists and Freebooters, Croom Helm, London
Stanley, M. (1980) 'The Geographical Distribution of Wealth in Medieval England', Journal of Historical Geography 6, 3, 315-320
Thorpe, H. (1973) 'The Lord and The Landscape', in D.R.

Mills English Rural Communities: The Impact of a Specialist Economy, Macmillan, London, 31-82

Thrift, N.J. (1983) 'On the Determination of Social Action in Space and Time', Environment and Planning, D., Society and Space, 1, 23-57

Tosh, J. (1984) The Pursuit of History, Longman, Harlow, Essex

Turner, M. (1980) English Parliamentary Enclosure, Dawson, Folkestone

Turner, M. (1981) 'Arable in England and Wales: Estimates from the 1801 Crop Return', Journal of Historical Geography 7, 3, 291-302

Urry, J. (1985) 'Social Relations, Space and Time', in D. Gregory and J. Urry eds.), Social Relations and Spatial Structures, 20-48

Wagstaff, J.M. (1978) 'A Possible Interpretation of Settlement Pattern Evolution Evaluation in Terms of 'Catastrophe Theory'', Transactions, Institute of British Geographers, N.S. 3, 2, 165-178

Wagstaff, J.M. (1979a) 'Settlement Pattern Evolution and Catastrophe Theory: A Reply', Transactions, Institute of British Geographers, N.S. 4, 3, 438-444

Wagstaff, J.M. (1979b) 'Dialectical Materialism, Geography and Catastrophe Theory', Area 11, 4, 326-332

Wallerstein, I. (1974) The Modern World-System. I, Academic Press, New York

Wallerstein, I. (1980) The Modern World-System. II, Academic Press, New York

Ward, D. (ed.) (1979) Geographic Perspectives on America's Past. Readings on the Historical Geography of the United States, Oxford University Press, New York

Williams, M. (1974) The Making of the South Australian Landscape, Academic Press, London

Williams, M. (1975) 'More is Smaller and Better: Australian Settlement, 1788-1914', in J.M. Powell and M. Williams (eds.), Australian Space, Australian Time, Oxford University Press, London, 61-103

Wishart, D. (1979) The Fur Trade of the American West 1807-1840, Croom Helm, London

Wolf, E.R. (1982) Europe and the People without History, University of California Press, Berkeley and Los Angeles

Wrigley, E.A. and Schofield, R.S. (1982) The Population History of England, 1541-1871, Arnold, London

Wynn, G. (1977) 'Discovering the Antipodes, A Review of Historical Geography in Australia and New Zealand, 1969-75, with a Bibliography', Journal of Historical Geography 3, 3, 251-266

Chapter Two

DATA SOURCES IN HISTORICAL GEOGRAPHY

J.D. Hamshere

INTRODUCTION

The range of data sources available to the historical geographer is so extensive as to defy even the simplest of
descriptions in a chapter of this nature. In the United
Kingdom alone, the tradition of preserving written records
containing an overt spatial element within them dates back
some 1,400 years to the foundation charters of Anglo-Saxon
monasteries. As one approaches the present, the amount of
documentation produced and preserved grows to alarming
proportions. Indeed, central government of the United
Kingdom today produces over one hundred shelf miles of
documentation each year. Of this, less than one per cent is
preserved and not all of that is available for public
inspection. Curiously, the British mania for keeping records
is matched by an equal penchant for restricting access to
them. Hence we have one hundred year rules on manuscript
census material, thirty year rules on Cabinet papers, and the
near impossibility of obtaining records on anything pertaining
to individual income or wealth. At present the most recent
national register of land ownership available for public
inspection is Domesday Book, and undoubtedly studies of the
nineteenth and twentieth centuries are considerably hampered
by the unwillingness of authorities to release in bulk even
such mundane documents as death certificates. This apparent
sensitivity to individual privacy is not so widely appreciated
in many other European countries, where, notably in Sweden,
social data on individuals are readily available for con-
sultation. Similarly in the United States, freedom of information acts theoretically allows access to governmental
documentation, even including the Watergate papers!

Despite the irritations and irrationalities of official
secrecy, it would be impossible for any British historical
geographer to complain about the general availability of data
sources. Archival skills have a long tradition and are expressed throughout a hierarchy extending outwards from the

DATA SOURCES IN HISTORICAL GEOGRAPHY

Public Record Office in London to the County Record Offices and now to the many excellent local history libraries that have recently been established. Few other European countries are so well served, but it is to be hoped current restrictions upon both local and central government expenditure will not inhibit the advance of record handling techniques. In France, a system is in train whereby it will be possible to access centrally by computer the holdings of all departmental record offices. Similarly, the development in the United States of Geographical Intelligence Systems and Artificial Intelligence (Blakemore, 1985) means that, increasingly, data sources can be both stored and accessed by computer. All of this seems a long way from the traditional journey to the archive with its leisurely pace and often gentlemanly hours of opening. Whilst this will no doubt continue for some years yet, the sheer bulk of data now being generated, together with an increasing backlog of unclassified material, will force the introduction of new technology into the storing of historical records. The opportunity of handling the original document may well become a pleasure vouchsafed to very few, and both historians and historical geographers will have to adapt to entirely new modes of data acquisition. The advent in the very near future of machine-readable record texts is but a first step in this process of the evolution and change of primary data sources.

In the past, historical geographers have had recourse to a wider range of data sources than documentary material derived directly from private or public archives. In the period when landscape evolution featured more prominently than at present in the studies of historical geographers, the employment of topographical and field work data was commonly integrated with the more formal documentary sources. It is probably true to say that it was the example of H.C. Darby's (1954-79) extensive study of a single source, Domesday Book, that did much to encourage British historical geographers to centre their studies upon primary documentary materials along lines traditionally employed by historians. Since that time it is noticeable that it has been largely in Britain that historical geographers have produced guides to the use of particular records. The series of monographs initiated by the British Historical Geography Research Group (1979), or Glasscock's (1976) Lay Subsidies of 1334 and Prince and Kain's (1985) guide to Tithe Records are but a few examples of what is becoming a large and growing literature. There are signs that a similar concern for broadcasting the primary sources available for study by historical geographers is being expressed in the United States, with the recent publication of a symposium on new sources for rural history (Parker, 1984).

In France and Germany the 'reflexive critique' that Baker (Baker and Gregory, 1984) feels has powered British historical geography has perhaps been less apparent. As

DATA SOURCES IN HISTORICAL GEOGRAPHY

Claval (1984) and Jäger (1972) demonstrate, historical geographers have tended to employ historical explanation for the purposes of elucidating modern problems. Whilst the close links between history and geography in the training of continental geographers, particularly in France (Clout, 1985) have given human geographers considerable facility in the handling of primary record material, their studies, by and large, have been less overtly source-oriented than has been common in British or North American historical geography.

Historical geographers have generally been less concerned about the distinction between primary and secondary sources than have many historians, yet relatively few works by historical geographers these days are not based around some primary record source. It is unlikely now that the work of historical geographers could be rather dismissively assessed as interesting surveys of the secondary (printed) sources, as has happened in the past. The 'purity' of the source expressed as an absolute insistence on the original rather than a printed or translated version, however, has not been a particularly important issue within historical geography. Historical geographers are generally concerned with the manipulation of material derived from employment of the source rather than an analysis of the source itself. Harley (1982) feels this to be a weakness within the subject and proposes a specific modelling of source materials based upon concepts drawn from linguistics. In this, the code (the source) is distinguished in its context as a message between the addresser (the constructer of the source) and the addressee (the receiver). By this means, it is argued that it would be possible "to perceive and describe the internal structures and tendencies of the evidence we employ". By and large, however, this 'structuralism' of the source material has been taken as little more than the implication that the historical geographer should "read as widely as possible about the source to be exploited, concerning the administration and collection of data and the purposes for which it was intended" (Baker, Hamshere, Langton, 1970).

Clearly, understanding the code of the source is of great importance to its successful exploitation. As Cheney (1973) wisely remarked, "Records, like the little children of long ago, only speak when they are spoken to, and they will not talk to strangers". Thus the language of the source has to be comprehended as do the methods, purposes and motivation of the compiler. It is in pursuit of this language of the source that the historical geographer has to remove himself to some degree from the spatially orientated world of geography into that of textual criticism. Many would no doubt feel that the specific modelling of such processes would represent a departure from the essential nature of historical geography - the acquisition of the language of the source being only a step, albeit a vital one, on the way to posing questions of the source. It is, however, worth developing upon Harley's ideas in order to outline the complexities presented by documentary source materials to the historical geographer. It is for this purpose that the source 'model' shown in Figure 2.1 has been constructed.

Prior to the extraction of data from the source, certain basic questions have to be asked concerning that source's construction. In some cases these questions can only be fully answered once analysis of the source itself has been made, demonstrating the highly interlinked nature of Figure 2.1 The addresser is the person or persons who undertook the original construction of the source. Questions need to be answered concerning their original purpose in making the record, as this is found to proscribe the ultimate contents of the finished document. Clearly we also need to know by what methods the material was collected, whether by structured survey, personal observation or the recording of orally given evidence. Finally, it is useful to know something of the persons concerned, their position within society and the general ideological framework within which they operated. Sometimes such information can be obtained from the many guides to record sources that exist, particularly in Britain, but more often than not, this aspect of research requires enquiry into the general history of the period being studied and the organisational and ideological ethos of its society.

Figure 2.1: Source Model for Historical Geography.

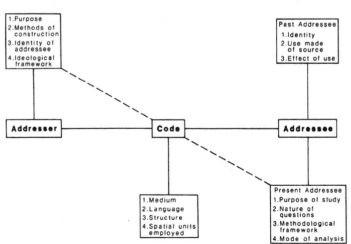

DATA SOURCES IN HISTORICAL GEOGRAPHY

The source material itself (the code) needs to be studied not merely for its data content, but also for the manner in which those contents are transmitted. The medium, be it parchment roll or computer tape, handwritten diary or printed parliamentary report, can affect the message that is being transmitted. Whilst we might not entirely accept McCluhan's stricture that the medium is the message, it is true that the medium does affect both the construction and receipt of the message. Insight can be gained into not only what is included, but also what has been left out. Historians have expended considerable scholastic effort in such studies, and it would be instructive for historical geographers to give closer attention to some of this work, for example to that extensively studied document, Domesday Book (HMSO, 1954, Rumble, 1985). Consideration of the language of the document extends beyond that relating to the problems of translation from, say, medieval Latin or French. Many documents are written in a form of shorthand, as are most medieval documents, which needs to be extended. Even more recent documents written in English are subject to changes in the conventional use of the language or to peculiarities of accounting and book keeping methods. In fact, the actual manner of expression within the document has been widely studied through the agency of content analysis to explore the motives, attitudes and aspirations of the document's compiler. Both the medium and the language are linked to the way in which the information is structured, which is, in turn, linked back to the purposes of the compiler. We need to understand the logic behind this structure in order to gain insight both into the relative importance of categories of information as perceived by the compiler and also the method employed in the data collection itself. In this latter respect it is in the spatial units within which the data are framed that is of great importance to the historical geographer. In a surprisingly wide range of documentary sources these spatial units are never explicitly stated and have to be inferred and reconstructed from internal evidence and other supplementary sources. More will be made of this point later, suffice it to say at this juncture that even amongst the most apparently spatially constructed sources, such as nineteenth century censuses, the reconstruction of the enumeration districts is often far from being straightforward (Lawton, 1982).

The role of the addresses in Figure 2.1 can be divided into two separate functions. Firstly, there is the original addressee, that is, the person or persons for whom the source was actually intended. This clearly is linked to the purpose of the construction of the source but is often not one and the same thing. Many sources, particularly those prepared for government, were never employed for the purposes for which they are originally intended. It is often important for the context of historical studies to understand by whom and how the source material was ultimately used and what effect it might have had in framing subsequent policy. The second category of the addressee is provided by the present user of the source. Whilst it is not the intention to suggest that our studies of the past must be entirely source led, it is nonetheless true that the purpose of our studies must be compatible with the source material. There is no point in asking questions of the source which it cannot answer, nor of invoking analytical methods that require such distortion of the source as to invalidate the answers. The modern addressee's view of the source and of the questions to be posed will, in turn, be coloured by his general methodological or philosophical standpoint, be it aligned anywhere along the spectrum from positivist empiricism to humanistic Marxism. With due attention to all these facets and the intricate system of feedbacks between them, the historical geographer then has to arrive at an appropriate mode of analysis that will reflect not only the intrinsic qualities of the source, but also the needs of his particular study. Clearly, this can be a complex and difficult task, but it is one that supplies much of the enduring fascination of the use of historical data sources.

It has been the nature of the questions that historical geographers should pose of their sources that has been the subject of recent methodological debate. The content of that debate is not the subject of this chapter, save to point out that it cannot be effectively conducted without some reference to the data sources which underpin any historical study. Change in methodology can, however, affect not only the questions asked of the sources, but also the selection of the source itself. One such change occurring in the past has already been alluded to; the change in emphasis from landscape evolution to the reconstruction of past geographies. This led to a change in emphasis from the use of topographical evidence to that of primary documentary sources. The question now remains as to what extent the debate on methodology in historical geography conducted over the past fifteen years has been reflected in changes to the types of source material employed.

DATA SOURCES IN HISTORICAL GEOGRAPHY 1950-1985

In order to ascertain the data sources employed in historical geography, some datum point has to be established. In this case the collection of papers found in Geographical Interpretations of Historical Sources (Baker, Hamshere and Langton, 1970) is used as a starting point. Before 1970, the work of historical geographers was scattered amongst books and journals of a historical, geographical and antiquarian nature, and this publication marked the first attempt to draw

DATA SOURCES IN HISTORICAL GEOGRAPHY

together a corpus of the works of British historical geographers. The advent, in 1975, of a specialist journal devoted
to historical geography, The Journal of Historical Geography,
allows comparison to be made of publication of articles over
some 35 years, spanning the period 1950-1985. Admittedly,
such a review is limited to articles largely derived from the
English-speaking world, but it is felt that, given the impossibility of encompassing all works produced by historical geographers, it does allow some sort of comparative framework.

The twenty articles reprinted in Geographical Interpretations spanned an original publication period of 1950-69
and represented an attempt by the editors to select works
illustrative of the use of British historical sources by
historical geographers. In this sense, the articles were not
necessarily representative of British historical geography in
the period, let alone the works of the continental or North
American scholars. At the time, however, it was noted how
closely the articles reflected the methodological ethos of
British historical geography of the previous twenty years. It
was appropriate, for instance, that the first article reprinted
utilised Domesday Book as its main source (Darby, 1950), for
it was Professor Darby's work that did more than most to
influence the research methods of historical geographers in
Britain. The data sources employed by the various authors in
Geographical Interpretations and in the Journal of Historical
Geography have been grouped into a broad classificatory
system for purposes of comparison in Tables 2.1 and 2.2. The
typology is dependent partly on the date of source material
and partly upon its provenance and purpose to which it has
been put. The system is thus not one that would be recognised by archivists.

Only one article in Table 2.1 made use of relict landscape features as a principal source, although in percentage
terms this probably under-represented the popularity of this
type of study in the period before 1970. Characteristically,
this type of study utilised early maps, particularly the first
editions of the Ordnance Survey one inch and six inch maps,
as a source from which distributions of relict features could
be established. Given prominent attention were settlement
morphologies (particularly green villages), place names,
moated sites, boundaries and field evidence of such features
as ridge and furrow as aspects of the morphogenesis of the
landscape. The basic data sources were thus removed in time
from the landscape that was being recreated and recourse was
made to such contemporary documentation as existed in order
to check such things as the early spelling of place names or
the population structure in Domesday Book. Such studies
were paralleled on the continent, particularly in France and
Germany (Smith, 1967), although they were rare in North
America and Australasia. In Sweden, beginning in the 1960s,
there has been a tradition of placing such studies within the
context of model construction or simulated processes of dif-

Table 2.1: Data Sources in Geographical Interpretations.

	Source Type	Date Range	Study Areas
1.	Topographical/Archaeological Relict features, O.S. maps	13th–14th centuries	England
2.	Medieval Domesday Book, Hundred Rolls Subsidies, Taxations	1086–1342	England, Warwickshire E. Anglia
3.	Parish Registers, Indentures, etc.	1538–1837	Colyton (Devon)
4.	Surveys and Estate Records Tithe Awards, Agricultural Reports Farm Accounts, Crop Returns	1794–1900	Chilterns, Wales, Leeds
5.	Local Government and Court Records Guild Rolls, Probate Inventories	1540–1682	Preston, E. Worcestershire
6.	Central Government Records Customs Registers, Taxation (property, income and rates), Parliamentary Reports, Boundary Commissioners Reports, Factory Inspectorate	1735–1900	England, S. Lincs., Kent, E. Midlands, Lancs., Chilterns
7.	Company Records Account Books	1690–1717	Midlands
8.	Directories	1828–9	Wales
9.	Census Enumerators Books	1851	Liverpool

fusion and colonisation (Byland, 1960). In more recent years morphogenetic studies of the landscape have continued to flourish with the addition of techniques such as metrology and have become particularly associated with the work of the Conference Europeenne Permanente pour l'Etude du Paysage Rural (Roberts and Glasscock, 1983). The interlocking of topographical evidence with archaeological findings supported by relevant documentary evidence has now become very much an inter-disciplinary pursuit, with contributions from archaeologists, historians and place-name experts as well as historical geographers, as is revealed by such recent publications as that on Medieval Villages (Hooke, 1985).

Medieval source materials were widely exploited by historical geographers in the period before 1970, with an emphasis on those sources that yielded quantitative rather than qualitative data. Thus Domesday Book, Hundred Rolls and taxation records tended to be more widely used than more qualitative sources such as manorial court records or chronicles. This was no doubt partially due to the palaeographic and linguistic problems associated with much qualitative medieval documentation, which undoubtedly restricted their widespread employment by historical geographers. The problems associated with the interpretation and exploitation of medieval source material in general resulted in the inclusion of a specific discussion of the source material itself as an important part of most studies. Such discussion was less common in studies based upon more recent data sources, although a similar focus of interest upon population and agricultural resources was apparent. Another common feature was that most data sources stemmed from the public domain of central government. Although the records employed were usually national in coverage, they were predominantly analysed at a regional or local level. As with medieval sources, the majority of the later records were quantitative in form, with the statistical content of the documents structured within identifiable spatial units, which were most commonly comprised of ecclesiastical parishes (parish registers, crop returns) or enumeration districts (census). As can be seen in Table 2.1, private records were rarely used, Johnson's (1981) use of early company records for a study of the charcoal iron industry being exceptional at the time.

The inbuilt spatial aspect of so many of these record types lent itself to the construction of distribution maps, and it is no surprise that the mapping of the historical data derived from the sources formed a major element of most studies. The 20 articles comprising Table 2.1 included 120 maps; only one article based upon parish registers (Wrigley, 1966) did not employ any maps as part of the analysis. Clearly, in the 1950s and '60s, historical geographers viewed the construction of distribution maps as the main means of exploiting their data sources - a methodology that lent itself to the reconstruction of past geographers. Geographical change over time was largely conducted by comparison of a series of distribution maps constructed for different time periods.

Whilst critics of the empiricism of such methods have perhaps under-estimated the amount of attention devoted to temporal change (e.g. Harley, 1958), it is true that the statistical analysis of the data was largely limited to the methods commonly employed in cartographic construct. Pie charts, proportional circles, bar graphs, outweigh any other method of analysis in the frequency of their employment. Time series graphs, histograms and tables were utilised to depict temporal change but, overall, statistical methods remained simple in what was a largely pre-computer period. D. Harvey's (1963) use of regression analysis on hop excise data was exceptional in historical geography, although by 1970 quantification and model building were already widely employed within human and economic geography.

The above discussion has been able to give some impression of the data sources employed in British historical geography in the period 1950-1969. The years since 1970 have witnessed a more widespread application of computers, an increase in the availability of late nineteenth century sources, notable in access to census material, and an increasing level of debate on the methodology of historical geography. All of these have implications for the type of source materials used in historical geography as well as for the methods employed for their analysis. The Journal of Historical Geography, launched in 1975, provides a useful source for comparison with earlier work. It must be borne in mind, however, that the sample is much larger and the articles encompass a much larger field of international scholars than appear in Table 2.1. The original transatlantic nature of the Journal has been progressively expanded to incorporate the work of other European, Australasian and even Chinese and Japanese scholars. It is thus possible to gain a reasonable impression of the range of data sources selected for study within a broad field of historical geographers.

The far larger sample provided by the Journal of Historical Geography yields a more extensive range of source materials with a wider international provenance than characterised Geographical Interpretations. Even so, it is noticeable that those sources rich in quantitative data, such as census, survey and taxation records, continue to be widely used. A significant change, however, has been the increased use of more qualitative sources; diaries, correspondence, literature, sales catalogues and landscape paintings have been increasingly exploited as principal sources and not merely as descriptive adjuncts to a more quantitative source. Some of this can be ascribed to the Journal's transatlantic influence, as such sources have long been employed amongst North American historical geographers to explore both the real and

Table 2.2: Data Sources in Journal of Historical Geography 1975-85.

Source Type	Date Range	Study Area
1. Topographical/Archaeological Place names, relict features, field evidence, pollen analysis	3rd century-19th century (predominantly early medieval) (No. of studies - 12)	Kent, Sussex, I.O.W., Ireland, Warwickshire, Notts., Mongolia, England & Wales, Cotswolds, Peru, Florence, En Sweden
2. Medieval Anglo-Saxon Charters, Law Codes, Domesday Book, Cartularies, Lay Subsidies, Muster Rolls, Calendars, Inquisitions Post Mortem, Poll Taxes	400-1550 A.D. (No. of studies - 10)	Sussex, Cotswolds, England & Wales, E.Anglia, Notts., W.Midlands, Odenwals, Ireland, N.E.Polard
3. Parish Registers, Indentures Apprentice lists, Migration registers, Immigration doc., Military conscription records	1486-1920 (predominantly 18th-19th centuries) (No. of studies - 17)	London, Cambridge, England, Kent, E.Anglia Laggon, Dundee, Perth, Stirling, Sweden, N.W.Iceland, France, Loir et Cher, N.Brabent, Minnesota, Upper Canada, Fall River (U.S.A.)
4. Colonial Records Reports, Surveys, Dispatches, Consulate Reports, Treaties and Statutes	1580-1940 (predominantly 16th-18th centuries) (No of studies - 12)	Ontario, New England, Caribbean, N.E.Brazil, Yucaton, Guatemala, Mexico, Guyana, S.Africa, Palestine, India, Zimbabwe

5. Surveys and Estate Documents
Tithe Awards, Cadastre, Hacienda Records, Farm Accounts, Leases, Enclosure Awards, Terriers

1598-1900 (predominantly 17th-18th centuries) (No. of studies - 20)

England, Kent, Cirencester, Holderness, Huddersfield, N.England, Oxon, London, Vale of Evesham, Grampians, Edzell, Edinburgh, Wales, Ontario, New England, Mexico, Scania

6. Travel and Exploration Documents
Itineraries, Accounts, Correspondence, Timetables, Guides, Turnpike trusts, Stagecoach services

Roman - 1922 (predominantly 18th-19th centuries) (No. of studies - 13)

England & Wales, S.Hants., New England, Mid West U.S.A., Great Plains, West U.S.A., Virginia, Jamaica, Mali, Jerusalem, Pannonia

7. Local Government and Court Records
Probate Inventories, Criminal Courts, Boards of Supervisors, Police Commissioners Records, Land Registration Documents, Officers & Board of Health Records, Corporation Records, Boards of Works, Assessment & Rate Records, Building Applications, Loan & Mortgage Documents, Land Archives

1580-1960 (predominantly 19th century) (No. of studies - 15)

Liverpool, London, Kensington, Leeds, Northampton, Watford, Birmingham (U.K.), San Francisco, Kingston (Ontario), Guadelejara, Peru, Palestine

Table 2.2: (continued)

Source Type	Date Range	Study Area
8. <u>Central Government Records</u> Royal Commissions, General Surveys, Senate Executive Documents & Immigration Reports, Board of Trade Reports, Federal Commissions, Parliamentary Debates & Reports, Annual Statistical Reports, Excise Records, Factory Inspectorate, Social Statistical Reports, Forestry Commission Reports, Crown Surveys, Taxation Records	1560-1938 (predominantly 19th century) (No. of studies - 18)	England, Lancashire, Herts. & Shropshire, S.Wales, U.S.A., Midwest, Fall River, Herault (Fr.), Ireland, N.E.Poland, U.S.S.R., Nelson Colony (N.Z.), New Zealand
9. <u>Census</u> Reports, Enumerators Books, etc.	1780-1961 (predominantly 19th century) (No. of studies - 30)	U.K., England & Wales, W.Yorkshire, Liverpool, Cambs., Huddersfield, Leeds, London, Dundee, Perth, Stirling, Grampians, Loir et Cher, U.S.A., Baltimore, Washington, Boston, Charleston, Minnesota, Kansas, Virginia, N.E. U.S.A., Mid-West, Philadelphia, Guyana, Ecuador, Victoria (Aust.), S.Africa

10. Directories, Newspapers Sales Catalogues

1769-1984 (predominantly 19th century) (No. of studies - 15)

U.K., England, England & Wales, Manchester, S.Hants., E.Anglia, Cirencester, Lancs., Oxon., Herts. & Northampton, Fall River (U.S.A.), Charleston, Milwaukee, Wisconsin, New England, Venezuela, Nelson Colony (N.Z.), Palestine

11. Company Records and other Private Organisations
Reports, Correspondence, Orange Order, Church Records, Cotton Spinners Operatives, Chamber of Commerce, Syndrical Bulletins

1652-1970 (predominantly 19th century) (No. of studies - 16)

U.K., England & Wales, London, Lancashire, Scotland, Argylshire, U.S.A., Mid-West, Fall River, Utah, Hudson Bay, Canada, Ontario, Gualdajona, Preague, Cape Colony France

12. Literature, Private Correspondence Diaries, Landscape Paintings

1606-1984 (predominantly 19th century)

England, Weald, Leeds., Wessex, N.W.England, U.S.A., Wn U.S.A., California, Virginia, Iowa, Great Plains, Upper Canada, Jamaica, Patagonia, Nelson Colony (N.Z.), Palestine, Venice, Mali imagined worlds of the frontier. This type of study has been extended to Australasia and to the old Spanish and Portuguese colonies in Middle and South America. In the latter cases, the rich archives located on both sides of the Atlantic have only just begun to be exploited. The use of narrative, descriptive and even pictorial sources is neither new to historical geography nor limited to studies of colonisation. Their use has, however, been extended in recent years by the introduction of humanistic-phenomenological approaches to perception studies of the past. If these paths continue to be explored, then we can no doubt look forward to the employment of 'a panoply of sources, which embraces all cultural symbols surviving from the past, including for example drawings and dance, literature and landscape, music and memory, paintings and poems, rituals and relics' (Baker, 1977).

Amongst historical geographers in the United States there has been a somewhat different practice in the use of primary source material than has characterised their British counterparts. The National Archive for the United States was opened as late as 1935, and only ten years ago Herman Friis (1975) still felt it necessary to extol the virtues of discovering 'the original in preference to accepting on faith the published record'. Whilst American historical geographers writing in the Journal of Historical Geography employ a wide range of primary or manuscript sources, their studies are often set within the context of an extensive survey of the secondary literature. There is thus a greater tendency for the analysis stemming from the primary source to be placed within some theoretical construct and related to a larger spatial setting than many equivalent British or European studies. On the whole, British and European historical geographers have continued to study more narrowly defined 'laboratory regions', as A.H. Clark (1972) called them, in which the analysis of a primary source, or set of sources, forms the major part of the study. In this sense the work of many British historical geographers has remained more 'source' orientated than their transatlantic colleagues.

The relative frequency of use of data sources employed in Geographical Interpretations or the Journal of Historical Geography is shown in Figure 2.2. The source materials have been arranged into chronological groupings and displayed as a percentage of all source materials used in the two bodies of work.

Despite the difference in size between the two samples, the general trend of a relative increase in the use of nineteenth and twentieth century source materials can be confirmed by other publications such as the Register of Research in Historical Geography (Wild, 1980; Whyte, 1984), or registers of theses presented and in preparation. The relative decline in the use of medieval and earlier sources no doubt

DATA SOURCES IN HISTORICAL GEOGRAPHY

Figure 2.2: Comparison of Sources Used in <u>Geographical Interpretation</u> and <u>Journal of Historical Geography</u>.

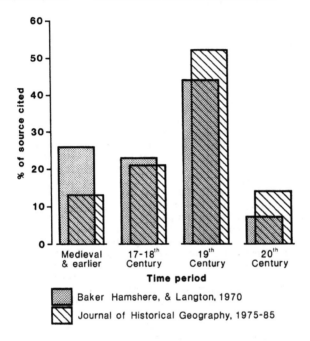

partially reflects the decline in the teaching of Latin in schools and the increasingly multi-disciplinary nature of studies using archaeological and topographical sources. Much of the increase in nineteenth century studies has been generated by the use of manuscript census returns which form the single most extensively used source in the <u>Journal of Historical Geography</u>. No doubt, analysis of census material has been greatly facilitated in recent years by the increased accessibility of computers. The increase in the study of twentieth century source materials has largely been generated by the exploitation of local government and private company archives. In terms of overall balance, this has resulted in a change in emphasis from concern for rural society and land-scape, a dominant theme of historical geography before 1970, to one based upon urban and industrial society. Concomitant with this has been the search for theoretical frameworks, often derived from urban sociology. The reconstruction of past geographies, however, still retains a very significant presence.

DATA SOURCES IN HISTORICAL GEOGRAPHY

MAPPING HISTORICAL DATA

It was noted in the analysis of Geographical Interpretations that mapping of historical data was the principal method employed in exploiting the source material. The articles averaged six maps each, with only one having no maps at all. Since 1970 it has been suggested that there has been a 'retreat from maps' (Balchin, 1972) in geography in general, and to some extent this is discernible in historical geography. In the 64 articles using primary source material in Journal of Historical Geography between 1982-85, an average of just under 4 maps was employed per article. The incidence of articles using no maps, however, rose to 18 per cent and a further 20 per cent used only one map, often simply to locate the study area. The construction of maps from historical data has thus become more polarised, although the majority (62% of articles) still regard it as a vital aspect of the exploitation of their source materials.

In recent years some attention has been focussed on the process of cartographic communication (Board and Taylor, 1977), in which the relationships between the constructer, the viewer of the map and the 'real world' have been explored. In terms of the history of cartography these ideas have been developed by Blakemore and Harley (1980), but little attention has been devoted either to the manner in which historical geographers construct their maps or to the perception of the past that these maps portray. Generally, the construction of maps from historical data sources poses problems over and above those associated with most forms of cartographic representation. In addition to the usual problems of the suitability of symbols, shading and class intervals, there are specialist problems associated with particular data sources being studied. Some of these have been identified in the source model in Figure 2.1. Of particular importance for mapping historical data is the identification of the spatial units within which the source material was originally collected. On many occasions this poses great difficulties of reconstruction and may not even be standard throughout the source material. A well-known example is provided by Domesday Book with its great variety of spatial units, none of which can be directly reconstructed from the source itself. Such problems are not limited to medieval source materials, for Lawton (1982) has pointed to the difficulties of integrating nineteenth century demographic data into a single special framework. The historical geographer thus has to produce spatial units that reflect the data source and yet produce meaningful spatial variations. This task is often a difficult one and should always be carefully explained, for the eventual distributions produced are predictated by the choice of the spatial units within which the data have been analysed.

DATA SOURCES IN HISTORICAL GEOGRAPHY

A further problem for the historical cartographer is the lack of standardisation of units of measurement employed in many historical sources. Whilst this problem is most pronounced in medieval documentation, it is by no means limited to it, as those who have struggled with 'baskets' or 'cartloads' as measures of industrial or agricultural production will testify. Considerable ingenuity has often to be applied in converting these measures into a cartographic form. A prime example of such ingenuity is provided by Darby's (1977) map of Domesday Woodland, where great skill has been used to produce a meaningful visual image from difficult and seemingly intractable data. The different woodland measures employed by the Domesday Commissioners are not strictly comparable, yet Darby has produced a map that not only retains the integrity of the original data, but also conveys an excellent impression of the distribution of woodland; in 1086. It is, however, still only an impression and not the actual distribution that is being displayed. In this sense, Darby's woodland map, along with most other examples of historical cartography, provides maps of data derived from particular sources rather than of 'real world' historical situations. Once again the pervasive influence of the source materials is emphasised.

It is surprising that in the methodological debates current in historical geography virtually no attention has been given to the mapping of historical data. As has been demonstrated, it is the most commonly used mechanism for displaying and analysing data and has powerfully affected the manner in which we visualise the past. A limited attempt to categorise 'historical-geographical maps' has been made by Simms (1982). Two broad categories are identified; the synchronic or temporally cross-sectional type of maps and the diachronic, which display change through time. This provides only a very simplistic typology more appropriate to the rather dated debate in historical geography concerning the cross-sectional recreation of past geographies as against the study of change through time. It does little justice to the vast range of historical cartography extending from attempts to recreate, cartographically, past landscapes in their topographical entirety (Harvey, 1985) to schematic representations of San Francisco following the 1905 disaster (Bowden, 1982). There is neither time nor space here to develop the theme of the role of historical cartography, but clearly there are important questions to be answered concerning the various styles of cartography employed by historical geographers and the impact these have had on our perception of the past. One suspects that in British historical geography, at least, the style of historical cartography employed in recent years has been considerably influenced by the work of Darby and Versey (1952-77).

DATA SOURCES IN HISTORICAL GEOGRAPHY

Before leaving the subject of the mapping of historical data, mention must be made of the historical geographical atlas. Apart from one or two notable exceptions (for example, the Atlas of Islam (Brice, 1981)), little interest has been shown by British historical geographers in producing maps of historical data in atlas form, and the work has been largely undertaken by archaeologists or historians. Amongst the better examples of these, for instance in the Atlas of Anglo-Saxon England (Hill, 1981), many of the maps produced are stylistically very similar to those produced by historical geographers. In Britain there are signs that historical geographers are beginning to revive interest in the atlas with the planning of an atlas of the Industrial Revolution (Withers, 1985). In both North America and on the continent, historical geographers have played a more active role in the production of historical atlases. In the United States a project for the mapping of the North American Plains is being undertaken at the Center for Great Plains Studies at the University of Nebraska - Lincoln, from which the Atlas of the Lewis and Clark Expedition has already begun publication (Moulton, 1983). Another recent example of regional historical atlases is that of California Patterns (Hornbeck et al., 1983). In Canada, the problems of producing maps of historical data for the Fifth edition of the National Atlas of Canada have been outlined by A.W. Wilson and D.S. McKay (1978). In Europe, German historical geographers have been prominent in the production of regional atlases of the Federal Republic (Mortensen et al., 1968) and in the proposed Atlas of Ireland (Simms, 1982). It is indicative that most of these projects are interdisciplinary ones in which historical geographers are playing an important role.

STATISTICAL ANALYSIS AND COMPUTING OF HISTORICAL DATA SOURCES

Writing in 1970, it was possible to look forward to a time when statistical procedures such as vigorously formulated sampling designs, variance and regression analysis, significance tests and various types of numerical taxonomy would be widely employed within historical geography (Baker, Hamshere and Langton, 1970, 18-19). To some extent this expectation has been realised, as examples of the use of all the above techniques and more can be found within historical geography. It is still, however, only a relatively small minority of historical geographers who employ such statistical techniques. The reasons for this can only partly be assigned to the limitations imposed by the data sources. Certainly some historical data do restrict the range of statistical techniques that can be effectively employed. The use of parametric techniques, for example, requires normally distributed data and accurate measures of central tendency and dispersion - a condition not found in many historical data sources. Similarly, the methods employed in the construction of many medieval sources have resulted in a degree of multicolinearity in the data that severely restricts the significance of such techniques as multiple-regression (Hamshere and Blakemore, 1976). Deficiencies of the data source can offer only limited explanation for the restricted employment of more advanced statistical techniques and reasons must be sought in the methodological preferences of the practitioners of historical geography. This can be expressed as a preference for small-scale aerial studies rather than large, a preference for the more empirical 'real world' historical reconstruction rather than 'timeless' comparison in geographical space. The increasing usage of more qualitative data sources revealed in Table 2.2 has also had an impact in restricting the employment of statistical techniques, but even here there are few signs of the adoption of the more rigorous forms of qualitative analysis found elsewhere in the social sciences.

The progress of computerisation has similarly been rather hesitant and patchy. The widespread access to computers from micros to mainframe found in most academic institutions has, as yet, failed to realise a large crop of computerised studies of historical data. The use of computers in small-scale studies by the application of standard statistical packages to small data sets, however, is probably more common than a review of historical geography literature at first sight suggests. Such procedures are relatively common in geographical research in general and as such might well fail to find mention in the text of the more statistical of historical geographical articles. Thus, most studies employing statistical procedures such as correlation, regression and Monte Carlo simulations, have probably employed this type of technology. This, however, is simply employing a computer as a sophisticated form of calculator and is very different from utilising the full range of computer technology to store large data sets, analyse them and to generate computer maps. Outside the work of the Cambridge Centre for Population Research, it is difficult to think of any project involving historical geographers that has utilised computers for data storage on such a large scale. Most uses made of computers involve the selection of data from a historical record and imply some loss of material in the process - usually, the statistical data elements of the source being retained at the expense of the qualitative. There are, however, indications that this is about to change. It is anticipated that in the near future Professor Warren Hollister's research team at the University of Santa Barbara will produce a machine-readable text of Domesday Book. This will offer many advantages not only in analysing the statistics, but also the general content of the text by the use of general purpose computer packages such as the Oxford Concordance Program (Hockney and Marriot, 1982). As Palmer (1985) has pointed out, 'in future, only masochists will approach Domesday's statistics without the aid of a computer'.

The success of such schemes to produce machine-readable records is going to be in part dependent upon the development of computer packages enabling access to the data base. In the future it is possible to envisage central computer holdings of manuscript census returns which will be accessed by the development of packages similar to SASPAC, which is currently employed for modern census returns. Whilst this might be in the future, there are already in existence software packages enabling the storage and manipulation of historical data. One such package deals with nineteenth century enumerator books and allows simple sorting and analysis of the data (Mills, 1985). At present such packages have largely been developed for teaching purposes and have only a restricted research potential. Historical geographers are well placed to aid and benefit by such developments and an increase in commitment in these directions is going to be necessary if we are to avoid technological obsolescence.

CONCLUSION

The foregoing discussion has only been able to highlight certain aspects of data sources in historical geography and has concentrated on the English-speaking world. It has been possible to demonstrate that since 1970 important changes have occurred in the type of data sources employed and in the methods used to analyse them. Nineteenth and twentieth century sources have been increasingly exploited as have the more qualitative types of source material. There has also been a refreshing 'internationalisation' of interest within the subject as linkages between sources derived from different countries are being increasingly forged. Despite these changes, the main stream of historical geographical research continues along lines already well established by 1970. Both the type of data source and the mode of its exploitation remain very similar to those found in works of the period 1950-1970. Methodological debate would therefore seem to have had only a limited impact on the relationship between historical geographers and their source material. There are some signs, however, that this relationship is about to be disturbed, not so much by methodological debate within historical geography, but by forces external to it. Technological change, restricted research funding and low levels of recruitment of young research workers will inevitably lead to a pooling of resources with other like-minded academic disciplines. Historical geographical research could thus become part of large-scale projects that are both multidisciplinary and

DATA SOURCES IN HISTORICAL GEOGRAPHY

spread across several academic institutions similar to the type
of research organisations established in France (Clout, 1985).
Indeed, there are already signs of this occurring with the
establishment of interdisciplinary Centres for Anglo-Saxon and
Nineteenth Century Studies in some British universities.
All such change carries in train something of a Faustian
bargain. The submission of the independence of personal
research to the discipline of team research will not be to the
taste of many. Similarly, there is a danger that the distinctiveness of historical geography will be swallowed up in the
interests of wider historical research. Herein lies perhaps the
greatest challenge of historical geography. Is it possible to
develop a methodological framework that continues to allow the
subject a distinctive identity, yet is not so esoteric that the
skills of its practitioners are of no use to multidisciplinary
team research projects?

REFERENCES

Baker, A.R.H. (1977) Conference Report, Journal of Histori-
    cal Geography 3, 3, 301-5
Baker, A.R.H. and Gregory, D. (1984) (eds.) Explorations
    in Historical Geography: Interpretive Essays, Cambridge
Baker, A.R.H., Hamshere, J.D. and Langton, J. (1970)
    Geographical Interpretation of Historical Sources, Newton
    Abbot
Blakemore, M. (1985) 'Cartography and Geographical Infor-
    mation Systems', Progress in Human Geography 9, 4,
    566-574
Blakemore, M.J. and Harley, J.B. (1980) 'Concepts in the
    History of Cartography', Cartographica, 17, 1-120
Board, C. and Taylor, R.M. (1977) 'Perception and Maps:
    Human Factors in Map Design and Interpretation', Trans-
    actions, Institute of British Geographers 2, 1, 19-36
Brice, W.C. (ed.) (1981) An Historical Atlas of Islam, Leiden
British Historical Geography Research Group (1979) Historical
    Geography Research Series
Bylund, E. 'Theoretical Considerations Regarding the Distri-
    bution of Settlement in Inner North Sweden', Geografiska
    Annaler, 42, 225-31
Cheney, C.R. (1973) Medieval Texts and Studies, Oxford,
    p. 8
Clark, A.H. (1972) 'Historical Geography in North America' in
    A.R.H. Baker (ed.), Progress in Historical Geography,
    Newton Abbot
Claval, P. (1984) 'The Historical Dimensions of French Geo-
    graphy', Journal of Historical Geography, 10, 229-45
Clout, H. (1985) 'French Geography in the 1980s', Progress
    in Human Geography 9, 4, 474-490

DATA SOURCES IN HISTORICAL GEOGRAPHY

Darby, H.C. (1950) 'Domesday Woodland', Economic History Review, 3, 21-43
Darby, H.C. (1977) Domesday England, Cambridge, 183-185
Darby, H.C. and Versey, G.R. (1952-77) Domesday Geography of England 7 vols. Cambridge
Friis, H. (1975) 'Original and Published Sources in Research in Historical Geography a Comparison', in R.C. Ehrenberg (ed.), Pattern and Process - Research in Historical Geography, Washington, 1975, 139-159
Glasscock, R.E. (ed.) (1975) The Lay Subsidy of 1334, London
Hamshere, J.D. and Blakemore, M.J. (1976) 'Computerising Domesday Book', Area 8, 4, 289-294
Harley, J.B. (1958) 'Population Trends and Agricultural Developments from The Warwickshire Hundred Rolls of 1279', Economic History Review, 11, 8-18
Harley, J.B. (1982) 'Historical Geography and Its Evidence: Reflections on Modelling Sources', in A.R.H. Baker and M. Billinge (eds.), Period and Place, Cambridge, pp. 261-273
Harvey, D.W. (1963) 'Location Change in the Kentish Hop Industry and the Analysis of Land Use Patterns', Transactions of the Institute of British Geographers, 33, 123-44
Harvey, P.D.A. (1985) 'Mapping the Village: The Historical Evidence', in D. Hooke (ed.), Medieval Villages, Oxford 47-60
Hill, D. (1981) An Atlas of Anglo-Saxon England, Oxford
HMSO (1954) Domesday Book Rebound, HMSO for the Public Record Office, London
Hockney, S. and Marriot, I. (1982) Oxford Concordance Program: A User Guide, Oxford University Computing Service
Hooke, D. (ed.) (1985) Medieval Villages, Oxford University Committee for Archaeology, 5, Oxford
Hornbeck, D., Kane, P. and Fuller, D.L. (1983) Californian Patterns: A Geographical and Historical Atlas, Palo Alto, California
Jäger, H. (1972) 'Historical Geography in Germany, Austria and Switzerland', in A.R.H. Baker (ed.), Progress in Historical Geography, Newton Abbot, pp. 45-62
Johnson, B.L.C. (1951) 'The Foley Partnership: The Iron Industry at the End of the Charcoal Era', Economic History Review, 4, 322-40
Lawton, R. (1982) 'Questions of Scale in the Study of Population in Nineteenth Century Britain', p.99-113, in A.R.H. Baker and M. Billinge (eds.), Period and Place, Cambridge, pp. 99-113
Mills, J., Mills, J. and Mills D. (1985) Analysis of Nineteenth Century Census, Mills Historical and Computing, Lincoln
Mortensen, H., Mortensen, G., Wenskus, R. and Jäger, H.

(1968) 'Historical-Geographischer Atlas des Preusen Landes Wiesbaden, 1968-

Moulton, G.E. (1983) Atlas of Lewis and Clark Expedition, Lincoln, Nebraska

Palmer, J. (1985) 'Domesday Book and the Computer', in P. Sawyer (ed.), Domesday Book: A Reassessment, London, p. 167

Parker, W.N. (ed.) (1984) 'New Sources for Rural History', Agricultural History, 58, 105-57

Prince, H.C. and Kain, R.J.P. (1985) The Tithe Surveys of England and Wales, Cambridge

Roberts, B.K. and Glasscock, R.E. (1983) 'Villages, Farms and Frontiers: Studies in European Rural Settlements in the Medieval and Early Modern Periods', Oxford, British Archaeological Reports, International Series, 185

Rumble, A.R. (1985) 'The Palaeography of the Domesday Manuscripts', in P. Sawyer (ed.), Domesday Book: A Reassessment, London, pp. 28-49

Simms, A. (1982) 'Cartographica Representation of Diachronic Analyses; the Example of the Origin of Towns', in A.R.H. Baker and M. Billinge (eds.), Period and Place, Cambridge, pp. 289-300

Smith, C.T. (1967) An Historical Geography of Western Europe Before 1800, Cambridge

Whyte, K.A. (1984) 'Register of Research in Historical Geography', Historical Geography Research Series, 14

Wild, M.T. (1980) 'Register of Research in Historical Geography', Historical Geography Research Series, 4

Wilson, A.W. and McKay, D.S. (1978) 'Mapping Canada's History', Canadian Cartographer 15, 1, 13-22

Withers, C.T. (1985) 'Mapping the Industrial Revolution', Journal of Historical Geography 11, 2, 196-7

Wrigley, E.A. (1966) 'Family Limitation in Pre-Industrial England', Economic History Review, 19, 82-109

Chapter Three

THE DARK AGES

G. Whittington

In answering the question 'what progress has been made recently in the historical geography of the Dark Ages?' it would be possible to answer that there has been none. While that statement might be a slight exaggeration, it is nevertheless very close to the truth. At that point, therefore, this chapter might be concluded; a glance at the contents page will, however, have indicated that that is not the case. That would leave unanswered, among others, such questions as why has there been no progress, should there have been any progress and is there a foundation from which progress could have been made? The main thrust of this contribution will be an attempt to survey and provide possible answers to those questions. Before so doing, something must be said as to the extent of time and space to be considered within the chapter's title. The Dark Ages is a specifically European concept relating to the period between the demise of the Roman Empire and the emergence of a coherent body of written evidence which allows a reconstruction of the many aspects of economy and society. The actual time span varies from area to area but it generally covers the middle to late centuries of the first millennium of the Christian era. The main feature of this period, as far as the scholar is concerned, is his reliance upon relict features as his source of data. That being so, it would be possible to undertake a review of the historical geography of all areas of the world for the period up to the variously dated emergences of a consistent and full written record. Such a task is plainly impossible, both from the point of view of its sheer enormity and the problem of not only covering but also discovering what has been written. From the investigations undertaken, however, the overall, let alone Dark Ages condition of historical geography is as dismal in Latin America and Africa as it was when surveyed in 1971 (Baker). The treatment of the Dark Ages here will be restricted to a European context while recognising that other societies in other areas do merit similar consideration.

HISTORICAL GEOGRAPHY AND THE DARK AGES:
THE FOUNDATIONS

While the Middle Ages and subsequent periods have been
dominant in the historical geographical literature, isolated
examples can be found of an early involvement with the Dark
Ages. A paper in Geographische Zeitschrift (Gradmann, 1901)
considered the landscape of central Europe for the period of
folk-wandering which followed the decline of Roman
supremacy. Such efforts stand out because of their unusual
nature as does a further major and perhaps more stimulating
work, An Historical Geography of England before 1800
(Darby, 1936). That collection of papers taking as its
starting point the prehistoric period of Southern Britain, set
the scene for the nature of historical geography in Britain for
many years. It also included the first attempt to deal with
aspects of Dark Ages landscape development over a wide area
of England. This work clearly drew on existing interests and
attitudes. Its author, Wooldridge (1936), was extending
temporally an investigation he had undertaken for south east
England (Wooldridge and Linton, 1933) and also spatially an
interest shown by Darby (1934) in the Anglo-Saxon period.
The Anglo-Saxon Settlement, as the chapter by Wooldridge
was entitled, can be used to shed light on the stuttering
nature of geographers' subsequent involvement in this period
of landscape development. In his introduction Wooldridge,
noting the contribution made by 'archaeologists and students
of place-names' to the work already undertaken by historians,
felt it necessary to write: 'even at the risk of being regarded
as an intruder in the field, the geographer cannot forgo the
task of bringing another technique, and still other facts, into
the discussion' (Wooldridge, 1936). There seemed to be a
conviction, perhaps born out of an underlying environmental
determinism, that the geographer must contribute but that it
could only be done apologetically.

It is difficult to assess the effect of this contribution in
stimulating historical geographers to further efforts in this
field. This stems from two factors. First there was the
interruption of scholarship caused by the Second World War.
The second reason is perhaps less clear cut but it arises from
the predominance of regional geography until well after the
European academic communities had got back into full stride
after the wartime hiatus. During that period it is possible to
find many more examples of geographical writings involved
with the period of the Dark Ages. They came about as a
result of genetic explorations of regional landscapes. Thus
the contributions by geographers to The Making of the
English Landscape series (e.g. Millward, 1955, Emery, 1974)
and similar works by Derruau (1949) in France, Aalen (1978)
for Ireland and Keuning (1979) in the Netherlands; at a lower
regional level the paper by Thorpe (1962) on Wormleighton

parish in Warwickshire comes into the same category. Explorations such as these were undertaken using a progressive chronological approach but interesting variations are also to be found which include a distinct methodological variation. In the many debates on the purpose and method of historical geography recourse is had to an attitude expressed by Hartshorne (1960):

> that in seeking the explanation of existing relationships of features, the normal procedure in geography is to start with the existing feature and trace back in time the process of its relations with other features.

This has been characteristic of many French workers who have considered aspects of the Dark Ages as part of a much broader exploration (e.g. Brunet, 1960) and examples are to be found among other European authors (e.g. Filip, 1972). Such writings seem to have a more positive driving force behind their excursions into the Dark Ages than those which were prompted by the forward chronological nature of the investigation; in many of the latter cases the Dark Ages appear to have just 'got in the way' and are dismissed quickly and one suspects thankfully that so little information about that period exists.

Contributions by geographers to the study of the Dark Ages can also be found which arise from quite different reasons. One develops out of Britain's position at the western margin of the European continent and the part that played during the continuing migration of people through Europe. The replacement of languages, cultures and economies has led to the superimposition of many different features in the landscape and to the existence of a Celtic fringe. These two features taken together led to the growth of a body of historical geographers who have been concerned with aspects of Dark Age geography. The origin of many members of this group is tied to the emergence of a strong school of geography in the University of Wales at Aberystwyth. Whether as a direct product of that school or due to conversion by one of its many missionaries, the interests developed and encouraged by Fleure, demonstrated in his paper on early population movement (Fleure and Whitehouse, 1916), led to a wide variety of studies being undertaken for the Dark Ages. Most influential, not only as authors, but also as teachers were E.G. Bowen and E. Estyn Evans. Interestingly and significantly their earlier published works appeared in non-geographical journals: for example, Evans published an historical geography of part of the Welsh Marches in Montgomeryshire Collections (1922-8); Bowen's first paper was in Geographical Teacher (1925-6) on rural settlement, a topic which was to occupy most of his research time, but he, too, turned to Montgomeryshire Collections in 1930. A study of

both his and Evans' subsequent writings reveals a pre-
dilection for placing their findings in non- geographical
outlets. Both workers had strong leanings towards archaeological and anthropological attitudes.

Evans' tenure of the geography chair at Queen's University, Belfast saw a spread of the Aberystwyth influence and
important papers on the Dark Ages came from his former
students, especially on rural settlement and land use (e.g.
Proudfoot, 1961, Gailey, 1962, Buchanan, 1970, McCourt,
1971). An encouraging re-emergence of a similar interest in
this period has come in the work of Barrett (1982) on the
spatial and temporal continuity of rural settlement in Ireland;
a work which makes use of statistical techniques. Bowen's
influence in the studies of this period come most notably from
his direct teaching or through the teaching of his former
pupils. In the latter case is a study of settlement in Scotland
which uses archaeological and place-name evidence
(Whittington, 1977). For the health of Dark Ages studies,
however, no other British geographer can match the contribution made by Jones in a stream of important, controversial
and thought-provoking articles. Although concerned with
rural settlement his work has carried modelling into Dark Age
studies following upon his concern with and investigation into
spatial organisation (Jones, 1961a, 1965, 1971). His involvement in this topic has not only led to publication in
archaeological journals (e.g. 1961b, 1962, 1963) but also into
conflicts with views held by an archaeologist (Alcock, 1962).
The Celtic fringe element is also present in the final contributing source to Dark Age studies by British geographers in
that it stems from the University of Aberdeen and the
interest of O'Dell (1939). This was furthered by Small (1968)
in a special concern with the Viking involvement in settlement
and landuse forms in northern Britain.

The most outstanding contributions to the study of the
geography of the Dark Ages have come from what might be
called the Swedish school. Workers involved here have used a
number of innovatory techniques and have had an impact
upon geographical work in other parts of Europe, although
more usually in a Middle Ages context. Helmfrid (1971) has
shown the great interest in the Dark Ages that has obtained
in Scandinavia from the time of work by Stenberger (1933)
right through to the vital part played by Hannerberg in the
1960s. The most striking contribution has been in the problem
of continuity in field and settlement structures. This type of
work has been undertaken in other parts of Europe but the
innovatory feature was the persistent pursuit by Hannerberg
of the metrological technique (Hannerberg, 1955, 1960). This
allowed him to identify the nature of measurements employed
in the layout of farms and their associated land for different
periods (Hannerberg, 1946). Since that date many more
articles have appeared on the same topic. To the metrological

approach has been added detailed morphometric analysis of farms and fields. Articles by Anderson (1959), Göranson (1958, 1971, 1982), Helmfrid (1962), Lindquist (1968), Sporrong (1971, 1982) and Widgren (1982), among others, have secured a lasting place for historical geography in the furthering of the understanding of Dark Age landscapes and the unravelling of landscape evolution. In 1976 Hannerberg brought much of this work together by publishing models of territorial organisation that he had developed and which he linked to those created by Jones for Wales.

The success of the Scandinavian contribution to Dark Age studies owes much to the conscious development of a programme of study adopted by the Department of Human Geography of the University of Stockholm. The work undertaken there and by Scandinavian historical geographers in general for this period is comparable with the success of the recent Japanese industrial effort. Important techniques were found in other disciplines and other countries and then, with vigour and skill, applied in an integrated fashion in the indigenous programme. The pioneering work of the Scandinavian and German botanists (e.g. Iversen, Faegri and Firbas) was capitalised upon. Working with the botanist M. Fries, Helmfrid (1959) introduced the pollen analytical technique into the historical geography of Scandinavian Dark Age study. A similar action saw the incorporation of phosphate analysis. The Swedish worker Arrhenius (1931) seems to have been the prime mover in exploiting the relationship he noted between the phosphate levels in soils and the effect of human habitation and exploitation of the land (Arrhenius, 1955) and Sporrong (1968) successfully applied it in a Dark Age study as an aid in the identification and differentiation of human and animal occupation in and around a habitation site. Before the exploration of such sites can be undertaken their location has to be identified and that is frequently a problem due to the lack of surface remains. A solution to this was demonstrated in 1924 by Crawford (trained as a geographer) when he exploited aerial photography for the location of deserted settlements and their field systems. This technique was also fitted into the overall strategy of the Scandinavian school as was the use of radiocarbon dating. The significance of carbon in the old land surfaces preserved under walls in abandoned sites was originally perceived by Niemeier (1958) in Germany. The exploitation of this technique was furthered by the Scandinavian workers following upon the work by Lindquist (1961, 1968) because they could see its potential for a much wider range of relict features.

It is this width of vision which has underpinned the success of the Scandinavian school. The pioneers had noted the endeavours and achievements of one worker in particular who was following a tradition built up over a long period by some illustrious predecessors. The work undertaken by

Gradmann as early as 1901 and furthered by the Mortensens (1937, 1938) culminated in one sense in the type of approach used in his research by Müller-Wille (1944). His analytical method allowed him to decompose field structures extant at the end of the eighteenth century and to suggest the different ages of their constituent parts. That study was taken further by the application of its findings to an area of northern Germany, allowing the tracing of field patterns back into the Dark Ages (Müller-Wille, 1965). Although that approach has been used by many other workers, they have confined themselves, both in Europe and Britain to the period of the Middle Ages (for the methodological importance and bibliography of this see Lienau and Uhlig, 1967). Notable exceptions to this have been the study by Backmann (1960) who traced the ownership of farms in the southern Germanic lands back into the sixth century and a study by Nitz (1961) of Frankish colonisation.

It is not surprising that the main thrust of historical geographers in Dark Ages study has been in the area of rural settlement, agrarian systems and territorial organisation because the landscapes of that period were predominantly rural. Urban beginnings had occurred before this period, however, and yet there are very few studies by historical geographers of the urban phenomenon or its continuity with earlier developments. Such as there are usually occur in a context of a seeking for urban origins and the Dark Ages receive little attention. Exceptions from this state of affairs include work by Millward (1972) on Leicester in the Saxon and Danish period, Gosselin et al. (1976) on the growth of Boulogne between the first and seventh centuries, Butlin (1977) on early town development in Ireland, Clout (1977) on early urban development in France and Giese (1980), much more narrowly oriented, discusses the structure of towns in the Islamic areas of the Soviet Union during the seventh and eighth centuries. The paucity of such studies, considering the geographers' strong interest in urban development and morphology, suggests that a major problem has been encountered here.

This survey of some representative studies undertaken by historical geographers for the period of the Dark Ages indicates a very uneven development. Foundations for such work were laid early in Britain and Germany. In the former area they seem not to have been capitalised upon and have only led to sporadic and largely unconnected studies since. In Germany an early interest was also encountered but has been developed more strongly in the realm of the historical geography of the Middle Ages or has been the stimulant behind the one undoubted area of success, that of the Scandinavian school. How is this to be judged in terms of progress? It is seen as vindicating the statement made at the beginning of this chapter that progress has been virtually non-existent,

both on a knowledge increment basis and in the development
of method and theory. A perusal of the registers of research
work in progress by historical geographers in Britain shows
little involvement in purely Dark Age problems. Consideration
of the work undertaken by the Scandinavians and Germans
encourages the belief that some progress has indeed been
made but that it may well have come to an end. As in
Britain, the workers who have contributed belong to an age
group which was trained in the philosophy of regional geography and its best representatives have adapted their
techniques, within that framework, to changes current in the
discipline as a whole. They are, however, of a generation
which has been active in geography for a long time. Innovations are usually initiated and developed by younger
workers. Where are they? Are there no successors?

DARK AGES STUDY IN OTHER DISCIPLINES

At this point it is important to ask a wider ranging question
with regard to progress in Dark Ages studies. The work and
overall contribution of the historical geographers needs to be
put into perspective by a survey of activity for this time
period by workers in other academic disciplines. Only when
this has been done will it be possible to provide the reasons
why historical geographers have had such limited success and
impact and why their future involvement seems to be so bleak
and meagre. In order to achieve this, the work undertaken in
Britain on the Dark Ages will be considered.

Interest in this period is of long standing with historians
actively involved from before the start of this century.
Milestones in the study are easily identified with the contributions by Seebohm (1883) and Maitland (1897) certainly
numbered among them. The debate their opposing views
engendered still continues to the undoubted benefit of historical studies of the Dark Ages. The publication by Collingwood
and Myres (1936) of a study of the relationship between
Roman Britain and the incoming Anglo-Saxons continued the
interest created by the earlier writers as did papers by Leeds
(1925, 1933). Works by Stephenson (1933) and Tait (1936)
encouraged an early interest in urban development. Thus the
historians, although starting sooner and within a wider
compass, shared with the geographers an early involvement
with the Dark Ages. The geographers' concern seems to have
petered out and yet they appeared to have aims such as
synthesis and the investigation of pattern and spatial organisation which might have been expected to lead them forward;
such aims, especially the latter two, did not bulk large in
historical methodology and so geographers might have been
expected to contribute in a complementary fashion.

It is not, however, in a growing and all absorbing interest in the Dark Ages by historians that a reason for a recession in geographers' involvement should be sought. The recent manifestation of the breaching of traditional academic discipline boundaries is a more potent cause. In a paper which has been much debated and written about since its publication, Darby (1953) considered the relations between history and geography. Such investigations proved difficult and somewhat inconclusive but how much more so would they be today. This, it is contended here, is due to major directional changes in two other areas of study, the orientation of which make a nonsense of the claim that geography can find its focus and raison d'être in spatial studies.

From as early a date as 1849 (Kemble) place-names had been perceived by historians as features of major importance in any consideration of the Anglo-Saxon presence in England. Before place-names can be employed for such a task it is essential that their earliest form be traced and meaning explained. This has been the task of philologists, often aided by historians (e.g. Mawer and Stenton, 1924). The availability of such studies enabled Wooldridge (1936) to consider the Anglo-Saxon settlement of England in relation to its three phases of entrance, expansion and consolidation. He mapped the distribution of place-name elements and analysed them in relation to the physical nature of the areas in which they occurred. His temporal model of settlement built upon the use of the successive toponymic elements of -ingas, -inga- (the entrance phase), -ing, -ingaham and -ington (the expansion phase) and -den, -ley and -field (the consolidation phase) and an analysis of the terrain in which they were located was needed because the place-name scholars showed little or no interest in such matters. Their attitude to their work, which summarises that commonly held during the first half of this century, was clearly stated by Wyld (1950) in his study of Lancashire place-names:

> place-names here are considered as elements of language and are treated as a purely linguistic problem. The work is not concerned with the question whether names fit the places to which they are attached or whether they ever did so.

Of course, there are exceptions to that statement but there has recently been a remarkable move away from such attitudes. Etymological problems are still dealt with but present-day place-name scholars seem to be as concerned with the context in which their subject matter occurs; geological, as well as topographical maps are now being used in their method of inquiry.

A result of this change of direction by many of the foremost place-name scholars working on the period of the

THE DARK AGES

Dark Ages has been that entirely new views on the settlement
of the Anglo-Saxons in England have emerged. As late as
1962 Ekwall was maintaining his original position regarding the
place-name element -ingas:

> at the time of the Anglo-Saxon invasion, and perhaps for
> some time afterwards, it was used to designate the
> people of a village by a collective name often formed with
> the suffix -ingas which frequently became the name of
> the village or entered into its name.

Myers had suggested as early as 1935 that place-names ending
in -ingas could represent a phase later than that of the
Saxon burials, since he had noted that the -ingas place-name
element and pagan burials were not spatially coincident in
Sussex. This problem is directly related to the method
employed by historical geographers in the use of place-names.
The scale at which distribution maps were produced led to an
apparent coincidence of the two remaining monuments to the
initial Saxon settlement. On closer inspection such a relationship did not exist and this posed a challenge to the Ekwall
hypothesis. Apologists, when confronted by the occurrence of
only one of the monuments instead of both together, evaded
the problem on the grounds that either the place-name had
not survived or that the burial still awaited discovery.

A publication by Dodgson (1966), perhaps the most
stimulating and revolutionary yet to be published on Anglo-
Saxon place-names, considered the problem of the non-
coincidence of the place-name element -ingas and the pagan
graves and suggested that it is simply resolved if they cease
to be considered as contemporaneous. Dodgson proposed that
the -ingas place-names resulted from a social development
contemporary with a colonisation phase but one which was
later than that of the immigration phase. If such were to be
the case then the settlement model examined by Wooldridge
(1936) would have to be revised. Furthermore, if the -ingas
element were not to be the earliest then a replacement for it
would have to be found. Interest focused on the element -ham
which was known to be of early occurrence, having an
association with the other early element -ingas in the form of
-ingaham. This same relationship exists in continental Europe
between the elements -heim and -ing. Progress in establishing
the early nature of the names containing -ham is hampered by
the difficulty in distinguishing it in its modern form from the
Old English -hamm.

To overcome this problem Gelling (1960) conducted a
systematic topographic survey of the occurrence of -hamm;
this in an attempt to distinguish it from -ham and also to
investigate its meaning. In this pioneer study, Gelling
suggested a northern limit (since revised further north,
Gelling, 1967) for the distribution of the element, corre-

sponding to the territory of the Saxons and Jutes while also concluding that 'enclosed plot on a promontory' was perhaps the comprehensive meaning of -hamm. Also in an examination of -hām and -hamm undertaken in Southern England Dodgson (1973) supported most of Gelling's findings but stressed that unless unequivocal Old English or Middle English spellings of -hamm or -hām are to be found, even the most striking topographical associations justify only a statement of probability. Even this position has been attacked (Sandred, 1976) on the grounds that, because a high proportion of major names occur in sharp river bends, the location there of the -hamm element has no significant topographical meaning; perhaps the place-name scholars' translation from etymology to locational analysis has still not been completely consummated.

Dodgson's original statement on the earliest settlement problem stimulated Kirk (1971), a historical geographer, to apply the technique of nearest neighbour analysis to the occurrence of the four categories of -ing place-names (-ingas, -inga-, -ing, and -ing-) in Kent, in an attempt to discuss whether Ekwall's (1962) or Dodgson's (1966) proposed sequence was the more valid. The analysis revealed a non-random placing of all four types of -ing place-names. This was followed by an investigation of the environmental surroundings of the -ingas and -inga- sites which showed that statistically a significant number occurred on 'inferior' sites and that they related to settlements which date from a period when no other land was available. Kirk's general conclusion was in accord with Dodgson's.

This work was followed by two other important investigations. Cox (1973) and Kuurman (1975) undertook a survey of the English place-name chronology over the Midlands and East Anglia, with special reference to the elements -hām, -ingahaam, and -ingas. Cox argues that the -ham element belongs to the pagan Anglo-Saxon period and he shows a strong relationship between it and the sites of Romano-British settlement, Anglo-Saxon pagan burials and the Roman road system. With these findings Kuurman is largely in agreement and the general conclusion is that the rare coincidences of -ingas and -inga- place-names with pagan burials indicate that the former are demonstrations of later settlements than those containing the element -hām.

Wooldridge (1936) also noted the occurrence of names derived from the pagan Saxons' gods such as Tiw, Thunor and Woden. He considered that their paucity precluded any conclusive statement from being made about their distribution but he nevertheless felt, by their very nature and from their location in areas of loam terrain as in west Surrey, that they must be early. Gelling (1961) in a persuasive article has reversed that suggestion, considering them to be late settlement creations.

The combined result of this work has been to overturn and virtually reverse the settlement model used by Wooldridge. That this should occur is not altogether surprising because, due to its inherent characteristics, inquiry is likely to result in drastic revisions of earlier findings. What is remarkable from the historical geographers' point of view is the means by which it has been achieved. The place-name scholar has abandoned the traditional stance and has embraced attitudes and techniques which the historical geographer has always assumed, perhaps complacently, were among the distinctive features of geography. In this context, however, the place-name scholar Watts (1976) has sounded an interesting warning. Commenting on the fact that some assumed early settlements occur on what are environmentally inferior sites he states:

> I suspect that the evidence of the one-inch drift maps may need a good deal of refinement. It would be interesting to hear a geographer's comment on the quality of the drift evidence for the kind of exercise place-name scholars have been conducting.... How delicate ought our geological information ideally to be? .... Is sand and gravel always superior to clay? A. Steensberg suggested soil fertility rather than its geological consistency as the crucial factor but how do we measure that?

Perhaps the most useful counter comment on that is that he should be considering consulting a pedologist rather than a geographer! Thomas (1974) in contemplating the use of place-names in landscape analysis, however, concludes that for their most efficient exploitation 'the co-operation of philologists, botanists, geographers and historians is required'.

Reference was made earlier to the infrequent examples that occur of historical geographers' incursions into urbanisation during the Dark Ages. While not on the same low scale, the contributions of historians is also inconsiderable, an exception being Loyn's (1971) thoughtful article on the towns of Anglo-Saxon England. Why should there be so small an involvement by two groups of workers who have long been interested in aspects of urban development? The answer to this probably lies in the difficulties of the data source; as Baker (1971) has observed:

> although the approach of the historical geographer towards a given theme may be influenced by the intellectual climate within which he is operating, it is nonetheless largely conditioned by the materials available.

Thus it is not surprising that the important advances in our knowledge of Dark Age towns and the urbanisation process have been made by workers who are active in archaeological excavation. Primary material has come from a variety of sites such as, for example, York (Addyman, 1974), Southampton (Addyman and Hill, 1968), Cirencester (Wacher, 1964), and Northampton (Williams, 1977a). Not content with being mere providers of data such workers have naturally gone on to analyse their information. Addyman and Hill (1968) considered the evidence that existed for a Saxon Southampton while Biddle (1973) looked at Winchester from the point of view of its function as an early capital. Williams (1977b) raised a vital problem concerning urban origin and development in the Dark Ages:

> Looking generally at the growth of urban centres, there is a danger of transforming a later urban character into an earlier site when the earlier site is of inferior status. For example, no one to my knowledge has suggested that West Stow (West, 1969) or Chalton (Addyman and Leigh, 1973) was a town, but if these sites had been found beneath a major medieval centre, I am sure the cry would have been one of urban continuity not of settlement continuity.

Further involvement, stemming from the availability of the primary data, has allowed synthesis of such investigations but in relation to the examination of particular problems perceived as being of vital importance to the period under consideration. The nature of the Saxon urbanisation process is one such issue and Biddle and Hill (1971) have looked at this within the planning context. Biddle (1976) also took up that theme in a research report produced for the Council for British Archaeology.

A second problem considered to be of a fundamental nature has also been addressed. In 1959 Finberg published a study on the settlement at Withington in the Gloucestershire Cotswolds in which he postulated continuity of occupation between the Roman and Saxon periods. The work suffered a sceptical reception but Finberg had raised an issue which refused to disappear (it is interesting to note that twenty years later, from the same stable, a further contribution on the same theme has appeared, Mackreth, 1978). The problem of what happened to the products of the urbanisation of England by the Romans, encapsulated in the theme continuity or catastrophe, has provided a driving force for work in Dark Age settlement studies. From their highly informed position as possessors of hard-won primary data several of the authors quoted above have pursued the urbanisation problem with the aim of shedding light on its 'Dark Ages' element. Notable in this is the article by Hill (1977) in which he considers the evidence for continuity. At present the general consensus on this problem seems to be that adduced by Hill and by Wacher (1975) that 'life in towns' rather than 'town life' appears to be the best conclusion at the moment.

The burden of this excursion into Dark Ages urbanisation seems to be that availability of and access to primary data along with a strong focus for any study appear to be vital for any advance in an understanding of this period. If this is the case why should it be that historical geographers have not carried back in time their considerable success in unravelling the nature of the field systems and settlements of the Middle Ages? In the case of primary data perhaps it is because of the availability of cadastral information for that period whereas its absence for earlier times has caused the historical geographer, map oriented by training and perhaps also by inclination, to shy away. This seems to be an inadequate reason and the answer to the question appears to lie in a further exploration of what lies behind the advances in the knowledge of Dark Ages urbanisation.

An obvious feature there is the activity of the archaeological excavator. It might, therefore, be concluded that advances in Dark Age knowledge should also be occurring in the rural zone, perhaps in quantity commensurate with the ratio of urban to rural studies apparent for the Middle Ages. When a survey of work in the rural areas for the Dark Ages is undertaken this proves to be true on both counts. The amount, range and treatment of information from the rural zone is not only considerable but again it reveals a deep penetration by the archaeological excavator. It will be convenient here to look at this work from three points of view: the consideration of rural settlement, the understanding of field systems and the attempts to unravel the organisation of space. Only when this has been undertaken will it be possible to examine further the failure of the historical geographer to contribute in significant amount to Dark Age study.

Our understanding of the settlement and exploitation of the rural area has been greatly increased in the last two decades by the publication of reports from archaeological excavations and an analysis of the conjunction of relict features in the landscape. Thus Morris (1966) has looked at the relationship between Roman roads and Saxon settlement in Hertfordshire, Thomas (1972) has considered the settlement of the Irish in western Britain for the post-Roman period and Ford (1976) has gone much further and produced a view of settlement patterns in Warwickshire, synthesising evidence obtained from a variety of approaches. Examples such as these could be multiplied many times.

Excavation reports have been arriving with an increased tempo, expanding the information on the nature and chronology of Anglo-Saxon settlement from a wide variety and distribution of sites. Among these are West Stow (West,

1969), Chalton (Addyman and Leigh, 1973), Catholme (Losco-Bradley, 1973), Walton, Aylesbury (Farley, 1976), Bishopstone (Bell, 1977), Old Down Farm, Andover (Davies, 1980) and Heybridge (Drury and Wickenden, 1982). Within this group it is possible to find values of differing kinds. Some (e.g. Catholme and Chalton) are mainly concerned with the excavation of buildings at ground level and reveal much concerning their structure and layout and the morphology of the settlement they produced. Others (e.g. Bishopstone and Walton) have in addition provided a different kind of evidence because they have involved the excavation from below ground level of buildings in the form of sunken huts. The Chalton site has proved to be important in a further different way. It has been demonstrated by the place-name scholars that the relationship between the Romano-British and the earliest Saxon settlers needed to be revised. Work undertaken at Chalton (Cunliffe, 1973) has shown that the upland areas of the Downs were the first to be settled by the Saxons and that their dwellings were in close proximity to Romano-British farms. Subsequently a centrifugal movement took place which carried settlement to lower elevations and on to different soil associations; such findings have been confirmed by work at Bishopstone. Progress is being made here not merely by providing additional information; it also occurs through the support being given to workers in the same field who, however, use a different approach.

Work on agrarian systems has not achieved the same level as that for the settlements. It is here that the archaeologist interested in such matters seems to be on common ground with the historical geographer; they are both confronted by an absence of information or so it would appear from the lack of work on Dark Age field systems. Yet the Scandinavians (e.g. Hannerberg, 1976) have shown that much can be achieved by the careful piecing together of information obtained from the use of a variety of techniques. Fowler (1976) has surveyed agriculture in the Dark Ages by scrutinising the ways in which our knowledge of it may be increased. Excavation, surviving field pattern evidence and extrapolation from estate boundaries are all considered as possible sources of information. Great hopes are pinned on the interpretation of animal bones from excavations and from palaeoenvironmental evidence because both should reveal what the farmers actually did. An attack on the whole problem is recommended in a two pronged form, going forwards from what we know about Romano-British agrarian practice while also working backwards from our much more detailed information about the early medieval period. That recommendation contrasts starkly with advice from Finberg (1972) who felt that to increase our understanding on this topic it is 'necessary to clear our minds of preconceptions' (mainly those ideas derived from the Domesday Book) and 'to work forwards from the beginning'. Fowler's approach seems the more sensible in that it must surely increase the chances of progress and, indeed, the Scandinavians have shown that such a method works. Until, however, much more effort is expended on the agrarian landscape, simple statements, such as those by Taylor (1974), will have to suffice; they basically amount to a confession of our ignorance. A conclusion that could be drawn from this is that the absence of information from excavations, largely due to the difficulty of identifying worthwhile excavation sites, has deterred the archaeologist from an involvement in agrarian matters to the same degree as has been displayed where the rural habitat is concerned.

One of the strengths of the historical geographers' contribution to Dark Ages studies in Britain lies in the work of Jones on the territorial organisation of space. In this context the archaeologists have made considerable advances, especially in taking up a topic which has been of longstanding interest among the historians, epitomised by the work of Grundy (e.g. 1919, 1920). Bonney (1966, 1972, 1976) has examined the emergence of estate and parish boundaries in areas of Wessex and, like Jones, in the case of the multiple estates, believes that such features have very early origins, even prehistoric. The relationship between pagan burials and boundaries is considered and once more the importance of the former features in assessing the development of the Anglo-Saxon settlement appears to be vital. Progress in this field has been made by the contribution of Arnold (1977) who used the evidence of place-names and cemeteries to define settlement territories, using a weapon from the 'new archaeology', the Thiessen polygon. A more recent study (Goodier, 1984) has taken up a point made by Hills (1979) that the results of the investigations of the relationship between boundaries and burials have been found wanting because of an absence of rigorous statistical testing. By using such an approach, rooted in hypothesis formulation, Goodier has not only revealed a more secure relationship but has also been able to show that it varies with time.

What can be learned from these excursions into the work of the archaeologists in relation to the settlement, field systems and boundary features of the Anglo-Saxon settlement? They are all elements which historical geographers have worked on and continue to study but, in the main, they only dose when medieval times are reached. Seeing that there has been such great progress for that period why has the earlier one not been tackled? A suggestion was made when considering the urban realm that it was due to the lack of primary data. That too might be concluded in relation to their lack of involvement in the morphology of the rural settlements. The techniques used by the archaeologist in other areas of the study are, however, available to the historical geographer and indeed are largely borrowed from the mainstream develop-

ments in that subject as a whole. Could it be that the
archaeologists have perceived the void in our knowledge
represented by the Dark Ages and have reacted to it in a
positive way? That may be the case but there must also be
reasons why the historical geographers did not react in the
same way. In concluding this review of the lack of progress
by historical geographers in the study of the Dark Ages some
reasons for that gap will be suggested.

WHY SO LITTLE PROGRESS?

It was stated earlier that major directional changes in two
areas of study had had a big impact on landscape studies.
One, in the realm of place-names, has already been considered; the second has to be addressed. It is related to the
realisation by archaeologists that their craft needed to be less
concerned with excavations as a means of merely obtaining
artefacts and information on structures and much more with
the distribution of features, patterns of movement and the
organisation of space that the artefacts and structures could
reveal. Their new perception of their activities took them
further and further away from being excavators only, towards
becoming proponents of a discipline in which the
Kulturlandschaft geographers would feel at home. The pursuit
of 'total landscape' archaeology, a theme enunciated in a
discussion of a Society of Antiquaries of London research
project, has as its aim 'to break away from the single-site
and single-period approach in an attempt to gain a fuller
understanding of the past' (Bowen and Cunliffe, 1973).
Embodied in the aim was to be an attempt to answer three
questions for an area: 'what size of population is involved,
how did it use the land, and, as a corollary, how does the
resolution of these questions throw light on the organisation,
technical capacity and thought-processes of the people
involved?' With that aim in mind and also the memory of
Dymond's (1974) Archaeology and History: a Plea for Reconciliation, it is perhaps worth reviving a comment made by
Kirk (1963): indeed it could be argued that the logical,
ultimate end of archaeology is to become geography.

It is not only the subject matter and aim of the archaeologists that ally their study so closely to that of the
historical geographer but also the penetration of their
discipline by methods derived from the experience of the
geographer. The publication of works with titles such as
Models in Archaeology (Clarke, 1972), Mathematics and
Computers in Archaeology (Doran and Hodson, 1974), Spatial
Archaeology (Clarke, 1977) and Simulation Studies in Archaeology (Hodder, 1978) reveal a similar concern over methods of
explanation when dealing with distributions over space and

through time. It appears possible to agree with Kirk that basically there is little difference between the aims and methods of historical geographers and archaeologists, with the caveat that the latter number excavation among their techniques while the geographer does not. If Darby (1953) were pondering the position of historical geography today, it might well be that the relationship between history and geography only would not be sufficient.

The adoption into archaeology of what Bowen and Cunliffe (1973) call 'procedures of analysis familiar to the geographer' not only led to a closer alignment of the two disciplines but almost certainly led to the increasingly later chronological involvement in the analysis of past landscapes by the archaeologist and the retreat of the geographer. For most archaeologists the past is all; for the historical geographer that position is no longer seen to be respectable. Along with the Quantitative Revolution came an attitude that was to cause the same damage to historical geography as the Cultural Revolution did to progress in China. Just as the Chinese had a generation indoctrinated with a particular viewpoint so has the geographical world; the only real difference is that for Maoism it is necessary to substitute Relevance. Geography students in further education have been told that unless they undertake studies which are relevant to the needs (i.e. materialist) of society then they are recidivists who are doomed to incarceration in that most frightful of all prisons, the Ivory Tower. This provides the answer to the question posed earlier as to where the successors of the past and elderly present practitioners in the geography of the Dark Ages are to come from. They are unlikely to come at all because that area of historical geography is suffering more than most from a lost generation despite the efforts of workers like Barrett (1982), Hooke (1981, 1983) and Unwin (1983).

The attitudes prevalent and fashionable in the seats of higher learning are also reinforced by other important factors in this issue which result from the schooling of the students. Gone are the days when large numbers of university geography students would have studied, as a matter of course, History, French and Latin until they left school. An interesting recent survey (Phillips and Unwin, 1985) has shown that those subjects along with English, Mathematics, Economics and Physics, were the most usual taken at school by university teachers in historical geography. Cosgrove (1985) has pursued this topic further by considering recruitment into the teaching of historical geography. He finds the situation bleak and when it is looked at from the point of view of Dark Age studies it would appear that any substantial progress by geographers in furthering the understanding of that period is entirely remote.

In two other respects the geographer has lost ground. Other disciplines today are taking much more interest in spatial relations and spatial organisation, features which many geographers saw as distinctive to their discipline. In the past, too, the geographer saw as his over-riding task the pursuit of synthesis and, indeed, pleas have again been made recently that that aim is still the way forward, especially for the historical geographer (Moodie et al., 1974, Baker, 1982). Much synthesis in the past, especially when couched in a regional framework demanded the geographer as polymath. The nature of modern inquiry is, however, such that it is becoming ever more narrowly specialised; the polymath is a dying (dead?) species and it may be questioned whether the regional method did not lose favour not only because it appeared to lead nowhere but also because it made impossible demands upon its practitioners. In landscape development, the historical geographer indulged in synthesis within a long time span. The Making of the Landscape series is a classic example of this and one which clearly demonstrates the flaws of such an approach; the canvas is too broad and the artist does not have enough brushes and paint or the expertise to use them.

Changes stemming from a variety of causes have occurred in mainstream geography and these have led to a questioning of the aims and methods of historical geography and also to a decline in its teacher recruitment. This has made itself felt strongly in the study of the Dark Ages where the ground has been occupied by the archaeologist. Perhaps this is a good thing. Perhaps the results will be better. That is debatable, however, for there are already apparent in archaeology the same problems of which the geographer is only too painfully aware. A new methodology has brought into existence more debate than practitioners. Humanistic, behavioural, marxist and positivistic attitudes have their adherents. Perhaps most worrying of all is the frequent failure to face up to the practicalities of achieving the aims set. How is the challenge of 'total landscape' archaeology to be met? By archaeological polymaths? How even are the reports from major excavations to be presented? The principal excavator is aided by his team of specialists and usually produces a synthetic report based on his own site work and the reports on environmental and artefactual features. It is unlikely that the excavator will understand all aspects of the latter and thus be able to provide a truly explanatory synthesis. Perhaps the main consolation for the archaeologist is that Relevance is more likely to be assessed from an academic standpoint unless he is battling on the financial support front.

The historical geographer can and should be involved in Dark Ages study; the archaeologist would benefit from his different form of training. Achieving a meeting of minds will need, however, something more substantial than the launching of the new journal Siedlungsforschung - Archäologie,

THE DARK AGES

Geschichte, Geographie (Denecke, 1984). Perhaps what is needed is a further study involving geography and archaeology along the lines of the plea by Dymond (1974) for a reconciliation between history and archaeology.

REFERENCES

Aalen, F.H.A. (1978) Man and the Landscape in Ireland, Academic Press, London

Addyman, P.V. (1974) 'Excavations in York, 1972-73: First Interim Report', Antiquaries Journal, 54, 200-31

Addyman, P.V., and Hill, D. (1968) 'Saxon Southampton: a Review of the Evidence', Proc. Hants. F.C. Archaeological Society, 25, 61-93

Addyman, P.V., and Leigh, D. (1973) 'The Anglo-Saxon Village at Chalton, Hampshire: Second Interim Report', Medieval Archaeology, 17, 1-25

Alcock, L. (1962) 'Settlement Patterns in Celtic Britain', Antiq., 36, 51-4

Andersson, H. (1959) Parzellierung und Gemengelage, Studien über die ältere Kulturlandschaft in Schonen, Lund

Arnold, C.J. (1977) 'Early Anglo-Saxon Settlement Patterns in Southern Britain', Journal of Historical Geography, 3(4), 309-15

Arrhenius, O. (1931) 'Markanalysen i Arkeologiens Tjanst', Geol. Fören. Förhand., Stockholm

Arrhenius, O. (1955) 'The Iron Age Settlements on Götland and the Nature of the Soil' in M. Stenberger (ed.), Vallhagar II, Munksgard, Copenhagen, pp. 1053-64

Bachmann, H. (1960) 'Zur Methodik der Auswertung der Siedlungs- und Flurkarte für die siedlungsgeschichtliche Forschung', Zeits. Agrargeschichte Agrarsoziologie, 8, 1-13

Baker, A.R.H. (1971) Progress in Historical Geography, David and Charles, Newton Abbot

Baker, A.R.H. (1982) 'On Ideology and Historical Geography' in A.R.H. Baker and M. Billinge (eds.), Period and Place, Cambridge University Press, Cambridge, pp. 233-43

Barrett, G. (1982) 'Problems of Spatial and Temporal Continuity of Rural Settlement in Ireland, AD 400 to 1169', Journal of Historical Geography, 8(3), 227-44

Bell, M. (1977) 'Excavations at Bishopstone', Sussex Archaeo. Coll., p. 115

Biddle, M. (1973) 'Winchester: the Development of an Early Capital' in H. Jankuhn (ed.), Vor- unde Frühformen der Europäischen Stadt im Mittelalter, Göttingen, pp. 229-61

Biddle, M. (1976) 'The Evolution of Towns: Planned Towns before 1066', C.B.A. Research Report, 14, 19-32

Biddle, M., and Hill, D. (1971) 'Late Saxon Planned Towns',

Antiq. Journal, 51, 70-85

Bonney, D.J. (1966) 'Pagan Saxon Burials and Boundaries in Wiltshire', Wiltshire Archaeological Nat. Hist. Magazine, 61, 25-30

Bonney, D.J. (1972) 'Early Boundaries in Wessex' in P.J. Fowler (ed.), Archaeology and the Landscape: Essays for L.V. Grinsell, John Baker, London, pp. 168-86

Bonney, D.J. (1976) 'Early Boundaries and Estates in Southern England' in P.H. Sawyer (ed.), Medieval Settlement: Continuity and Change, Arnold, London, pp. 72-87

Bowen, E.G. (1925-6) 'A Study of Rural Settlements in South-West Wales', Geographical Teaching, 13, 317-26

Bowen, E.G. (1930) 'A Map of the Trehelig Common Fields', Montgom. Coll., 41, 163-8

Bowen, H.C., and Cunliffe, B. (1973) 'The Society's Research Projects: I The Evolution of the Landscape', Antiq. Journal, 53, 9-15

Brunet, R. (1960) Structure Agraire et Economique Rurale des Plateaux Tertiaires entre la Seine et l'Oise, Caen

Buchanan, R.H. (1970) 'Rural Settlement in Ireland' in N. Stephens and R.E. Glasscock (eds.), Irish Geographical Studies, Queen's University, Belfast, pp. 146-61

Butlin, R.A. (1977) 'Urban and Proto-Urban Settlements in Pre-Norman Ireland' in R.A. Butlin (ed.), The Development of the Irish Town, Croom Helm, London, pp. 11-27

Clarke, D. (1977) Spatial Archaeology, Academic Press, London

Clarke, D. (ed.) (1972) Models in Archaeology, 2nd edition, Methuen, London

Clout, H.D. (1977) 'Early Urban Development' in H.D. Clout (ed.), Themes in the Historical Geography of France, Academic Press, London, pp. 73-106

Collingwood, R.G., and Myers, J.N.L. (1936) Roman Britain and the English Settlement, Oxford University Press, Oxford

Cosgrove, D. (1985) 'Present Fears for the Future of the Past: Report of a Survey into Academic Recruitment of Historical Geographers', Area, 17(3), 243-6

Cox, B. (1973) 'The Significance of the Distribution of English Place-Names in -ham in the Midlands and East Anglia', English Place-Name Society Journal, 5, 15-73

Crawford, O.G.S. (1924) Air Survey and Archaeology, HMSO., London

Cunliffe, B. (1972) 'Saxon and Medieval Settlement Patterns in the Region of Chalton, Hampshire', Medieval Archaeology, 16, 1-12

Darby, H.C. (1934) 'The Fenland Frontier in Anglo-Saxon England', Antiq., 8, 185-201

Darby, H.C. (1953) 'On the Relations of Geography and History', Transactions, Institute British Geographers,

_19_, 1-11

Darby, H.C. (ed.) (1936) An Historical Geography of England before 1800, Cambridge University Press, Cambridge

Davies, S. (1980) 'Excavations at Old Down Farm, Andover. Part I. Saxon', Proc. Hants. F.C. Archaeological Society, 36, 161-80

Denecke, D. (1984) 'A New Journal on Settlement History in Germany', Journal of Historical Geography, 10(2), 201

Derruau, M. (1949) La Grande Limagne Auverngnate et Bourbonnaise, Clermont Ferrand

Dodgson, J.M. (1966) 'The Significance of the Distribution of the English Place-Name in -ingas, -inga- in South-East England', Medieval Archaeology, 10, 1-29

Dodgson, J.M. (1973) 'Place-Names in ham Distinguished from hamm Names in Relation to the Settlement of Kent, Surrey and Sussex', Anglo-Saxon England, 2, 1-50

Doran, J.E., and Hodson, F.R. (1974) Mathematics and Computers in Archaeology, Edinburgh University Press, Edinburgh

Drury, P.J., and Wickenden, N.P. (1982) 'An Early Saxon Settlement within the Romano-British Small Town at Heybridge', Medieval Archaeology, 26, 1-40

Dymond, D.P. (1974) Archaeology and History: A Plea for Reconciliation, Thames and Hudson, London

Ekwall, E. (1962) English Place-Names in -ing, 2nd edition, Lund

Emery, F.V. (1974) The Oxfordshire Landscape, Hodder and Stoughton, London

Evans, E.E. (1922-8) 'An Essay on the Historical Geography of the Shropshire-Montgomeryshire Borderlands', Montgom. Coll., 40

Farley, M. (1976) 'Saxon and Medieval Walton, Aylesbury', Records Bucks., 20

Filip, K-H. (1972) 'Fruhformen und Entwicklungsphasen sudwestdeutscher Altsiedellandschaften unter besonderer Berucksichtegung des Rieves und Lechfelds', Forschungen zur Deutschen Landeskunde, 202

Finberg, H.P.R. (1959) Roman and Saxon Withington: A Study in Continuity, Leicester University Press, Leicester

Finberg, H.P.R. (1972) 'Anglo-Saxon England to 1042' in H.P.R. Finberg (ed.), The Agrarian History of England and Wales, I, Cambridge University Press, Cambridge, pp. 385-525

Fleure, H.J., and Whitehouse, W.E. (1916) 'The Early Distribution and Valleyward Movement of Population in South Britain', Archaeol. Cambren., 20, 101-40

Ford, W.J. (1976) 'Some Settlement Patterns in the Central Region of the Warwickshire Avon' in P.H. Sawyer (ed.), Medieval Settlement: Continuity and Change, Arnold, London, pp. 274-94

Fowler, P.J. (1976) 'Agriculture and Rural Settlement' in
    D.M. Wilson (ed.), The Archaeology of Saxon England,
    Methuen, London
Gailey, R.A. (1962) 'The Evolution of Highland Rural Settle-
    ment', Scottish Studies, 6, 155-77
Gelling, M. (1960) 'The Element hamm in English Place-Names:
    A Topographical Investigation', Namn Och Bygd., 48,
    140-62
Gelling, M. (1961) 'Place-Names and Anglo-Saxon Paganism',
    University of Birmingham Historical Journal, 8(1), 7-25
Gelling, M. (1967) 'English Place-Names derived from the
    Compound wicham', Medieval Archaeology, 11, 87-104
Giese, E. (1980) 'Aufbau, Entwicklung und Genese der
    islamisch-orientalische   Stadt   in   Sowjet-Mittelasien',
    Erdkunde, 34(1), 46-60
Goodier, A. (1984) 'The Formation of Boundaries in Anglo-
    Saxon England: A Statistical Study', Medieval Archae-
    ology, 28, 1-21
Göransson, S. (1958) 'Field and Village on the Island of
    Oland. A Study of the Genetic Compound of an East
    Swedish Rural Landscape', Geografiska Annaler, 40
Göransson, S. (1971) 'Village Planning Patterns and Terri-
    torial Organisation. Studies in the Development of the
    Rural Landscape of Eastern Sweden (Oland)', Acta Univ.
    Upsal., 4
Göransson, S. (1982) 'Landuse and Settlement Patterns in the
    Mälar Area of Sweden before the Foundation of Villages'
    in A.R.H. Baker and M. Billinge (eds.), Period and
    Place, Cambridge University Press, Cambridge, pp.
    155-63
Gosselin, J-Y., Seillier, C.L., and Leclercq, P., 'Boulogne
    Antique: Essair de Topographie Urbaine', Septentrion,
    6, 5-15
Gradmann, R. (1901) 'Das Mittel Europäische Landschaftsbild
    nach Seiner Geschichtlichen Entwicklung', Geogr. Zeits.,
    7, 361-77 and 435-47
Grundy, G.B. (1919) 'The Saxon Land Charters of Wiltshire',
    Archaeology Journal, 76, 141-301
Grundy, G.B. (1920) 'The Saxon Land Charters of Wiltshire',
    Archaeological Journal, 77, 8-126
Hannerberg, D. (1946) 'Tunnland, Oresland, Utsäde och
    Tegskifte', Gothia, 6
Hannerberg, D. (1955) 'Die Älteren Skandinavischen Acker-
    Manse', Lund Stud. Geog., Ser. B. 12
Hannerberg, D. (1960) 'Schonische "Bolskiften"', Lund Stud.
    Geog., Ser. B. 20
Hannerberg, D. (1976) 'Models of Medieval and Pre-Medieval
    Territorial Organisation', Journal of Historical Geo-
    graphy, 2(1), 21-34
Hartshorne, R. (1960) Perspective on the Nature of Geo-
    graphy, John Murray, London

Helmfrid, S. (1959) 'Eine Pollenanalytische Untersuchung zur Geschichte der Kulturlandschaft im Westlichen Teil der Provinz Ostergötland, Schweden', Geografiska Annaler, 41

Helmfrid, S. (1962) 'Ostergotland "Västanstång". Studien uber die Altere Agrarlandschaft und ihre Genese', Geografiska Annaler, 44, 1-277

Helmfrid, S. (1971) 'Historical Geography in Scandinavia' in A.R.H. Baker (ed.), Progress in Historical Geography, David and Charles, Newton Abbot, pp. 63-89

Hill, D. (1977) 'Continuity from Roman to Medieval: Britain' in M.W. Barley (ed.), European Towns: Their Archaeology and Early History, Academic Press for C.B.A., London, pp. 293-302

Hills, C.M. (1979) 'The Archaeology of Anglo-Saxon England in the Pagan Period: A Review', Anglo-Saxon England, 8, 297-329

Hodder, I. (ed.) (1978) Simulation Studies in Archaeology, Cambridge University Press, Cambridge

Hooke, D. (1981) 'Anglo-Saxon Landscapes of the West Midlands: the Charter Evidence', British Archaeological Rep. Brit. Series, 95, Oxford

Hooke, D. (1982) 'Pre-Conquest Estates in the West Midlands: Preliminary Thoughts', Journal of Historical Geography, 8(3), 227-44

Jones, G.R.J. (1961a) 'Early Territorial Organisation in England and Wales', Geografiska Annaler, 43, 174-81

Jones, G.R.J. (1961b) 'Settlement Patterns in Anglo-Saxon England', Antiq., 35, 221-32

Jones, G.R.J. (1962) 'Settlement Patterns in Celtic Britain', Antiq., 36, 54-5

Jones, G.R.J. (1963) 'Early Settlement in Arfon: the Setting of Tre'r Ceiri', Trans. Caern. Historical Society, 24, 1-20

Jones, G.R.J. (1965) 'Early Territorial Organisation in North England and its Bearing on Scandinavian Settlement' in A. Small (ed.), The Fourth Viking Conference, Aberdeen University Studies 149, pp. 67-84

Jones, G.R.J. (1971) 'The Multiple Estate as a Model Framework for Tracing Early Stages in the Evolution of Rural Settlement' in F. Dussart (ed.), L'Habitat et les Paysages Ruraux d'Europe, Université de Liège, pp. 251-67

Kemble, J.M. (1839-48) Codex Diplomaticus Aevi Saxonici, English Historical Society, London

Keuning, H.J. (1979) Kaleidoscoop de Nederlansche Landschappen, Martinus Nijhoff, The Hague

Kirk, S. (1971) 'A Distribution Pattern: -ingas in Kent', Journal of English Place-Name Society, 4, 37-49

Kirk, W. (1963) 'Problems of Geography', Geography, 48, 357-71

Kuurman, J. (1975) 'An Examination of the -ingas, -inga- and -ingaham Place-Names in the English Midlands', Journal of English Place-Names Society, 7, 11-44

Leeds, E.T. (1925) 'The West Saxon Invasion and the Icknield Way', History, 10, 97-109

Leeds, E.T. (1933) 'The Early Saxon Penetration of the Upper Thames Area', Antiq. Journal, 13, 229-51

Lienau, C., and Uhlig, H. (1967) Flur und Flurformen, W. Schmitz, Giessen

Lindquist, S.-O. (1961) 'Some Investigations in Field-Wall Areas in Uppland and Ostergötland', Geografiska Annaler, 43, 205-20

Lindquist, S-O. (1968) 'Det Förhistorika Kulturlandskapet i östa Ostergötland. Acta Univ. Stockholm, Studies in North-European Archaeology, 2

Losco-Bradley, S. (1973) 'The Anglo-Saxon Settlement at Catholme', Trent Valley Archaeological Research Committee, 8, 1-35

Loyn, H. (1971) 'Towns in Anglo-Saxon England: the Evidence and Some Possible Lines of Enquiry' in P. Clemoes and K. Hughes (eds.), England Before the Conquest, Cambridge University Press, Cambridge, pp. 115-28

McCourt, D. (1971) 'The Dynamic Quality of Irish Rural Settlement' in R.H. Buchanan, E. Jones and D. McCourt (eds.), Man and His Habitat: Essays Presented to Estyn Evans, Routledge and Kegan Paul, London, pp. 126-64

Mackreth, D.F. (1978) 'Orton Hill Farm, Peterborough: a Roman and Saxon Settlement' in M. Todd (ed.), Studies in the Romano-British Village, Leicester University Press, Leicester, pp. 209-28

Maitland, F.W. (1897) Domesday Book and Beyond, Cambridge University Press, Cambridge

Mawer, A., and Stenton, F.M. (1924) 'Introduction to the Survey of English Place-Names', English Place-Name Society, Cambridge, 1

Millward, R. (1956) Lancashire: The Making of the English Landscape, Hodder and Stoughton

Millward, R. (1972) 'Saxon and Danish Leicester' in N. Pye (ed.), Leicester and its Region, Leicester University Press, Leicester, pp. 218-34

Moodie, D.W., Lehr, J.C., and Alwin, J.A. (1974) 'Zelinsky's Pursuit: Wild Goose or Canard?', Historical Geography Newsletter, 4, 18-21

Morris, C. (1966) 'Roman Roads and Saxon Settlements in Hertfordshire', Hertfordshire Past and Present, 6, 3-8

Mortensen, H., and Mortensen, G. (1937) 'I. Die Preusisch-Deutsche Siedlung am Westrand der Grossen Wildnis um 1400', Die Besiedlung des Nordöstlichen Ostpreussens bis zum Beginn des 17 Jahrhunderts, Leipzig

Mortensen, H., and Mortensen, G. (1938) 'II. Die Wildnis im

THE DARK AGES

östlichen Preussen, ihr Zustand um 1400 and ihre
Frühere Besiedlung', Die Besiedlung des Nordöstlichen
Ostpreussens bis zum Beginn des 17 Jahrhunderts,
Leipzig
Müller-Wille, W. (1944) 'Langstreifenflur und Drubbel', Deut.
Archiv. Land u. V. Forschung, 8, 9-44
Müller-Wille, W. (1965) Eisenzeitliche Fluren in den
festländischen Nordseegebieten, Münster
Myers, J.N.L. (1935) 'Britain in the Dark Ages', Antiq., 9,
455-64
Niemeier, G. (1958) 'Erste Ergebnisse von c^{14}-Datierungen in
der Kulturlandschaftsgeschichte Nordwestdeutschlands',
Ber. z. Deut. Landfsk., 1
Nitz, H.J. (1961) 'Regelmässige Langstreifenfluren und
fränkische Staatskolonisation', Geographische Rundschau,
13, 350-65
O'Dell, A.C. (1939) The Historical Geography of the Shetland
Isles, T. and J. Manson, Lerwick
Phillips, M., and Unwin, T. (1985) 'British Historical Geo-
graphy: Places and People', Area, 17(2), 155-65
Proudfoot, V.B. (1961) 'The Economy of the Irish Rath',
Medieval Archaeology, 5, 94-122
Sandred, K. (1976) 'The Element hamm in English Place-
Names: a Linguistic Investigation', Namn Och Bygd., 64,
69-87
Seebolm, F. (1883) The English Village Community, Longman,
London
Small, A. (1968) 'The Historical Geography of the Norse
Viking Colonisation of the Scottish Highlands', Norsk
Geogr. Tids., 22, 1-16
Sporrong, U. (1968) 'Phosphatkartierung und Siedlungs-
analyse', Geografiska Annaler, 50, 62-74
Sporrong, U. (1971) 'Kolonisation, Bebyggelseutvechlig och
Administration', Medd. f. Kulturgeog., Inst. Stockholm
Universit.
Sporrong, U. (1982) 'Individualistic features in a Communal
Landscape: Some Comments on the Spatial Organisation of
a Rural Society' in A.R.H. Baker and M. Billinge
(eds.), Period and Place, Cambridge University Press,
Cambridge, pp. 145-54
Stenberger, M. (1933) Öland Under Aldre Jarnalder,
Stockholm
Stephenson, C. (1933) Borough and Town: A Study of Urban
Origins in England, Monog. Med. Acad. Amer. 7,
Cambridge, Massachusetts
Tait, J. (1936) The Medieval English Burgh, Pub. Univ. Man.
Histo. Ser. 70, Manchester
Taylor, C.C. (1974) 'The Anglo-Saxon Countryside' in T.
Rowley (ed.), Anglo-Saxon Settlement and Landscape,
British Archaeo. Rep., 6, 5-15
Thomas, C. (1972) 'The Irish Settlements in Post-Roman

Western Britain: A Survey of the Evidence', Journal Royal Inst. Cornwall, 6, 251-74

Thomas, C. (1974) 'Place-Name Analysis in the Geographical Study of Landscape', Stud. Celt., 8-9, 299-318

Thorpe, H. (1962) 'The Lord and the Landscape, Illustrated through the Changing Fortunes of a Warwickshire Parish, Wormleighton', Transactions, Birmingham Archaeo. Society, 30, 38-77

Unwin, T. (1983) 'Township and Early Fields in North Nottinghamshire', Journal of Historical Geography, 9(4), 341-6

Wacher, J. (1964) 'Cirencester, 1963. Fourth Interim Report', Antiq. Journal., 44, 9-18

Wacher, J. (1975) The Towns of Roman Britain, Batsford, London

Watts, V.E. (1976) 'Comment on "the Evidence of Place-Names by M. Gelling"' in P.H. Sawyer (ed.), Medieval Settlement: Continuity and Change, Arnold, London, pp. 212-22

West, S.E. (1969) 'The Anglo-Saxon Village of West Stow', Medieval Archaeology, 13, 1-20

Whittington, G. (1977) 'Place-Names and the Settlement Pattern of Dark Age Scotland', Proc. Society of Antiq. of Scotland, 106, 99-110

Widgren, M. (1982) 'Field Evidence in Historical Geography: A Negative Example?' in A.R.H. Baker and M. Billinge (eds.), Period and Place, Cambridge University Press, Cambridge, pp. 303-11

Williams, J. (1977a) 'Excavations in St. Peter's Street, Northampton', Northampton Development Corp. Archaeo. Monog., 2

Williams, J. (1977b) 'The Early Development of the Town of Northampton' in A. Dornier (ed.), Mercian Studies, Leicester University Press, Leicester, pp. 130-50

Wooldridge, S.W. (1936) 'The Anglo-Saxon Settlement' in H.C. Darby (ed.), An Historical Geography of England before 1800, Cambridge University Press, Cambridge, pp. 88-132

Wooldridge, S.W., and Linton, D.L. (1933) 'The Loam Terrains of South-East England in their Relation to its Early History', Antiq., 7, 297-303

Wyld, H.C. (1950) The Place-Names of Lancashire, English Place-Name Society, London

Chapter Four

MEDIEVAL ECONOMY AND SOCIETY

I.D. Whyte

INTRODUCTION

The term 'medieval' can be interpreted in various ways.
Writing this chapter in England during 1985, the 500th
anniversary of the Battle of Bosworth, the conventional end
of the Middle Ages in school textbooks, with the 900th
anniversary of the compilation of the Domesday Book falling in
1986, it is tempting to choose such clear-cut limits for this
review. When, however, one is dealing with the slow-moving
rhythms of agrarian life, with economic and demographic
cycles which may last for centuries, the choice of political
events as boundaries can be meaningless. Nevertheless, since
this chapter, relating to a period rather than to a systematic
theme requires a chronological framework, one which begins
in the mid 11th century and ends in the early 16th may sound
more convincing than 1066 to 1485. At the beginning of this
period Europe was settling down after the Viking diaspora
and was about to enter a phase of expansion which peaked
during the later 13th century before the onset of catastrophic
decline during the 14th century. The next major upswing did
not begin to gather momentum until the early 16th century by
which time a variety of new influences, conventionally labelled
'the Renaissance', were starting to alter social and economic
conditions. Even so, some backtracking into pre-Norman times
will be necessary in considering some themes while others
spill over into the later sixteenth century.

Over how wide an area the term 'medieval' has validity is
also debatable. Unfortunately it is impossible within a chapter
such as this to pursue the term to its geographical limits. To
achieve some depth of treatment, this review will focus on the
British Isles with comparative glances at continental Europe.
Even within these limits the task is formidable. Selectivity has
been essential and this has been accomplished by concentrating on a range of problems which have been the specific
concern of historical geographers, and to which they have
made significant contributions. Some topics, such as medieval

rural settlement and urban morphology, are considered within a systematic framework elsewhere in this volume and have not been treated in detail here. Emphasis is placed mainly on work which has appeared within the last ten years.

Geographers interested in the medieval period have, to an even more marked degree than historical geographers in general, ignored the new approaches which have begun to penetrate the subject within the last decade. Perhaps inevitably, the Quantitative Revolution had only a limited impact on the work of historical geographers in this area as relatively few medieval sources offered scope for large-scale number crunching. The potential of computer search-and-sort packages is perhaps exciting. This approach has already been applied to the Domesday Book (Hamshere and Blakemore, 1976) and charters in the Scottish Register of the Great Seal (Doherty and Gibson, 1983). Nevertheless, this only represents technical advances in data processing within a traditional positivist approach. While historical geographers, working in a more numerate and computer-literate milieu than most medieval historians, may gain temporary kudos from demonstrating expertise in such techniques, historians will soon catch up as archaeologists have done already. There has been little attempt to re-assess the fundamental analytical frameworks within this branch of historical geography and the emphasis of published work is still often descriptive rather than analytical. The debates which have been launched within historical geography on such themes as structuralism, Marxism, humanism and phenomenology have had little impact as yet on the study of medieval economy and society. The evaluation of data sources still remains the focus of attention for many workers rather than consideration of the framework within which the sources are interpreted. Many ideas which historical scholarship has absorbed in recent years from sociology, anthropology and philosophy have barely influenced historical geographers. To be fair, many medieval historians also appear to assume that their approaches are value-free, but historians have in general been more receptive to external influences than their geographical colleagues.

A recent survey of research in historical geography, mainly within the British Isles (Whyte, 1984), shows that only a tenth of the research projects listed were concerned specifically with the Middle Ages. This has two important implications: first, with so few geographers active in this field, it is unlikely that they will make a major contribution to medieval studies. This is borne out by the present review. While the work of geographers has been highlighted here, in most areas, with the possible exception of the study of rural settlement and field systems, the frontiers of knowledge are being pushed forward mainly by historians. Second, with work on the medieval period being a minority interest, this area of research is unlikely to represent the cutting edge of innovation within historical geography. Indeed, almost the reverse is true, such work reflecting in miniature the traditional empirical and positivist approaches which have dominated historical geography until very recently (Butlin, 1982). In Whyte's register over half the projects are concerned with rural settlement, field systems and landscape evolution. There is comparatively little interest in topics such as population, social structure or conceptual issues like the transition from feudalism to capitalism.

THE MEDIEVAL LANDSCAPE

Historical geographers have traditionally been concerned with interpreting past landscapes and studying their evolution: landscape research is still prominent in historical geography's contribution to medieval studies. W.G. Hoskins' 'The Making of the English Landscape', published in 1955, was a milestone in the development of landscape research, a book which was at once scholarly, stylish and readable. In this respect it still stands alone. Despite the advances in landscape studies in the last 30 years Hoskins' book remains the best general introduction to its theme. Hoskins also initiated the 'Making of the English Landscape' series of county volumes which have appeared steadily since the late 1950s. The approach taken by the various authors has generally followed Hoskins' model. The quality of the series has improved through time and the more recent examples (Allison, 1977; Bigmore, 1979; Reed, 1979) demonstrate some of the ways in which knowledge of medieval landscapes has developed in the last 30 years.

Recent advances in the study of English medieval landscapes have been reviewed by Aston (1982-3) and Taylor (1980-1). It has been established that the origins of man's impact on the English landscape lie much further back, and were more complex, than Hoskins thought. Hoskins saw the origins of the modern landscape in the Saxon and medieval periods, attaching little importance to the contributions of prehistoric and Roman times. Now, however, much initial settlement and woodland clearance which was thought to have been the work of medieval man has been shown to have occurred during pre-medieval, even pre-Roman times. It is now considered that during the first millennium B.C. population was higher and the landscape more intensively exploited than at the time of the Domesday Survey. This has necessitated a re-appraisal of the significance of medieval landscape development. The idea that nucleated villages and open field systems were introduced in early Saxon times has been shown to be incorrect. There is growing evidence that rural settlement was dispersed in early and mid-Saxon England and that the change to nucleated settlement only took place later. The indications of high levels of population and intense landscape exploitation in earlier times have diminished the scale of medieval settlement expansion and land clearance. It is now appreciated that instead of recording the start of a major phase of expansion, the Domesday Book may come near the end of a process rooted far back in prehistory. It is also clear that there was more change and less stability in the medieval landscape than was once thought. The process of settlement nucleation in late Saxon times had barely begun before new influences were encouraging a further phase of dispersion into isolated farms, hamlets and daughter villages. This evidence for rapid change should make us wary of assuming that early forms, such as field and settlement layouts, necessarily persisted with little alteration into post-medieval times. Long-term stability of this kind may well have been the exception rather than the rule. If so many elements in the English medieval landscape had their origins in prehistory one might expect this tendency to be more marked in Scotland, Wales and especially Ireland where there was more direct cultural continuity from prehistoric to medieval times. This has been reflected in a number of studies, notably those of Jones (1985) for Wales and Smyth (1982) for Leinster. Jaeger (1982) has also demonstrated interesting parallels between Ireland and Prussia both beyond the margins of the Roman Empire, where tribal territorial structures survived into medieval times.

In view of the steady flow of new ideas appearing in the journals it is remarkable that so few general surveys of medieval landscapes and their evolution have appeared since Hoskins' time. Cantor's volume (1982) is an unexciting survey which fails to emphasise new developments. Reed's collection of essays, while conveying something of the excitement generated by new research, lacks coherence and fails to define the objectives of landscape research. There is a need for a review of recent developments in our understanding of the English medieval landscape which considers techniques of landscape study, interpretations and theoretical approaches.

An encouraging development in recent years has been the greater attention which has been paid to the evolution of Scottish landscapes, a neglected theme compared with England. Parry and Slater's collection of essays (1980) marks a move away from the idea that the Scottish rural landscape was almost entirely the product of 18th and 19th century changes. Dodgshon (1981) has brought his many papers on medieval and post-medieval Scotland together into a major study of land and society. If progress in Scotland has been limited until recently, that in Ireland has been even less impressive. Aalen (1978) is more at home with Irish pre-history and his chapters on medieval Ireland are mostly reworkings of earlier published material with few new insights.

MEDIEVAL ECONOMY AND SOCIETY

The problem with research on medieval landscapes has been its highly empirical nature and its lack of any generally accepted theoretical and methodological framework. At their worst landscape studies can be purely descriptive, with effort being directed towards source evaluation and fieldwork rather than the processes underlying landscape change. It is easy to explain such changes by simplified influences such as population pressure without questioning the nature of man's activities and how they were reflected in the landscape. One problem with the medieval period is that documentation is rarely adequate enough to provide insights into individual motives and behaviour. We can only imagine, for example, how individual peasants saw the fields which they cultivated, and their motives in maintaining or altering their farming systems. Sometimes one can identify the work of an individual landowner in shaping the landscape; more frequently the nebulous concept of 'landlord control' is advanced to explain changes.

1977 saw the appearance of the final volume of H.C. Darby's Domesday Geography, a project which has been described as one of the most remarkably scholarly enterprises of the age. The five regional Domesday Geographies, the gazetteer and the summary volume have come to epitomise the traditional cross-sectional approach in historical geography. It may be unfortunate if some medieval historians view this as the main paradigm within historical geography. Indeed, it is important to appreciate that Domesday Book lends itself uniquely to this sort of treatment. Nevertheless, the Domesday Geographies have become standard reference works whose meticulous scholarship shows how much can be achieved by the painstaking analysis of a single major source. This does not imply, however, that the last word has been said on the geography of Domesday England or even of the Domesday Book. Recent experiments in computerising Domesday data (Hamshere and Blakemore, 1976) have shown that it is possible to wring new information and examine new sets of relationships as a result of the computer's ability to handle large bodies of information. If Darby's work has sometimes been criticised as source-bound empiricism, it is nevertheless an essential stepping stone to further analysis.

THE RURAL ECONOMY

Historical geographers studying the medieval period have continued to be preoccupied with open field systems; their origins, evolution, regional and local diversity, and eventual removal. The contribution of historical geographers to the debate on medieval field systems during the 1950s and 1960s lay mainly in the provision of case studies highlighting the regional variety of field systems rather than with more theoretical aspects of their origin and nature. Within the last 10 years or so, however, geographers have made important contributions to these broader issues. Baker and Butlin's (1973) survey was an important landmark. Described by Baker (1979) as a stocktaking exercise, it synthesised previous research on a regional basis. Its focus was on spatial variation and the evolution of the different types of field system described. It brought out some of the short-comings of existing work and pointed the way forward by showing lacunae in spatial and temporal coverage.

Developments in the study of field systems since 1973 have been surveyed by Baker (1979; 1983). The most important single contribution to the debate on the origins of open field systems has been that of Dodgshon. Initially he examined Scottish infield-outfield farming (1973; 1975a), stressing the importance of differences in tenure between the two main components of the system. He saw the distinction between infield and outfield as being due to a contrast between an original core of assessed land and later, un-assessed intakes from the waste (1975b,c; 1977). Subsequently he extended this work to an analysis of the origins of British field systems (1975d,e; 1978; 1980). Dodgshon believed that field systems were not designed, that the term 'system' is misleading, implying a coherence which may never have existed. Field systems were not a conscious response to particular sets of conditions, but more makeshift in their origins and development, an amalgamation of responses to various influences which were not necessarily directly related to agriculture. He stressed that field systems should be considered in the context of the total history of the communities that worked them. Field systems were not so much the deliberate creation of agrarian communities as influences on the nature of these communities. He concluded that sub-divided fields arose essentially from piecemeal colonisation and the sharing out of assarts. Communal farming in turn arose as a response to the logistical problems of working subdivided fields. He saw shareholding as being important in producing subdivided fields though influences such as population growth, partible inheritance and the operation of an active land market might also have encouraged fragmentation.

McCloskey (1975; 1976; 1979) has presented a case for open fields being the result of risk avoidance by the medieval peasantry, the scattering of parcels being an insurance policy reducing losses from natural hazards. Fragmentation of land within an open field system gave cultivators diversified parcels of plots with different types of soil so that no matter what weather conditions prevailed, something would be produced for basic subsistence. Over time, as agricultural technology and yields improved, the need for dispersal was reduced and the incentives grew to consolidate and enclose land. While there is no direct evidence that medieval farmers actually thought this way, McCloskey has developed and clarified ideas which have been implicit in much of the earlier literature on field systems. Baker (1979) has suggested that risk avoidance was a secondary consideration, an incidental advantage once the fragmentation of plots had been produced for other reasons.

Campbell (1980; 1981a,b) has also made an important contribution to the study of field systems. He has pointed out that despite the considerable body of writing on regional variations the precise nature and geographical distribution of different types of field system in England are still imperfectly known, while no generally accepted set of criteria exists for differentiating field systems. To remedy this, he has produced a functional classification of open fields within the arable-dominated parts of England which, on the basis of combinations of attributes, allows a division into five types of field system with several sub-types. In searching for the origins of regional variations in the degree to which open fields became subject to communal regulation, Campbell proposes an alternative to Thirsk's model which influenced much of the writing on field systems in the 1960s and 1970s. Thirsk explained the development of common rights and regulations over subdivided fields as a response to pressures caused by population growth and the resulting problems created in communities where the arable area had expanded to such a degree that there was a shortage of pasture. Campbell suggests that common field systems would not have developed in periods of population pressure because it was unlikely that the cultivators would have agreed to the restrictions involved. The adoption of such regulations would have fossilised the structure of field systems into a rigid pattern whose very existence would have worked against population growth. Growing population pressure could, Campbell believes, have been countered within the structure of an irregular common field system by advances in agricultural technology such as the introduction of legumes and other fodder crops, by the replacement of oxen by horses for ploughing, or by the intensification of existing practices. His work on field systems in east Norfolk (1983) confirms that such innovations were indeed introduced. The rigidity of regular common field systems was, he argues, more likely to have developed during periods of population stagnation or decline as a response to the need to make efficient use of scarce labour resources. Under such circumstances opposition to the institution of a rigid system is likely to have been less than during a period of population growth. Lastly, Campbell has examined the direct and indirect influence of lordship in the creation of regular field systems, pointing out that the areas in which regular common field systems dominated, central and southern England, were characterised in medieval times by a high incidence of strong, undivided lordship. His

comments on the relationships between the organisation of
field systems, lordship, and the development of peasant
communities as corporate entities demonstrate how recent
research on field systems has been concerned with interpreting them in their socio-economic context. This approach
points the way forward for further work on the inter-
relationships between field systems and the societies which
created them.

An example of recent progress in the study of field
systems at a regional scale is the work of Harvey (1980;
1982; 1983; 1984) on Holderness, the Yorkshire Wolds. She
has uncovered evidence of a widespread regularity in the
layout of open fields, and the way in which land was
apportioned within them. In these areas in the 16th century
open fields were characterised by simple divisions into long
parallel strips with few furlongs. Each furlong had a similar
number of strips or lands. The oxgangs which formed the
tenurial basis of the townships were organised so that each
one had a land in the same position in every furlong. Their
layout showed evidence of planning contrasting with the
complexity and irregularity of field systems in other parts of
England.

The occurrence of such regular fields over so large an
area was interpreted as indicating that large-scale planning
had been enforced from above by landlord control. In a few
townships it was possible to show that this system had
existed in the mid 13th century. The problem was to pinpoint
and explain its origins and account for its survival. As the
system had originated before detailed documentary sources
were available one difficulty was that indirect evidence, such
as the Domesday Book and archaeological data, had to be
used to indicate the presence or absence of factors which
might have encouraged such large-scale reorganisation. The
problems of this line of reasoning have been discussed by
Austin (1985). The emphasis on morphology distracts attention
from the processes which created and perpetuated particular
forms, and the ways in which they functioned. Absence of
evidence of change can easily be interpreted as evidence of
stability. Yet the more the origins of features in the medieval
landscape are sought in pre-Norman times the more it becomes
necessary to use evidence which is either specific to particular sites, such as archaeological data, or extremely generalised, as with the interpretation of settlement processes, and
population change derived from place name evidence. One
result of pushing the origins of regular field systems back
before the 11th century is that the cause and effect linkages
used to explain their establishment and spread become crude
and oversimplified. A second result is that chronology
becomes increasingly flexible, ending up with only a choice of
possible periods when the field systems might have been laid
out and little hope of definitive proof.

These difficulties emerge in Harvey's work. She suggests that two important factors which might have encouraged large-scale reorganisation of field systems in the areas she studied were unity of landlord control and phases of wide-spread disruption possibly accompanied by temporary depopulation. A favourite period in which to place changes in the organisation of settlements and field systems in Northern England has been immediately after the Norman Conquest, in the wake of the Scandinavian invasion and William I's campaign of 1069-70. Unfortunately, for Holderness and the Wolds the supposed pre-requisite of unity of landownership did not occur at this time.

This caused Harvey to push the search for the origin of regular open fields back to the period following the Scandinavian invasions of the late 9th century when there was tenuous evidence for unity of control in this area. Similarities between the layout of holdings within the fields in this area and the Scandinavian system of sokskifte encouraged this choice. A problem with seeking such early origins was that the laying out of the new field systems had evidently been undertaken in a landscape which was already densely settled and fully exploited; the limits of the fields extended to the township boundaries in some cases and the fields themselves were sufficiently large to require no further intakes by assarting during the build up of population in the 12th and 13th centuries. However, current interpretations favour the idea that much of England, including East Yorkshire, was fully exploited but with a dispersed settlement pattern which was very different from the nucleated villages which were intimately associated with the field systems. An origin before late Saxon times thus seems unlikely. Harvey suggested that such systems had been more widespread in England at one time but that they vanished as increasing population pressure in medieval times led to the subdivision of existing fields and the assarting of more land.

Support for the mid-late Saxon origin of field systems like these is likely to come mainly from excavation. Taylor (1981) has discussed the difficulties of the archaeological interpretation of remains of field systems. They are difficult to date; their occurrence under other datable features only provides a crude chronology and by their nature fields are disturbed annually by cultivation so that early forms are destroyed by the creation of later ones. Nevertheless, instances have been discovered where medieval furlong boundaries overlie Roman ditches and appear to have been aligned using them. Hall (1981) has suggested that archaeological evidence can be used to reconstruct the early forms of field systems by mapping furlong boundaries which tend to be more permanent and prominent than the ridge and furrow within them. By establishing a relative chronology of furlong boundaries in some Northamptonshire parishes he has reconstructed an early arrangement of long strips, similar to Holderness and the Wolds. Again, he places the origins of these systems in later Saxon times.

The potential of early documentary sources, such as pre-Conquest charters, for providing information on the early landscapes, especially arable land, has been shown by Hooke (1978-9; 1981a,b) in her study of the West Midlands. The boundary clauses in late Saxon charters, analysed systematically, allow spatial differences in the intensity of cultivation, and something of its nature, to be reconstructed and explained. Her work has suggested that in some areas the high level of intensity of agriculture in late Saxon times reflects a pattern of development which was already evident during the Roman period.

Many problems associated with the origins and development of open field systems mirror those of settlement morphology. It has often been too readily assumed that regularity could only have been imposed from above, not generated from below. It has also been taken for granted that a single landowner had to dominate an area for regular forms to be widespread. The possible role of fashion, imitation and the diffusion of ideas has often been ignored. Perhaps the greatest problem with recent work on field systems, as Dodgshon (1979) has emphasised, is that research has produced increasing numbers of detailed case studies, and now a variety of theoretical standpoints, without making much real progress. There has been a proliferation of ideas instead of a convergence of viewpoints. This is evident in the essays edited by Rowley (1981). Rowley at least is optimistic that despite the lack of consensus, a good deal of important new research is being undertaken. Nevertheless, it is clear that some of the theoretical work of the last ten years, notably that of Dodgshon himself, has encouraged consideration of the origins, evolution and diversity of field systems in a broader context, so that there is now a more active dialogue between the more abstract and the more empirical research in this area.

Medieval historians have long accepted that short runs of severe weather conditions had large-scale effects on society causing crop failures, livestock mortality and famines. Geographers and economic historians have continued to study the effects of weather on medieval agriculture (Dury, 1981; Hallam, 1984). Historians have been less convinced of the influence of longer-term climatic changes on medieval society. This has been due partly to the lack of detailed, accurate chronologies of climatic variations during medieval times and a failure by many historians to understand the mechanisms which produced them. In addition, the deterministic interpretations of the effects of climate on man put forward by some geographers earlier this century have helped to discredit this approach. The problem with relating climatic

variations to socio-economic changes is to establish clear
cause-and-effect linkages which filter out other influences.

Parry (1975; 1976a,b; 1978) has made an important
contribution to solving these problems. His argument has been
that it is necessary to focus on areas of marginal settlement
and agriculture where the effects of climatic shifts, amplified
with small increases in altitude in the maritime environments
of N.W. Europe, were likely to be particularly significant and
easy to identify. Using a case study from South East
Scotland, he demonstrated that remains of cultivation and
associated settlements occurred at altitudes which, under
present climatic conditions, would not allow even the hardiest
of crops to ripen in most years. Using cartographic and
documentary evidence Parry produced a chronology of settlement and land abandonment in this area from medieval times
onwards. The medieval climatic optimum coincided with an
expansion of cultivation to high altitudes. Parry showed that
during this optimum, temperature, exposure and soil moisture
conditions would have allowed subsistence crops of oats, with
an acceptable probability of failure, in all but the highest
parts of this area. The subsequent late-medieval climatic
deterioration was accompanied by a retreat of settlement and
cultivation. Parry did not claim that climatic change was the
sole cause of this, but he did suggest that the increasing
probability of harvest failure at a given altitude was an
important background factor encouraging permanent abandonment when other proximal causes made cultivation more
difficult. This restrained and cautious interpretation has
pointed the way forward for work on settlement changes in
marginal areas and Parry has refined his theoretical
approaches to the study of the links between climatic change
and human events (1981). Other studies have used behavioural approaches to identify structural weaknesses within
medieval societies to determine whether they allowed adapta-
bility or caused inertia in the face of climatic stress. A
notable contribution in this field has been McGovern's (1981)
analysis of the failure of the medieval Norse Greenland
colony.

POPULATION AND RURAL SOCIETY

The secular patterns of population change during medieval
times are reasonably clear for Western Europe as a whole, but
at a smaller scale there are problems in reconstructing population distribution, trends and characteristics at regional and
even national levels. Population figures for individual
countries are only estimates based on non demographic data
like the Domesday Book, the Hundred Rolls or the 14th
century Lay Subsidies. If estimates of the population of a
comparatively well-documented country like England can vary

MEDIEVAL ECONOMY AND SOCIETY

by 100% using the same source (Dodgshon, 1980), the figures
for more poorly recorded countries like Scotland or Ireland
are pure guesswork, based largely on backward projections
from later periods. Although for England absolute population
densities are uncertain, relative variations can be established
from a number of sources though problems of interpretation
may cast doubt on the scale of such differences; for example
between the south and east of England and the north and
west in 1086 (Dodgshon, 1980). The vital importance of
population fluctuations as a major influence on the economy
has long been appreciated and this tradition has been
continued in studies such as that of Hatcher (1977) on population change and the English economy in the later Middle
Ages. The role of epidemic disease is, inevitably, prominent
in studies of late medieval demography and the crisis of the
Black Death has continued to exert a grim fascination,
whether in general studies such as that of Gottfried (1984),
or regional ones like that of Shirk's (1981) for Aragon.

While surveys like the Domesday Book and the lay
subsidies can provide indications of relative, though less
certainly of absolute, spatial variations in population density,
it is more difficult to determine the extent to which regional
and local trends departed from the national picture. In the
period before the later 16th century when parish registers
first become available in some quantity, sources for smaller-
scale demographic studies become even more scanty and
indirect than for the early-modern period. Manorial court rolls
provide surrogate data on population changes at the level of
the village community but regional patterns are more difficult
to establish.

Medieval surveys can also show spatial and temporal
changes in the geographical distribution of wealth. This has
been attempted for England by Darby et al. (1979) using
three cross sections; 1086, 1334 and 1524-5 with a framework
of small geographical units. The resulting maps allow changes
in the pattern of wealth between the three surveys to be
assessed. The patterns of change for the two periods are
markedly different. Between 1086 and 1334 the most substantial increases in wealth occurred in the Fenlands, those
parts of Northern England which had been devastated by
William I in 1069-70, and areas of late colonisation of marsh,
woodland and upland. Between 1334 and 1525 the growth of
wealth mainly reflected the expansion of the textile industries
in areas like the South West, Essex, Suffolk, and parts of
the West Riding. Areas which had a steady increase in wealth
throughout the two periods included London and its environs,
larger towns like Bristol and Coventry and some rural areas
like the Weald. This research has produced a framework for
further inquiry at a more detailed level but it has been
criticised by Stanley (1980) who warns of the dangers of

using three such widely-spaced sources produced under different circumstances, for different purposes.

One of the most important recent advances in the study of medieval demography and society has been the growth of detailed reconstructions of small communities using nominal linkage analysis of manorial records. This has occurred at a time when links between historical demography, sociology and anthropology have become closer, and the need to link population characteristics to their socio-economic context is increasingly appreciated, along with the realisation that technical sophistication in handling population data needs to be matched by the use of appropriate theoretical approaches.

Within the last 20 years detailed work on the nature of early-modern European society has focused on themes such as family and household structures, the distribution of wealth, inheritance customs, community relations and geographical and social mobility. These themes have been pushed back into medieval times, adapting to different problems and sources in the process. Few documentary sources offer the opportunities for detailed community reconstruction like the material used by Le Roy Ladurie in his study of Montaillou in the Pyrenees (1976) but increasing use of more generally available sources has opened up many new possibilities. Macfarlane (1977) has shown that the amount of information which can be assembled about ordinary people in early-modern times is surprisingly great. It is now appreciated that medieval peasant societies are also better recorded than has often been appreciated, particularly in England. Manorial court rolls provide information on many aspects of community relations including demography, kinship and neighbourliness, debt and credit, crime and violence, and population mobility. Earlier studies of medieval communities used manorial court rolls in a generalised, anecdotal way. More recently there has been a veritable explosion of research and writing on medieval peasant societies integrating manorial records at the level of the village or estate. Much of this work has drawn on sociological and anthropological theory as well as quantitative analysis. Marxist historians have made an important contribution, notably Hilton's work on the peasantry of the west Midlands (1966; 1975).

Another major focus of research has been the Pontifical Institute of Medieval Studies at Toronto. Under the leadership of Raftis a number of community studies have been undertaken, notably by Britton (1978), Dewindt (1972) and Raftis himself (1974). Other community studies have been those of Harvey (1977) for the Westminster Abbey estates, Howell (1983) on Kibworth Harcourt, Dyer (1980) on the estates of the Bishopric of Worcester, and Razi (1980) on Halesowen. An important contribution has been that of Smith (1979; 1983). Other aspects of medieval society which have been examined include birth control (Biller, 1982), family structures

(Herlihy, 1983), and household composition (Hammel, 1980; Ring, 1979). Goody et al. (1976) have edited an important collection of studies on family and inheritance in Western Europe from medieval times onwards. There have also been notable advances in the study of geographical mobility using evidence such as place-name surnames (McClure, 1978-9). Earlier ideas that medieval rural society was relatively static have been replaced by the realisation that considerable mobility occurred. Field (1983) has examined this theme using an English case study, MacPherson (1984) using one from Scotland and Angers (1979) a French example.

In recent years historical geography has been influenced by Marxist theories of social and economic change. In the context of the medieval period Marxist research has focused on the structure of relations within feudal society and the nature of the transition from feudalism to capitalism. There has been considerable debate among Marxist historians over the nature of this transition. Marx himself was principally concerned with analysing capitalism and only sketched out his ideas regarding the structure of feudal society and the ways in which the transition to capitalism occurred leaving plenty of scope for debate. The early work of historians on the transition has been summarised by Hilton (1976). More recently, contrasting views on the transition have been put forward by Hindess and Hirst (1975) and Martin (1983).

Traditional interpretations saw the genesis of capitalism taking place within the towns. Many Marxist historians believe, however, that capitalism arose in the countryside out of forces operating within the feudal system rather than from influences imposed from outside rural society. Marx considered England to be the classic example of the development of agrarian capitalism and the debate has been widened recently within a European context by Brenner (1976; 1982). Brenner suggests that capitalism developed out of agrarian society in England because of the lack of a large class of peasant landowners and the existence of strong landlord control over rents. In other parts of Europe, notably France, such a transition was inhibited by the importance of peasant owner occupiers and inheritance strategies which produced fragmentation of ownership.

Brenner's arguments were designed to counter not only the traditional Pirenne model which stressed the importance of towns and trade in promoting capitalism but also models based on demographic influences which explained economic trends in terms of the external, independent variables of fertility and mortality. Brenner's first paper provoked a lively debate with contributions from Postan and Hatcher (1978) and Le Roy Ladurie (1978) who were among the principal exponents of the demographic theory. Brenner saw the emergence of the classic landlord-capitalist, tenant, wage labourer structure in England as the key to the transformation of agriculture and rural society, encouraging investment and innovation by both landlords and tenants with the creation of a rural proletariat divorced from the land. Much of the recent research on the socio-economic structure of late medieval and early-modern Britain tends to support Brenner's thesis. On the other hand it has been suggested that landlord-tenant-wage labourer structures existed elsewhere without producing the same effects; in parts of France for example (Cooper, 1978).

The strains which existed within feudal society, especially conflict between landlord and peasant, have continued to be a focus of attention. At a macro-scale medieval peasant revolts have been studied by Hilton (1973) and Hare (1982) for England, Blickle (1979) for Germany and by Fourquin (1978) and Francois (1974-5) at a European scale. Charlesworth (1983) has provided a geographical perspective on rural rebellions in Britain at the end of the Middle Ages. He has examined early 16th century rural protests and the degree to which they can be considered as part of the traditional feudal antagonism between lord and peasant, or as a reaction to the rise of agrarian capitalism. More recently attention has turned to the study of medieval crime at a smaller scale. Stone (1983) has considered the changing nature of violence in English society from the 14th century onward. Urban crime has also been analysed by Hammer (1978) for Oxford, Cohn (1980-1) for Florence and Blanshei (1982-3) for Bologna.

TRANSPORT AND COMMUNICATIONS

As Hindle (1982) has pointed out, much has been written about medieval trade, but comparatively little attention has been given to roads and road transport. The better documentation relating to international and seaborne commerce has lured economic historians and historical geographers away from the more mundane, but arguably more important, patterns of regional and local trade. The growth of the medieval economy and the increase in population between the 11th and 14th centuries must have greatly increased traffic on existing roads while the pioneer colonisation or re-settlement of marginal areas must have led to the creation of new road networks. Much of what has been written on roads has been anecdotal, concentrating on how badly they were maintained, and the difficulties and dangers involved in travelling them. Labarge's (1982) study of medieval travellers continues this tradition. This is largely due to lack of evidence and as a result many histories of road transport treat the medieval period as a brief, unimportant interlude between the Roman period, supported by abundant archaeological evidence, and early-modern times, when better documentation becomes available. Taylor (1979) has provided a guide to the physical remains of medieval roads. Other archaeological evidence can be used to establish the antiquity of a route as Hooke (1976-7) and Steane (1983) have shown. The problem is that medieval roads were worn by use rather than being constructed and so are difficult to date as visible features on the ground may have developed over a long timespan, though some roads in areas like the Fens were deliberately constructed. Nevertheless, landscape evidence only provides information on limited, disjointed stretches of road, often in association with other kinds of features such as deserted settlements.

For the reconstruction of road networks on a larger scale, documentary sources are essential. Hindle has made an important contribution to the study of medieval road networks at a national scale using cartographic sources (1976; 1980; 1982). The legacy of the Roman road system is clear with some 40% of the routes shown on the Gough map of c1360 being on the line of Roman roads. At a more detailed level he has demonstrated the potential of combining documentary material with field evidence in studies of medieval road networks in Cumbria (1977; 1984) and Cheshire (1982). The difficulty of reconstructing medieval road networks at this more detailed scale emerges from Hindle's study of roads in medieval Cumbria (1984). All that the documentary evidence allows one to do in many cases is to locate the centres of power and authority, such as castles and monastic houses, and to postulate the links which might have existed between them and their dependent settlements. Although roads are mentioned in charters and other documents such references are generally sporadic and are of little use in reconstructing more general patterns of movement, although they may allow one to establish the line of a medieval route at a specific locality. In the study of medieval transport, he has also examined the seasonality of movement (1978), while Langdon (1984) has made important claims for a revolution in vehicle transport in the 12th and 13th century with a change from oxen to horses.

TOWNS

Since developments in the study of urban morphology will be considered in Chapter 9, this section reviews recent work on other aspects of medieval towns. Despite the growth of urban history in recent years, less attention has been paid to economic and social aspects of medieval towns in Britain than for later periods. Many studies have focused on individual towns and have lacked the broader, comparative perspective which has been characteristic of recent work on early-modern towns in Britain.

The origins of medieval urban development have been a fruitful area of investigation, encouraged by growing numbers of excavations of medieval town centres since the 1960s (Clarke, 1977; Brooks, 1977). On a European scale the best recent survey of medieval urban origins is Barley (1977) in which the question of continuity of urban functions from Roman to medieval times is considered in the context of Britain, France and the Western Mediterranean. French (1983) has also discussed urban origins in medieval Russia while Carter (1983) has provided an introduction to the diffusion of urbanisation in medieval Europe reviewing the various theories of urban origins. The shadow of Henri Pirenne still looms large over much of this work. His influence is discussed by Hodges (1982) who has developed a new approach based partly on recent archaeological discoveries. Instead of being a period of economic stagnation as a result of declining long-distance trade, Hodges re-interprets the 9th and 10th centuries as a time of urban growth stimulated by exchanges at regional and local rather than international levels. His approach to the place of towns in the early medieval economy derives much of its inspiration from anthropology and provides a refreshing new perspective on an established theme. In England the question of continuity of urban functions from Roman to early medieval times has been long debated but greater emphasis is now being laid on the late Saxon period as an important formative phase in urban development, helping to establish a framework around which the later expansion of urban centres in medieval times was organised. The importance of this period has been considered by Aston and Bond (1976), Reynolds (1977) and Radford (1978).

In Scotland, without Roman towns, large Viking trading centres, or planned Saxon burghs, it is still acknowledged that the introduction of planned towns in the twelfth and thirteenth centuries by incoming Anglo-Norman landholders represented a revolution in settlement and economic activity. Nevertheless, current research now places greater stress on the existence of early forms of urban life before the 12th century, providing a basis for the more dramatic and sweeping development of medieval burghs (Dicks, 1983). In Ireland comparatively little research has been carried out on medieval urban origins but recent work has again emphasised the importance of Norse coastal trading centres from the 9th century, as well as the proto-urban nuclei provided by the principal Early-Christian monastic foundations. The origins of urbanisation in medieval Ireland have been discussed by Butlin (1977) while Graham (1979) has analysed the distribution of medieval Irish boroughs with a model showing how urbanisation occurred.

Although much of the work of historical geographers relating to medieval towns has concentrated on urban morphology, there has been continuing interest in the development of urban hierarchies. This is exemplified by Slater

(1985) who has used the early 14th century lay subsidies to reconstruct the urban hierarchy of Staffordshire. By combining different subsidies he has shown the discrepancies which occurred between rankings by wealth and population size. He has also investigated the potential of the subsidies for studying migration patterns and urban occupations using surname evidence. Fox (1981) has used the taxation lists of the Scottish royal burghs to show how the urban hierarchy developed from a primitive system in the 14th century to a more sophisticated one three centuries later. The role of smaller medieval market centres has been considered by Unwin (1981) who draws interesting parallels between periodic market systems in medieval Nottinghamshire and modern underdeveloped countries in order to explore the functional role of the former. The reasons behind the proliferation of market centres in the 13th and 14th centuries have been considered by Britnell (1978; 1981) while Hilton (1984) has examined the social structure of small medieval market centres in England. Surveys for Scandinavia, the Slav countries, Russia and North West Europe are given in Barley (1977).

Recent developments in the study of medieval urban society in England have been reviewed by Platt (1976) and Reynolds (1977). Research on the social structure of medieval towns has tended to focus on the period from the 14th century onwards when documentation becomes more abundant. Among economic historians there has been a vigorous debate concerning the existence, nature and scale of a late medieval urban decline in England caused by a fall in population and consequent economic contraction, as well as changes in patterns of economic activity such as the movement of the textile industries from the towns to the countryside on a significant scale. The decline has been dismissed by Bridbury (1975; 1981) who takes an optimistic view based on statistics derived from the lay subsidies of 1334 and 1524-5 which demonstrate a steady growth of urban wealth. Bridbury's interpretation of the subsidies and the validity of his statistics have been questioned by Rigby (1979). The opposite view, that urban decline was widespread in England at this time, has been vigorously expressed by Pythian-Adams, who has produced an analysis of conditions in late-medieval Coventry (1979a) and a general review of the problem (1978) in which he suggests that the decline began as early as 1300 and continued with little interruption into the 1570s. Dyer (1979) has attacked Pythian-Adams' theory of general decline, casting doubt on his interpretation of many types of evidence suggesting that towns were experiencing economic and social problems on a grand scale. While acknowledging that some towns such as Coventry undoubtedly underwent a long-term decline, Dyer suggests that the general pattern was one of periods of short-term difficulty for individual towns, often balanced by economic growth elsewhere. In a reply to this critique, Pythian-Adams (1979b) has accused Dyer of pursuing a 'neutralist textbook orthodoxy' and remains unrepentant. Much of the fuel for this debate has been generated by a lack of agreement over basic terminology, in particular what is meant by 'decline' and 'decay' in this context. The linkages between population change, economic fluctuations and urban societies have often been assumed rather than demonstrated. Growth of population and of urban wealth were not necessarily positively correlated, nor was economic growth in the countryside and in the towns always closely linked although Dyer (1979) believes that given the small proportion of the population living in towns at this period, it is unlikely that a general urban crisis could have occurred when the indicators for the economy as a whole were fairly healthy. Reynolds (1980) has suggested that a more rigorous framework of hypothesis testing is needed to evaluate the various claims and counter claims. Many of the generalisations which have been made about urban decline have been based on a limited number of case studies. Amongst the more notable of these have been Palliser's study of York (1978), Saul's of Great Yarmouth (1982) and Rigby's of Grimsby (1984). Both sides of the debate have been reviewed by Reynolds (1977; 1980). More detailed studies of individual towns are required with careful evaluation of the statistical sources as well as more qualitative evidence for decline.

Historical geographers have directed comparatively little attention to the study of urban social patterns in medieval times by testing the contrasting models of residential location in pre-industrial cities suggested by Sjoberg and Vance. The validity of these models has been discussed by Langton (1975) in the context of a case study of Newcastle. Langton concluded that the social geography of Newcastle was more complicated than either model, resembling in some respects a hybrid of the two. The scope for extending this work further back into medieval times is considerable. Overall, there has been an unfortunate gap in urban history between medievalists and early-modern specialists, each using different sources and considering different problems. Although the sources for reconstructing urban societies and economies undoubtedly improve through time, it is really in the 14th rather than the 16th century that records first become fairly abundant (Reynolds, 1977); more detailed studies of change in individual towns spanning the late medieval and early modern periods would be advantageous. An example of how much information can be assembled about an individual, even at a fairly early date, is Donnelly's (1980) study of Thomas of Coldingham, a merchant of Berwick upon Tweed in the late 13th century.

Research into medieval urban development in other parts of Britain continues to lag behind that in England and has

MEDIEVAL ECONOMY AND SOCIETY

concentrated more upon morphological aspects. This can be
seen by the lack of new perspectives on medieval burgh
society in Adams' survey of Scottish urban development
(1977). The treatment of medieval towns in Gordon and Dicks'
(1983) volume on Scottish urban history also concentrates on
the physical character of the medieval burgh rather than its
socio-economic structures. Equally, Butlin's survey of Irish
towns (1977) demonstrates the need for more detailed research
on urban society and for broader, more comparative
perspectives.

CONCLUSION

Within the limited space available it has only been possible to
review certain aspects of recent progress in the study of the
medieval economy and society. Attention has been focused on
some of the areas where historical geographers have made a
significant contribution within the last decade, or where it is
considered that they may be able to bring their expertise to
bear on particular problems in the future. The indications are
that geographers studying the Middle Ages are slowly beginning to respond to paradigm shifts within geography, moving
away from mere description and a narrow preoccupation with
source evaluation towards more theoretically-based approaches. There will always be a place for the positivist
tradition. Nevertheless, the way ahead is clearly for historical
geographers to recognise that their work cannot be value free
and that blind unquestioning adherence to a particular
approach is limiting, denying them the excitement and the
challenge of a range of alternative theoretical standpoints.

REFERENCES

Aalen, F.H.A. (1978) Man and the Landscape in Ireland,
     London
Adams, I.H. (1977) The Making of Urban Scotland, London
Allison, K.J. (1977) The East Riding of Yorkshire Landscape,
     London
Angers, D. (1979) 'Mobilite de la Population et Pauvrete dans
     une Vicomte Normande de la Fin du Moyen Age', Journal
     of Medieval History, 5, 233-48
Aston, M. (1982-3) 'The Making of the English Landscape;
     the Next 25 Years', Local Historian, 15, 323-32
Aston, M., and Bond, J. (1976) The Landscape of Towns,
     London
Austin, D. (1985) 'Doubts about Morphogenesis', Journal of
     Historical Geography, 11, 201-9
Baker, A.R.H. (1979) 'Observations on the Open Fields; the
     Present Position of Studies in British Field Systems',

Journal of Historical Geography, 5, 315-26

Baker, A.R.H. (1983) 'Discourses on British Field Systems', Agricultural History Review, 31, 149-55

Baker, A.R.H., and Butlin, R.A. (1973) (eds.), Studies of Field Systems in the British Isles, Cambridge

Barley, M.W. (1977) (ed.), European Towns: Their Archaeology and Early History, London

Bigmore, P. (1979) The Bedfordshire and Huntingdonshire Landscape, London

Biller, D. (1982) 'Birth Control in the West in the 13th and Early 14th Centuries', Past and Present, 94, 3-26

Blanshei, S.R. (1982-3) 'Crime and Law Enforcement in Medieval Bologna', Journal of Social History, 16, 121-38

Blickle, P. (1979) 'Peasant Revolts in the German Empire in the Late Middle Ages', Social History, 4, 223-40

Brenner, R. (1976) 'Agrarian Class Structure and Economic Development in Pre-Industrial Europe', Past and Present, 70, 30-75

Brenner, R. (1982) 'Agrarian Class Structure and Economic Development in Pre-Industrial Europe. The Agrarian Origins of European Capitalism', Past and Present, 97, 16-113

Bridbury, A. (1975) Economic Growth: England in the Later Middle Ages, London

Bridbury, A. (1981) 'English Provincial Towns in the Later Middle Ages', Economic History Review, 34, 1-24

Britnell, R.H. (1978) 'English Markets and Royal Administration before 1200', Economic History Review, 31 183-96

Britnell, R.H. (1981) 'The Proliferation of Markets in England 1200-1349', Economic History Review, 34, 209-21

Britton, E. (1977) The Community of the Village. A Study in the History of Family and Village Life in 14th Century England, Toronto.

Brooks, N.P. (1977) 'Urban Archaeology in Scotland' in M.W. Barley (ed.), pp. 19-33

Butlin, R.A. (1982) 'Developments in Historical Geography in Britain in the 1970s' in A.R.H. Baker and M. Billinge (eds.), Period and Place, Cambridge, pp. 10-18

Butlin, R.A. (1978) (ed), The Development of the Irish Town, London

Campbell, B.M.S. (1980) 'Population Change and the Genesis of Common Fields on a Norfolk Manor', Economic History Review, 33, 173-92

Campbell, B.M.S. (1981a) 'Commonfield Origins: the Regional Dimension' in T. Rowley (ed.), pp. 112-29

Campbell, B.M.S. (1981b) 'The Regional Uniqueness of English Field Systems? Some Evidence from East Norfolk', Agricultural History Review, 29, 17-28

Campbell, B.M.S. (1983) 'Agricultural progress in medieval England: Some Evidence from Eastern Norfolk', Economic History Review, 34, 26-46

Cantor, L.M. (1982) The English Medieval Landscape, London
Carter, H. (1983) An Introduction to Urban Historical Geography, London
Charlesworth, A. (1983) An Atlas of Rural Protest in Britain 1548-1900, London
Clarke, M. (1984) The Archaeology of Medieval England, London
Cohn, S. (1980-1) 'Criminality and the State in Renaissance Florence 1344-1466', Journal of Social History, 14, 211-34
Cooper, P. (1975) 'In Search of Agrarian Capitalism', Past and Present, 80, 20-65
Darby, H.C. (1977) Domesday England, Cambridge
Darby, H.C., Glasscock, R.E., Sheail, J., and Versey, G.R. (1979) 'The Changing Geographical Distribution of Wealth in England 1066-1334-1525', Journal of Historical Geography, 5, 247-62
Dewindt, E.B. (1972) Land and People in Holywell cum Needleworth. Structures of Tenure and Patterns of Social Organisation in an East Midlands Village, Toronto
Dicks, B. (1983) 'The Scottish Medieval Town: A Search for Origins' in G. Gordon and B. Dicks (eds.), Scottish Urban History, pp. 23-51
Dodgshon, R.A. (1973) 'The Nature and Development of Infield-Outfield in Scotland', Transactions, Institute of British Geographers, 59, 1-23
Dodgshon, R.A. (1975a) 'Towards an Understanding and Definition of Runrig. The Evidence for Roxburghshire and Berwickshire', Transactions, Institute of British Geographers, 64, 15-33
Dodgshon, R.A. (1975b) 'Scandinavian Solskifte and the Sunwise Division of Land in Eastern Scotland', Scottish Studies, 19, 1-14
Dodgshon, R.A. (1975c) 'Runrig and the Communal Origins of Property in Land', Juridicial Review, 20, 189-208
Dodgshon, R.A. (1975d) 'Infield-Outfield and the Territorial Expansion of the English Township' Journal Historical Geography, 1, 327-36
Dodgshon, R.A. (1975e) 'The Landholding Foundations of the Open Field System', Past and Present, 67, 3-29
Dodgshon, R.A. (1977) 'Changes in Scottish Township Organisation during the Medieval and Early Modern Periods', Geografiska Annaler, 58B, 51-65
Dodgshon, R.A. (1978) 'The Origins of the Two Field and Three Field System in England: A New Perspective', Geographia Polonica, 38, 49-63
Dodgshon, R.A. (1980a) The Origins of British Field Systems: An Interpretation, London
Dodgshon, R.A. (1980b) 'The Early Middle Ages' in R.A. Dodgshon and R.A. Butlin (eds.), An Historical Geography of England and Wales, London, pp. 81-117
Dodgshon, R.A. (1981) Land and Society in Early Scotland Oxford

MEDIEVAL ECONOMY AND SOCIETY

Doherty, J., and Gibson, A. (1983) 'Computer-Assisted Data Handling in Historical Geography', Area, 15, 257-60

Donnelly, J. (1980) 'Thomas of Coldingham, Merchant and Burgess of Berwick upon Tweed (died 1316)', Scottish History Review, 59, 105-25

Dury, G. (1981) 'Climate and Settlement in Late Medieval Central England' in C. Delano Smith and M.L. Parry (eds.), Causes of Climatic Change, Nottingham, pp. 40-53

Dury, G. (1984) 'Crop Failures on the Winchester Manors 1232-1349', Transactions, Institute of British Geographers, 9, 401-18

Dyer, A. (1979) 'Growth and Decay in English Towns 1500-1700', Urban History Yearbook, pp. 60-72

Dyer, C. (1980) Lords and Peasants in a Changing Society: The Estates of the Bishopric of Worcester 680-1540, Cambridge

Field, K.P. (1983) 'Migration in the Later Middle Ages: The Case of the Hampton Lovett Villeins', Midland History, 8, 29-48

Fourquin, G. (1978) The Anatomy of Popular Rebellion in the Middle Ages, Amsterdam

Fox, R.C. (1981) 'The Burghs of Scotland 1377, 1601, 1677', Area, 13, 161-7

Francois, M.E. (1974-5) 'Revolts in Late Medieval and Early-Modern Europe: A Spiral Model', Journal of Int. History, 5, 19-44

French, R.A. (1983) 'The Early and Medieval Russian Town' in R.A. French (ed.), Studies in Russian Historical Geography, London, pp. 249-78

Goody, J., Thirsk, J., and Thompson, E.P. (1976) Family and Inheritance: Rural Society in Western Europe 1200-1800, Cambridge

Gordon, G., and Dicks, B. (eds.) (1983) Scottish Urban History, Aberdeen

Gottfried, R.S. (1984) The Black Death, London

Graham, B. (1979) 'The Evolution of Urbanisation in Medieval Ireland', Journal of Historical Geography, 5, 111-26

Hall, D. (1981) 'The Origins of Open Field Agriculture: The Archaeological Fieldwork Evidence' in T. Rowley (ed.), The Origins of Open Field Agriculture, pp. 22-38

Hallam, H.E. (1984) 'The Climate of Eastern England 1250-1350', Agricultural History Review, 32, 124-32

Hammel, E.A. (1980) 'Household Structures in 14th Century Macedonia', Journal of Family History, 5, 247-73

Hammer, C.I. (1978) 'Patterns of Homicide in a Medieval University Town: 14th Century Oxford', Past and Present, 78, 3-23

Hamshere, J.D., and Blakemore, M.J. (1976) 'Computerising Domesday Book', Area, 8, 289-94

Hare, J.M. (1982) 'The Wiltshire Risings of 1450: Political and

118

Economic Discontent in Mid 15th Century England',
Southern History, 4, 13-32
Harvey, B. (1977) Westminster Abbey and its Estates in the
Middle Ages, London
Harvey, M. (1980) 'Regular (field) and Tenurial Arrangements
in Holderness, Yorkshire', Journal of Historical Geo-
graphy, 6, 3-16
Harvey, M. (1982) 'Regular Open Field Systems on the
Yorkshire Wolds', Landscape History, 4, 29-39
Harvey, M. (1983) 'Planned Field Systems in Eastern
Yorkshire: Some Thoughts on Their Origin', Agricultural
History Review, 31, 91-103
Harvey, M. (1984) 'Open Field Structures and Landholding
Arrangements in Eastern Yorkshire', Transactions,
Institute of British Geographers, 9, 60-74
Hatcher, J. (1977) Plague, Population and the English
Economy 1348-1530, London
Hatcher, J., and Postan, M.M. (1978) 'Agrarian Class Struc-
ture and Economic Development in Pre-Capitalist Europe:
Population and Class Relations in Feudal Society', Past
and Present, 78, 24-55
Herlihy, D. (1983) 'The Making of the Medieval Family:
Symmetry, Structure and Sentiment', Journal of Family
History, 8, 116-30
Hilton, R.H. (1966) A Medieval Society: The West Midlands at
the End of the 13th Century, London
Hilton, R.H. (1973) Bondmen Made Free: Medieval Peasant
Movements and the Risings of 1381, London
Hilton, R.H. (1975) The English Peasantry in the Middle
Ages, Oxford
Hilton, R.H. (1976) The Transition from Feudalism to Capital-
ism, London
Hilton, R.H. (1984) 'Small Town Society in England before the
Black Death', Past and Present, 105, 53-78
Hindess, B., and Hirst, P. (1975) Pre-Capitalist Modes of
Production, London
Hindle, B.P. (1976) 'The Road Network of Medieval England
and Wales', Journal of Historical Geography, 2, 207-33
Hindle, B.P. (1977) 'Medieval Roads in the Diocese of
Carlisle', Transactions, Cumberland and Westmorland
Society, 77, 83-95
Hindle, B.P. (1978) 'Seasonal Variations in Travel in Medieval
England', Journal of Transport History, 4, 170-8
Hindle, B.P. (1980) 'The Towns and Roads of the Gough Map
(c1360)', Manchester Geographer, 1, 35-49
Hindle, B.P. (1982) 'Medieval Roads and Tracks' in L.M.
Cantor (ed.), The English Medieval Landscape, pp.
192-217
Hindle, B.P. (1984) Roads and Trackways of the Lake
District, Ashborne
Hodges, R. (1982) Dark Age Economics: The Origins of

Towns and Trade A.D. 600-1000, London
Hooke, D. (1976-7) 'The Reconstruction of Ancient Route-
    ways', Local Historian, 12, 202-20
Hooke, D. (1978-9) 'Anglo Saxon Landscapes of the West
    Midlands', Journal of English Place Name Society, 11,
    3-23
Hooke, D. (1981a) 'Open Field Agriculture: The Evidence
    from the Pre-Conquest Charters of the West Midlands' in
    T. Rowley (ed.), The Origins of Open Field Agriculture,
    pp. 39-63
Hooke, D. (1981b) 'Anglo-Saxon Landscapes of the West
    Midlands: The Charter Evidence', British Arch. Reports,
    British Series, Oxford, 95
Howell, C. (1983) Land, Family and Inheritance in Transition.
    Kibworth Harcourt 1280-1700, Cambridge
Jaeger, H. (1982) 'Reconstructing Old Prussian Landscapes
    with Special Reference to Spatial Organisation' in A.R.H.
    Baker and M. Billinge (eds.), Period and Place,
    Cambridge, pp. 44-50
Jones, M.L. (1984) 'Society and Settlement in Wales and the
    Marches 500 B.C. to A.D. 1000', British Arch. Reports,
    Oxford, 121
Labarge, M.W. (1982) Medieval Travellers, London
Langdon, J. (1984) 'Horse Hauling: A Revolution in Vehicle
    Transport in 12th and 13th Century England', Past and
    Present, 103, 37-66
Langton, J. (1975) 'Residential Patterns in Pre-Industrial
    Cities: Some Case Studies from 17th Century Britain',
    Transactions, Institute of British Geographers, 65, 1-27
Le Roy Ladurie, E. (1976) Montaillou, London
Le Roy Ladurie, E. (1978) 'Agrarian Class Structure and
    Economic Development in Pre-Industrial Europe: A Reply
    to Professor Brenner', Past and Present, 79, 55-59
Martin, J. (1983) Feudalism to Capitalism: Peasant and
    Landlord in English Agrarian Development, London
McCloskey, D. N. (1975) 'The Persistence of English Common
    Fields' in W.N. Parker and E.L. Jones (eds.), European
    Peasants and Their Markets, pp. 71-119
McCloskey, D.N. (1976) 'English Open Fields as Behaviour
    towards Risk', Research in Economic History, 1, 124-70
McCloskey, D.N. (1979) 'Another Way of Observing the Open
    Fields', Journal of Historical Geography, 5, 426-9
McClure, P. (1979) 'Patterns of Migration in the Late Middle
    Ages: The Evidence of English Place-Name Surnames',
    Economic History Review, 32, 167-82
Macfarlane, A. (1977) Reconstructing Historical Communities,
    Cambridge
McGovern, T.H. (1981) 'The Economics of Extinction in Norse
    Greenland' in T.M. Wigley, M.J. Ingram and G. Farmer
    (eds.), Climate and History, Cambridge, pp. 404-33
McPherson, A.G. (1984) 'Migration Fields in a Traditional

Highland Community 1350-1850', Journal of Historical Geography, 10, 1-14

Palliser, D.M. (1978) 'A Crisis in English Towns? The Case of York 1460-1640', Northern History, 14, 108-25

Parry, M.L. (1974) 'Secular Climatic Change and Marginal Agriculture', Transactions, Institute of British Geographers, 64, 1-13

Parry, M.L. (1976a) 'The Abandonment of Upland Settlement in Southern Scotland', Scottish Geographical Magazine, 92, 50-60

Parry, M.L. (1976b) 'The Mapping of Abandoned Farmland in Upland Britain: An Exploratory Survey in S.E. Scotland', Geographical Journal, 142, 101-10

Parry, M.L. (1978) Climatic Change, Agriculture and Settlement, Folkestone

Parry, M.L. (1981) 'Climatic Change and the Agricultural Frontier: A Research Strategy' in T.M. Wigley, M.J. Ingram and G. Farmer (eds.), Climate and History, Cambridge, pp. 319-36

Parry, M.L., and Slater, T.R. (1980) The Making of the Scottish Countryside, London

Platt, C. (1976) The English Medieval Town, London

Pythian-Adams, C. (1978) 'Urban Decay' in Late Medieval England' in P.A. Abrams and E.A. Wrigley (eds.), Towns in Societies, London, pp. 159-86

Pythian-Adams, C. (1979a) Desolation of a City: Coventry and the Urban Crisis of the Late Middle Ages, Cambridge

Pythian-Adams, C. (1979b) 'Dr Dyer's Urban Undulations', Urban History Yearbook, 73-6

Radford, C.A.P. (1978) 'The Pre-Conquest Borough of England, 9th-11th Centuries', Proc. British Academy, 64, 131-54

Raftis, J.A. (1974) Warboys: 200 Years in the Life of an English Medieval Village, Toronto

Razi, Z. (1980) Marriage and Death in a Medieval Parish: Economy, Society and Demography in Halesowen 1270-1400, Cambridge

Reed, M. (1979) The Buckinghamshire Landscape, London

Reed, M. (1984) Discovering Past Landscapes, London

Reynolds, S. (1977) An Introduction to the History of the English Medieval Town, London

Reynolds, S. (1980) 'Decline and Decay in Late Medieval Towns: A Look at Some of the Concepts and Arguments', Urban History Yearbook, 76-8

Rigby, S. (1979) 'Urban Decline in the Later Middle Ages. Some Problems in Interpreting the Statistical Data', Urban History Yearbook, 46-59

Rigby, S. (1984) 'Urban Decline in the Later Middle Ages: The Reliability of the Non-Statistical Evidence', Urban History Yearbook, 45-60

Ring, R. (1979) 'Early Medieval Peasant Households in Central Italy', Journal of Family History, 4, 2-25

Rowley, T. (1981) (ed.), The Origins of Open Field Agriculture, London

Saul, A. (1982) 'English Towns in the Late Middle Ages: The Case of Great Yarmouth', Journal of Medieval History, 8, 75-88

Sharpe, J.A. (1985) 'The History of Violence in England: Some Observations', Past and Present, 106, 206-15

Shirk, M.V. (1981) 'The Black Death in Aragon', Journal of Medieval History, 7, 357-68

Slater, T.R. (1985) 'The Urban Hierarchy of Medieval Staffordshire', Journal of Historical Geography, 2, 3-20

Smith, R.M. (1979) 'Kin and Neighbours in a 13th Century Suffolk Community', Journal of Family History, 4, 219-56

Smith, R. (1983) 'Hypothèses sur la Nuptialite en Angleterre, 13th-14th Siecle', Annales, 38, 107-35

Smyth, A.P. (1982) Celtic Leinster: Towards an Historical Geography of Early Irish Civilisation A.D. 500-1600, Blackrock

Stanley, M.J. (1980) 'The Geographical Distribution of Wealth in Medieval England', Journal of Historical Geography, 6, 315-24

Steane, J. (1983) 'How Old is the Berkshire Ridgeway?', Antiquity, 57, 102-8

Stone, L. (1983) 'Interpersonal Violence in English Society 1300-1800', Past and Present, 101, 22-33

Taylor, C.C. (1979) Roads and Tracks of Britain, London

Taylor, C.C. (1980-1) 'The Making of the English Landscape: 25 Years On', Local Historian, 14, 195-201

Taylor, C.C. (1981) 'The Origins of Open Field Agriculture' in T. Rowley (ed.), The Origins of Open Field Agriculture, pp. 13-21

Unwin, T. (1981) 'Rural Marketing in Medieval Nottinghamshire', Journal of Historical Geography, 7, 231-51

Whyte, K.A. (1984) Register of Research in Historical Geography 1984, Norwich

Chapter Five

AGRICULTURE AND RURAL SOCIETY

J.R. Walton

My purpose is to present an interpretative survey of recent contributions to the historical geography of agriculture and agrarian society for the period 1500 to 1900. Practicalities and the inescapable consequences of my own academic socialisation dictate that the approach be Anglocentric. Had space allowed, it would have been possible to extend discussion beyond Britain or, more accurately, England, to embrace the rest of the world - a piece of conceptual neo-colonialism broadly in sympathy with one important strand in the recent historiography of historical geography and the related disciplines of economic and social history. The so-called world-system approach argues that from the sixteenth century, if not earlier, the global periphery was linked to the countries of the north-west European core, which were 'miraculously' spared the limitations upon growth intrinsic to the process of growth itself (Jones, 1981a), by steadily tightening bonds of dependency and subordination. In the 'pre-modern' world, trade between cultural regions was for the most part implemented by communities of resident aliens, who were constantly at risk of absorption into the host culture of the countries concerned (Curtin, 1984). There was a degree of equity in the reciprocal processes involved. In the world-system world, European military superiority and the gradual growth of commercial and industrial activity meant that the acquisition, on terms favourable to Europe, of global markets and global supplies of raw materials was both possible and imperative. Hegemony, achieved in a cultural and commercial, as well as political sense, eventually ensured that the destinies of all peripheral areas, and not only those of large-scale European settlement, were increasingly bound up with the core (Wallerstein, 1974; 1980; Wolf, 1982; Harvey, 1983). Take this line of reasoning, even denied some of its reductionist extravagance (Dodgshon, 1977), and a Europocentric view of the past stands legitimised by the very course of the events it surveys.

AGRICULTURE AND RURAL SOCIETY

THE GLOBAL CONTEXT

While my text is narrowly focused on Britain, it may be helpful to outline some of the concerns of toilers in other vineyards. As a framework for surveying this recent work, the world-system has many advantages. But it is not within the analytical context of the approach that most of this work has been produced. Amongst historical geographers generally, truffle hunters still comfortably outnumber parachutists, to borrow Le Roy Ladurie's metaphor (see Whyte, 1984). The overarching world view has increasing appeal, but it belongs to a realm of synthesis which is alien to the largely documentary tradition of research in the subject. There exists a strong and enduring preference for localised research, more and more informed by theory (Gregory, 1981; Butlin 1982), but laid on sound empirical foundations nonetheless. Most accretions to the existing corpus of knowledge and interpretation are based on painstaking documentary or field research. The quantities of data involved are alone sufficient to ensure that the focus is essentially local, albeit more of this work than formerly is now presented as case studies, illustrative of some broader theme or deeper truth.

Much the greatest volume of research on non-European areas of European colonisation or influence concerns North America. As befits a land where present national character was forged in conflicts between the indigenous and the adventitious, and in the interrelationships between different immigrant groups, the cultural tradition remains strong. The study of folk housing as a cultural form, pioneered by Kniffen (1979), lives on in numerous regional surveys. Some see the folk houses as a source of raw data providing access to the mental world of their builders (Glassie, 1975). Others are more concerned to trace particular elements in rural building styles or practice back to the colonial hearth areas of the New World or beyond that to the source areas of migration in the Old (Wonders, 1979; Jordan, 1980; Marshall, 1981; Hewes, 1981; Hewes and Jung, 1981). Other students have been more interested in immigrant culture from the standpoint of social or demographic rather than material evidence, and many studies are now available which reveal that the group identity of migrant communities was often persistent, even when migration levels from such communities were relatively high (Brunger, 1982; Ostergren, 1979; McQuillan, 1978; 1979). It has been suggested that, even on the frontier, pioneers of disparate provenance were not thrown together in the wide embrace of a grand, noble and difficult enterprise, but remained isolated in small, clustered groups which shared primary allegiances to particular countries, provinces, or, occasionally, parishes of origin (Rice, 1977). The frontier looks less likely to have fulfilled those functions in shaping American destiny which Turner attributed to it, even though the hypothesis continues to stimulate, fascinate and persuade (Walsh, 1981).

In many ways, it is the totality of interaction between cultures and between man and environment which is of greatest interest to North American historical geographers working in the cultural tradition. Only that totality can provide a full or acceptable explanation of the processes underlying the emergence of definable cultural regions, and cultural regions, or regional cultures have traditionally been a major objective in their work. But studies which focus on only a part of this totality have also gained favour and produced pathbreaking results. Several authors have explored the interactions between settlers and Indian society (Sheehan, 1980; Kupperman, 1980; Axtell, 1981). Others have examined the destruction or adaptation of aboriginal environments and the spread of agriculture which accompanied it. Many writers in this latter category are simply concerned to gain an understanding of the chronology and geography of these changes and of the mechanisms which produced them, but evolving personal reactions to the wilderness, its 'discovery' by east coast society, have also been examined at some length (Williams, 1982; Bogue, 1982; Clark, 1984; Nash, 1973; Stilgoe, 1982). It is notable that settler-environment interaction is also well represented as a theme in work on the historical geography of Australia (Jennings and Linge, 1980; Powell, 1981).

States of the periphery other than those of large-scale north European settlement have rather less historical geography on offer. So far as Latin America is concerned, much of the work accessible to English-speaking readers reflects the interest of modern-day academic conquistadores from Europe and, especially, North America. Themes which figure prominently include the agriculture and agricultural landscapes of the aboriginal world (Donkin, 1979; Gade, 1979), the nature of the interaction between indigenous and settler society (Newsom, 1976; Hennessy, 1978), production for global markets and its effects upon systems of cultivation (Albert, 1976; Donkin, 1977; Palacios, 1980; Hall, 1982; Galloway, 1982), and the failure of Latin American economies to develop during the nineteenth century in the same way as those of the North (Leff, 1982).

Africa has been less intensively cultivated than Latin America by historical geographers, if not by historians. Geographers working in African states which have gained their independence since the War are beset by other more pressing problems, and the region has attracted fewer overseas scholars than either Latin America or Asia. South African historical geographers, notably Christopher (1976; 1984), have explored the causes and consequences of the European drive into southern Africa from the late seventeenth century, focussing primarily on the landscape impact. Histori-

125

cal geographers are now conscious of the need to transcend 'the deliberate and dangerous mystification of a powerful racist ideology', to 'understand how South Africa's brutal landscapes were forged in subordination and struggle' (Crush, 1986), although it is an approach which promises more than it has yet delivered.

THE BRITISH RURAL LANDSCAPE: PARADIGMS LOST?

For reasons already outlined, the greater part of this chapter will be concerned with the dynamics of change in the British countryside, where much attention has recently been directed to the circumstances of the decay of feudalism and the gradual advent of new capitalistic systems and methods of production. These were shaped by innovation in technique, product and organisation, changes in market demand, changes in the distribution of ownership, and changes in the relationships between landowner, tenant and labourer. They were also underwritten by a new political economy which underpinned different perceptions of the land and its functions (Tribe, 1978). But not every important facet of recent work in the historical geography of the countryside is comfortably accommodated within this structure. This section surveys recent studies which lie outside it.

Those who completed their initial research training ten or more years ago will find the present landscape of historical geography both familiar and different. Some of the strands which were prominent in historical geography's theoretical and methodological fabric at that time are still there. But the fabric is now differently textured thanks to the novel materials which have been woven into it, many of them sympathetically matched to Baker's (1972) belief that future progress would be contingent upon historical geography being less insulated than it had been from concepts and methods developed elsewhere (Gregory, 1981; Baker and Gregory, 1984). Neither at that period nor this may we speak of historical geography being conducted within one unique methodological frame of reference. The sub-discipline could never be accused of rigid adherence to a single exclusive paradigm, and in this respect it seems to be no different either from geography as a whole, or from any of its other major branches (Stoddart, 1981; Johnston, 1984).

The process of conceptual and methodological development has been partly a matter of extension and importation, and partly a matter of reorientation and redefinition. Within this scheme, the concept of landscape fits much more into the latter category than the former, with both new and old approaches providing exemplars of work concerned with agriculture and rural society. 'Changing landscapes' was one of the central conceptual props of the historical geography of the thirties, forties, fifties and early sixties (Baker, 1972). It was itself supported by the twin notions that the present landscape is not only the end product of an extended process of historical evolution but also, ultimately, a mutual interest, perhaps the mutual interest of all geographers no matter what their specific specialist inclinations. Thus the approach was satisfactorily embedded within the conventionally accepted definitions of geography, from the point of view of both content and method. This was seen as a considerable virtue, inter-disciplinary interaction at that stage still being viewed with suspicion. One might say that contact with other disciplines was regarded not so much as cross-fertilisation but as miscegenation, with malformed and sterile hybrids the likely outcome. Landscape (the present day landscape, that is) served both as a prompt to the sorts of issues which the historical geographer ought to explore, and as a useful vehicle for structuring and organising the presentation of historical geography's substantive research findings. Thus, several academic generations were introduced to historical geography through introductory courses at degree level which had the present day landscape as their organisational focus (if not their only explanatory tool), and through volumes like Hoskins's highly influential Making of the English Landscape (1955), which specifically addressed the themes of landscape evolution and landscape history. Since agriculture has contributed more than any other economic activity to the appearance of the present-day cultural landscape, the approach gave considerable emphasis to this sector, thereby helping to forge a sub-discipline which was widely perceived as 'rustic at heart'.

The 'holistic' landscape approach lives on in the county volumes in the 'Making of the English Landscape' series, in the national volumes in the 'World's Landscapes' series, and in other studies which focus on particular themes or areas (for example, Darby, 1983). However, these contributions represent a much smaller proportion of the total output of historical geography than was once the case, especially as compared with the enduring strength of the landscape tradition in North American historical geography. Furthermore, although geographers are still well represented among the authors, it is also increasingly apparent that the field has considerable appeal to the practitioners of other disciplines and sub-disciplines, notably archaeology, economic and social history, architectural history, botanic history, toponymy, and literary and artistic history and criticism (Austin, 1985). The multi-disciplinary character of landscape history is a sign of strength rather than weakness. But it is surely unfortunate for historical geography, especially in view of the large contribution which historical geographers made to this subject area in the past, that mounting public interest in landscape history, prompted partly by increasing concern for the rate

AGRICULTURE AND RURAL SOCIETY

of erosion of historic landscapes, partly by the nostalgia
which has been an increasingly apparent part of eighties
middle-class culture and taste, should be satisfied in large
measure by authors who are not historical geographers, or
whose affiliations to the sub-discipline are well disguised.
Historical geography may well need its David Bellamy, but it
has yet to find him.

The 'holistic' interpretation of landscape is, in any case,
increasingly less favoured than a 'categoric' approach where
authors focus on those features of landscape history which
are specific to their own areas of specialism (Austin, 1985). A
number of these approaches incorporate work by historical
geographers or are of interest to historical geographers
concerned with agriculture and rural society.

First, we may note the large amount of attention which
historical geographers have devoted to the evolution of one of
the most prominent features of the present rural landscape,
its field systems. Although much of this work is medieval in
focus, and is therefore fully examined in Whyte's essay in
this volume, a great deal, even of the earlier work, uses
retrogressive methods, which treat large-scale maps and other
documentary sources, many of eighteenth century or later
date, as keys to unlocking earlier phases in the evolutionary
history of the field systems themselves (Baker and Butlin,
1973; Harvey, 1978; Sheppard, 1979). Increasing willingness
to analyse field systems in the context of the local economies
and societies which operated them has not only improved the
quality of interpretation offered but also helped to reinstate
the position of field system studies within agrarian historical
geography as a whole (Baker and Butlin, 1973; Dodgshon,
1980; 1981; Baker, 1983).

A second major approach is represented by the work of
students of vernacular architecture (the British equivalent of
American folk architecture), who have devoted a large proportion of their efforts to examining and cataloguing the
buildings of the countryside both in Britain (see, for
example, Smith, 1975; Mercer, 1975; Barley, 1961; 1985;
Harrison and Hutton, 1984; Fenton and Walker, 1981), and
France (Meirion-Jones, 1982). Vernacular buildings, constructed using methods and materials indigenous to their
localities, are much better represented in the countryside,
where polite influences took longer to penetrate, than in the
towns (Brunskill, 1971). Although the first concerns of most
published work in this area are description and classification,
the interpretative content of these studies provides material
of relevance to students of agrarian change. For example,
there is a general relationship between spatial and temporal
variations in the rate of new building or the reconstruction of
old buildings and similar variations in the prosperity of the
agricultural sector. Architectural evidence has not only made
an important contribution to debates about rural labourers'

living standards (Gauldie, 1981), but has also promoted a better understanding of the long-term chronology and geography of agricultural profits and investment. Observing the large representation of late sixteenth and early seventeenth century rural housing in the limestones and chalk scarplands of the Midlands and South East (see Wood-Jones, 1963; Portman, 1974), Hoskins (1953) developed the idea that a 'Great Rebuilding' occurred during those years. More recently, scholars have suggested that the experiences of the south-east were not those of the country as a whole. Evidence of large-scale rural rebuilding in north-west England and Scotland only becomes available some hundred years later with evidence of rising real living standards (Machin, 1977a; Marshall, 1980; Whyte, 1975). In any case, Machin (1977b) suggests rebuilding was not a single event but a recurrent response to cyclical fluctuations in agricultural profits.

The study of rural buildings has also cast some light on several other issues: rural family structure, physical evidence for yeoman farmhouses built on the 'unit system' (houses extended or divided, as in central or southern Europe, to provide separate but contiguous living space for the older generation) calling into question the demographer's orthodoxy that extended family forms were of little consequence in pre-industrial England (Gresham, 1971; Machin, 1975; Sandall, 1975; Carson, 1976; Barley, 1985); certain changes in methods of production or agricultural innovation, where these may be inferred from the functions new agricultural buildings were intended to fulfil (Peters, 1969; Wiliam, 1982); and evidence in the decay of the vernacular and the gradual penetration of 'polite' materials and styles of the advent of external influences, which had their counterpart in agriculture's increasing susceptibility to wider market forces, and the gradual predominance of agricultural techniques and materials originating outside the immediate region of production (Perkins, 1975).

Work on historical ecology presents a third group of studies which, while they are rooted in the history of the landscape, are of great relevance to the student of agrarian change. This category embraces a diversity of approaches and interests. Former land-uses have been reconstructed, using both botanical methods and historical sources, with particularly successful results in the context of the botanic and economic history of Britain's rapidly disappearing broadleaved woodlands (Rackham, 1980), although other residual or pressurised landscapes, like the wetlands, have been exposed to similar treatment (for example, Sheail and Wells, 1983). The merits and demerits of the so-called Hooper hypothesis, which holds that species diversity provides a simple linear measure of hedgerow age, continue to be debated (Pollard, Hooper and Moore, 1974; Cameron, 1984). More usefully still, studies in historical ecology have reminded agrarian historians that the agricultural pest problem in the past is not to be equated with present day experience. Rabbits provided a valuable source of protein with negligible opportunity costs as long as they remained confined to their warrens, not so once they had tasted freedom (Sheail, 1978). Pine-martens, squirrels, hedgehogs and several species of birds attracted generous bounties and were hunted close to extinction in some cases (Jones, 1972; 1981b). The scale of the slaughter denotes the scale of the problem, underlining the importance to agriculture of the twentieth-century pesticide revolution. But it also suggests that the human tendency to magnify a threat to the point of overkill is by no means new.

Some workers within the historico-ecological tradition have been particularly concerned with evidence for climatic change in the past. This very broad theme involves researchers in several countries, many of whom have become interested in climatic fluctuation for reasons other than historical ecology, although they often use botanical evidence to provide results and refine interpretations (see, for example, the papers collected in Rotberg and Rabb, 1981). Parry's studies of settlement and arable cultivation on the hill margins of southern Scotland (1975; 1976a; 1978) present an intriguing unfolding picture: an upper altitudinal limit of cereals cultivation as sensitive as any barometer, oscillating in exact phase with climatic fluctuations. Parry sees climate as a necessary and largely sufficient condition for upland settlement, without being too concerned about the other forces which created a latent demand for marginal land, a demand which could only be satisfied during periods of climatic amelioration. (See Grigg, 1982 for an explanatory survey of agrarian change which treats climatic change as one of several explanations). In the light of Wrigley and Schofield's (1981) convincing demonstration that the pre-industrial fertility regime was essentially Malthusian, with nuptiality and thus fertility showing a lagged response to changes in real living standards, there would be some value in a closer exploration of the interrelationships between climatic change, commodity prices, real incomes, demographic change and settlement on the physical margins, thus much extending the work of those like Hoskins (1964; 1968) and Harrison (1971) who have treated certain aspects of this nexus.

The research recently reviewed by Outhwaite (1986) suggests that late sixteenth and early seventeenth century population growth was associated with a veritable explosion of colonising settlement by small-scale producers on the upland margins of northern England, and on fen and forest edges in the Midlands and South (Porter, 1975; 1978; Skipp, 1970; 1978; Spufford, 1974). This was followed by a mid and late seventeenth century phase of population stagnation and decline which coincided with the bottom of the climatic trough. But that coincidence does not explain causality in respect of aggregate national statistics, much less the specific fate of the geographically marginal. Did their very marginality make them prime candidates for a classic if spatially uneven Malthusian crisis, which may have comprised positive as well as preventive components, mortality increase as well as delayed marriage and fertility decline? Appleby (1980), looking at the broader, aggregate picture, has stressed the relative success of English institutions in coping with the worst effects of the so-called 'Little Ice Age,' but it does not follow that the success was uniform or universal. Or did these areas, or at least certain favoured parts of them, find their salvation in a widening economic base of craft production (possibly as an agricultural by-employment, possibly not - compare Thirsk, 1961 and Sharp, 1980), which then led forward, via the much contested process of proto-industrialisation or some variant of it (Coleman, 1983; Houston and Snell, 1984; Clarkson, 1985), to a fully fledged, although spatially more focused landscape of factory industry?

I leave this an open question for the simple reason that present evidence offers suggestions rather than conclusions. What is clear from a number of studies (Levine, 1977; Mills, 1982; Rogers, 1981; Hudson, 1981; 1983) is that local variations in agricultural geography and in the patterning of social relationships within the agricultural community were critically important in determining the chronology, character and geography of industrial growth. The point is nowhere more elegantly demonstrated than in Hudson's (1983) contribution, which shows how woollen manufacture in the West Riding was associated with areas where fertile soils, strong manorial control, late enclosure and thus the persistence of small estates provided appropriate financial and labour conditions for the rise of master clothiers. The worsted sector, on the other hand, which used artisanal systems of industrial production from an early date, came to the fore in those places where poor soils, weak manorial control, earlier enclosure, and partible inheritance had brought into existence a large and landless labour force.

TRANSITIONS AND TRANSFORMATIONS IN POST MEDIEVAL BRITISH AGRICULTURE

In recent years, interest in the dynamics of change through time (which was always implicit in and potentially destructive of both landscape and cross-sectional traditions within historical geography) has become an all-consuming concern. The historical geographer is thus contributor to several broader historical debates, three of which focus on agriculture and rural society. I have identified these as the rise of agrarian capitalism, enclosure and the diffusion of innovations.

131

Although these themes all treat different aspects of the same problem - the transition from a less to more modern agricultural sector - they are often represented as contrasting, even incompatible or conflicting interpretations of agrarian change. Certainly, analyses of the origins and rise of agrarian capitalism, which all show an awareness of Marxian positions even if not explicitly Marxist in their own analytical frame of reference, contrast with the largely empirical, implicitly neo-classical content of work on technological innovation in agriculture, written in what has been characterised as the 'cows and ploughs' tradition (Butlin, 1982; Overton, 1984a). My own inclination is to emphasise compatibility rather than conflict. Most studies of agricultural innovation are not concerned with the evolution of rural landownership or social structures, or even with the way inequalities of rural wealth may have increased as a consequence of technological change. Instead, they treat the social distribution of landownership as one of a number of explanations of spatial and temporal variation in the progress of innovation. Unequal social and economic relations are therefore not denied so much as assumed.

A theme not specifically taken up under any of the three headings, and therefore examined briefly here is that of the agricultural revolution. Overton (1984a, p. 139) concludes his recent discussion with the observation that the phrase 'agricultural revolution' is 'beyond redemption', 'meaningless except as a pejorative label for a most unfortunate historiographical episode'. The point is well made. Ever since Ernle (1912) gave us an agricultural revolution shoehorned into the second half of the eighteenth century and strongly associated with parliamentary enclosure, and the works of the Norfolk improvers and their publicists, the term seems to have served to blinker and misdirect. Admittedly, it has also promoted a coherent image of agricultural history which has won the subject a popular audience it might otherwise have lacked. But 'imageability' is hardly a virtue if the image itself is false. What is more, the succession of claim and counter-claim which has characterised more recent treatment of the subject has not revealed a capacity for sustained dispassionate judgement by agricultural historians. Not only is there little evidence of consensus about what an agricultural revolution should comprise, but there is equally little evidence of participants' willingness to take a long-term view. This is true of the debate about the English agricultural revolution:- Kerridge's (1967; 1969b) belief in a sixteenth and seventeenth century agricultural revolution, Mingay's (1963) and Chambers and Mingay's (1966) attempts to rehabilitate the original claims of the late eighteenth and early nineteenth centuries (albeit stripped of the inflexibilities which characterised Ernle's original formulation), and the suggestion of revolutionary change in the second half of the nineteenth century to be found in the contributions of Thompson (1968) and Sturgess (1966).

It is also illustrated in more recent discussions about the agricultural revolution in Scotland. Presuming that the word 'revolution' implies that previous systems were overturned, Whittington (1975) suggests that the orthodoxy of a late eighteenth and early nineteenth century agricultural revolution (or 'Improving Movement' as it is more widely known north of the Border - see Turnock, 1982) be replaced by a concept of continuous evolution, beginning 'at least' in the seventeenth century. Contributors to the ensuing debate, who targeted their criticisms on each other almost as much as on the original, variously drew attention to the unwarranted rigidity which Whittington ascribes to the orthodox model (Parry, 1976b); the importance of the second half of the eighteenth century within the context of 'evolutionary' rather than 'revolutionary' change (Adams, 1978); the close association during the eighteenth century (unlike England) between enclosure and the implementation of long leases (Mills, 1976); and the fact that increasing numbers of Scottish tenants enjoyed the benefits of written leases from the early seventeenth century (Whyte, 1978). The last named has been particularly successful in demonstrating the progressive quality of Scottish agriculture and Scottish rural society during the seventeenth century (Whyte, 1979).

The problem of 'the agricultural revolution' is essentially a problem of reification. The term is a label. The label may be attached to particular phases of agricultural history, but it still remains a label, not an entity or thing. It does not confer any special quality on the historical events of those periods beyond that which is inherent in the events themselves. It implies coherent progressive change, thereby ignoring the conservative forces which were always a part of the agricultural sector, even during phases of relatively rapid change. And it tends to undermine the sensitivity and subtlety of interpretation, not enhance it.

THE ADVENT OF AGRARIAN CAPITALISM

Of course, similar criticisms have been applied to the idea of feudalism, which Elton (Fogel and Elton, 1983, p. 79) has characterised as 'a categorising concept invented to make discourse easier; it never existed in reality and cannot therefore have risen or declined.' In the event, most students of the rise of agrarian capitalism have been as little concerned as was Marx himself to elevate feudalism above the conceptual level. Marx's primary interest was the economic and social structures of capitalism, and it is against that background (searchingly explored by reference to nineteenth century sources) that the idea of a transition from some pre-existing principle of economic and social organisation originated. Subsequent researchers interested in the period under discussion have been more concerned to explore the features of the post transitional world than those of its predecessors. This is true, for example, of Dunford and Perrons' (1983) historical materialist interpretation of British post medieval history, which reviews contributions by historical geographers as well as those by economic and social historians. Work more directly or exclusively concerned with agriculture has focussed on the themes of changing landownership structures and tenurial relationships, the decay of communal forms of social and economic organisation in agriculture, pressures on customary use rights (partly as a consequence of a more strictly interpreted common law which attached greater importance to legal title), the fate of the 'peasantry', and the quality of relationships between different social groups or classes.

Much recent work represents continuing fall-out from the 'Brenner debate.' Briefly, Brenner's papers (1976; 1977) attempted to place class relationships at the forefront of the history of agrarian change. Specifically, Brenner argued that English and French agrarian history diverged during the later medieval and early modern periods because the two countries had a different experience of the operation of class power. Whereas in France the peasantry succeeded in curtailing the advances of the lords and therefore remained in control of substantial proportions of the total land area, in England they failed to gain freehold rights and were thus more readily dispossessed.

A large part of the ensuing debate questions Brenner's assumptions about the nature of medieval society and economy and the contrast between the English and French agrarian experience (Cooper, 1978). But there have also been moves to explore the history of post-medieval English agrarian society in the context of Brenner's arguments.

These began with Croot and Parker (1978), who drew attention to Brenner's neglect of the revisionist interpretation of 'the long sixteenth century', which emphasises the considerable security of tenure of small-scale producers (Kerridge, 1969b). The argument gains support from a number of detailed case-studies. For example, Bettey (1982) stresses the continuing strength of manorial custom in seventeenth-century Dorset. Zell (1984) draws attention to a manor in sixteenth-century Sussex where 'true' customary rights and obligations were formally defined at the insistence of the tenants, and to their advantage. Two students of landlord-tenant relationships in northern England (Hoyle, 1984; Spence, 1984) suggest that landlords were often triumphant, but that there was nothing inevitable about that outcome.

To date, it is in the context of Cumbria, that 'odd corner of the land', that the Brenner thesis has been most thoroughly tested. In a wide-ranging analysis of eighteenth-century Cumbrian estate material, Searle (1986) suggests that landlords attempted to step up the amount of surplus they extracted from their tenants by first replacing customary entry fines by arbitrarily determined fines, and, later in the century, when it was recognised that this strategy had failed, by trying, with equal lack of success, to suppress their customary access to use-rights of estate timber. In a less ambitious study which focuses on the Gilsland estate, Gregson (1984) suggests that the growth of the leasehold sector during the eighteenth century was largely accounted for by the enclosure of common and waste land on the upland margins of the parish (and thus took place within the constraints of the existing rules of the customary economy), and not by an all-out attack on customary tenure. Searle concludes that the 'consolidation of capitalist class relations was ... a complex and protracted process, which owed more to the volatility of commodity markets and the unintended consequences of inheritance practices, than it did to the Sturm und Drang of the class struggle.'

Clearly, there are problems in Brenner's analysis of English conditions. But the work of his critics is not completely problem free either. Leaving aside the question of whether or not Cumbrian evidence can offer much in the way of proof or disproof when ranged against an argument conceived on a Europe-wide scale, it is not clear that the Cumbrian customary tenantry conforms to a model of the 'peasantry' acceptable to Brenner or anyone else. I remain as unconvinced as Searle by Macfarlane's (1978; 1984) attempts to deny the peasantry a place in post-medieval England, and find the argument that a residual peasantry survived into the nineteenth century generally persuasive (Beckett, 1984; Reed, 1984). However, I am not persuaded of the authenticity of Searle's peasantry. The Cumbrian customary tenantry ranged from rich to desperately poor. Their number included individuals who were adept at defending their traditional rights by recourse to legal or other means, who produced for the market on a large scale, who frequently lived in substantial farmhouses, and whose better-off members were at least the equals of the lesser gentry in respect of their wealth, domestic comfort, and the educational advantages they managed to secure for their sons (Jones, 1962; Marshall, 1972; 1973; 1980; Beckett, 1982). In a strictly formal sense, the social formation of eighteenth-century Cumbria may have been 'feudal', as Searle claims. But it was a feudalism in which at least some of the supposedly subordinate had learned the essentially capitalistic message that he who buys cheap (or at least ensures that outgoings remain low) and sells dear (in a rising market for agricultural commodities) is likely to prosper. Some lords recognised that their best hope for extracting surplus lay in conceding full freehold rights in return for a single lump sum payment (Jones, 1962). If the Cumbrian case proves anything, it is surely the tendency of categories to mislead.

ENCLOSURE

The enclosure of subdivided fields and the reclamation of wastes made a substantial contribution to the appearance of present-day landscapes, and has thus been of considerable interest to practitioners of the landscape school, already discussed. In recent years, there has also been evidence of reawakening interest in the causes and consequences of enclosure among students of agrarian capitalism.

Much of the appeal of enclosure to scholars of both persuasions lies in its visibility. And of its various phases, none is more visible than that which was implemented by several thousand parliamentary acts during the eighteenth and nineteenth centuries. In England alone, these secured the enclosure of approximately 4.5 million acres of surviving open field arable, and 2.3 million acres of commons and waste, mostly on the fen and moorland margins (Turner, 1980; 1984a). In the upland counties of Wales, between 20 and 30 per cent of the total land area was affected by commons and waste enclosures (Williams, 1970). Throughout Britain, such enclosures were very much a feature of the nineteenth century. But much the largest proportion of open field enclosure took place between 1750 and 1830, with particularly intensive bursts of activity in the years 1775 to 1780, and 1810 to 1815 (Turner, 1980; 1984a).

Parliamentary enclosure of open field arable is highly visible in the landscape for several reasons. New fields and roads were set out by a relatively small number of very busy commissioners and surveyors according to rules laid down by parliament itself. There is therefore a certain uniformity about a parliamentary enclosure landscape, which is generally recognisable as such no matter where it happens to be. Furthermore, the distribution of such landscapes is highly uneven, being heavily concentrated in a belt which stretches from the Yorkshire coast southward through the east and south-east Midland counties to end at Salisbury Plain.

So far as students of agrarian capitalism are concerned, the visibility of parliamentary enclosure lay not so much in its physical presence, but in the chronology and character of the process itself. Parliamentary enclosure coincided with the growth of the industrial economy. Chronology alone suggested that it might have served as a means of driving a dispossessed peasantry into wage labour, both in agriculture and in factory industry. The conspiracy theory gained ground as scholars reflected on the likely role of parliament in the whole affair. It seemed inconceivable that a body comprised of larger landowners, enacting bills which had been submitted by or on behalf of large landowners (the rules, such as they were, required that a bill needed the consent of certain fixed majority proportions of owners by value, not by number, and therefore provided no automatic protection for the smaller man), would have taken the initiative in upholding the interests of smallholders or those without land and no formal legal title to common rights which were enjoyed by custom. During the late nineteenth and early twentieth centuries, writers like Marx, Hasbach, Slater, the Webbs, and, most notably, the Hammonds argued the case for dispossession and conspiracy, in works which were often more notable for polemical vigour than for sober scrutiny of the available evidence. As other scholars turned their attention to sources like the land tax, petitions and counter petitions to parliament, or data relating to poverty and labour migration, so it became possible to argue that the case against enclosure was at best unproven, at worst grossly exaggerated (see Mingay, 1968 for a summary).

The existence of this revisionist movement is well known. Since some of the work dates back to the twenties or earlier, it hardly qualifies for inclusion in such a survey as this. I mention it here only because it may be less well known that recent years have seen the growth of another revisionism which promises to rehabilitate some of the conclusions of the Hammonds and their co-religionists, if not the premises on which the conclusions were based. The new revisionism arises from a willingness to submit the available documentary sources to much fuller critical scrutiny. For instance, studies which treat the land tax as a source susceptible to analysis by nominal linkage methods have shown that while earlier workers may have been correct in claiming that enclosure did not bear responsibility for the single-handed destruction of small landowners as a group, it was clearly associated with the disappearance of individual owners and occupiers, and that these effects were not confined to the smaller producers (Turner, 1975; Walton, 1975). Work by Martin (1977; 1979a; 1984) on a variety of Warwickshire sources, including land tax assessments, parish registers, enclosure awards, and manorial documents, has identified a large sub-stratum of small landowners and tradesmen who lost the rights of access to land, to their considerable distress and disadvantage. Innovative research by Snell (1985) using poor law material suggests that increased poverty and higher levels of seasonal unemployment were both features of the post enclosure regime. Martin (1979b) has also drawn attention to close connections between leading petitioners and the small groups of M.Ps who steered each bill through the Commons. Further local studies like Chapman's (1982) on Horsham, Sussex, will doubtless show how the decision to seek an enclosure reflected complex underlying alliances and animosities in relationships between powerful interests. It is already clear that in this process the less influential were not only excluded, but that their interests were often sacrificed.

In a superb study of Northamptonshire, Neeson (1984) shows that an absence of counter petitions to parliament or of references to riots in the Home Office papers cannot be taken as positive evidence of popular acquiescence. Though rarely on a scale likely to attract the attention of the central authorities of the state, grass-roots opposition was nonetheless real. Predominantly local in character, its visibility to present-day historians depends upon the chance survival of local ephemera. 'The Raunds riot', Neeson (1984, p. 136) remarks, 'is known only because a vicar from a neighbouring parish wrote a poem about it many years after the event; much fence-breaking evidence turns up only because its victims advertised rewards in the vainest of hopes of enticing an informer; the Staverton counter-petition is known only because it was advertised for sale (and then sold without further trace) in an antiquarian bookseller's list in the 1970s.'

Parliamentary enclosure is now increasingly seen as one of the mechanisms which helped secure the destruction of the customary economy and its associated culture. If land which had traditionally been set aside for popular recreation was included within an enclosure proposal, then this often sparked off, or acted as a focus for popular protest (Malcolmson, 1981; Bushaway, 1982). However, it was more tangible and economically devastating losses stemming from the suppression of traditional rights like gleaning or gathering fuel which had the severest impact on the poor. So far as the progenitors of enclosure were concerned, these rights were marginal. But they were far from marginal to the marginal themselves. A central part of their hereditary rights and property was removed (Thompson, 1976). The process thus helped extend the notion that only a clearly defined legal title legitimised access to the fruits of the soil. A sequence of events initiated some hundred years previously by a parliament increasingly defensive of property rights (Hay, 1975), and subsequently by the implementation of the Game Laws (Munsche, 1981), achieved its apogee in the criminal sub-culture of the nineteenth century poacher (Jones, 1979; Hopkins, 1985). And the concept of the moral economy, the defensive appeal by the poor to the sanctity of their traditional rights, and a reminder of the customary obligations of the rich, which had long been a part of the food rioter's stock-in-trade, began to look even more threadbare and unconvincing, although it continued to be invoked (Thompson, 1971; Thwaites, 1985; Storch, 1982).

Recently, it has been argued that in North America land and labour were treated as commodities in exactly the same way as in England (Lemon, 1980; contrast Harris, 1977 and

Henretta, 1978). Yet, emigrants' letters home also reveal that this was not yet a society where the definition of private property had become so refined that access to use-rights was denied. The New World presented to its newcomers an environment of low demographic pressure and relatively open social networks. Freedom of access to plentiful supplies of firing or to unimagined quantities of game was a source of wonderment to letter writers (Snell, 1985). The contrast with home was barely credible.

Just as there has been a recent revival of interest in parliamentary enclosure, so, too, has much attention been devoted to the theme of pre-parliamentary enclosure. To some extent this represents a spin-off from work on the parliamentary phase. It had long been appreciated that enclosure before the 1730s, by agreement or by Chancery decree, must have affected quite substantial acreages, and a number of local and national studies were available which drew attention to this fact (Hodgshon, 1979; Yelling, 1977; 1978; Butlin, 1979; Beresford, 1979). However, it was only with the publication of Turner's (1978) edition of Tate's Domesday of English Enclosures that a reasonably accurate global estimate could be made of the acreages involved (see Chapman (1978) and Chapman and Harris (1982) for problems in the accuracy of estimates of parliamentary enclosure acreages). Wordie (1983) suggested that approximately 45 per cent of the total land area of England was enclosed before 1500, that very little was added to this total during the sixteenth century, but that 24 per cent was enclosed during the seventeenth century, as compared with approximately 20 per cent during the entire parliamentary enclosure phase. In other words, the visibility of parliamentary enclosure is illusory to the extent that it was preceded by a more concentrated phase of enclosing activity which accounted for a greater proportion of the nation's total land area.

These figures prompt two questions, neither particularly easy to answer. First, did seventeenth-century enclosure have a significant social cost, as did parliamentary enclosure? Second, was seventeenth-century enclosure associated with significant improvements in crop yield and agricultural output generally? The first question returns us to the heart of the Brenner debate, and its essential presumption (founded on particular interpretations of the English evidence) that the more substantial owners benefited at the expense of the smaller owners during the sixteenth and seventeenth centuries. Brenner's hypothesis does not require that enclosure actually played a causal role in bringing about these changes. Since most enclosure took place piecemeal, by the agreement of the parties concerned, it should perhaps been seen as contingent rather than causal. By the same token, it seems unlikely that large numbers of unwitting pauperised victims were trampled under in the process. There is something to be said for the argument that seventeenth-century enclosure mopped up many areas of subdivided field where dissent or conflict of interest were not a problem, leaving the more contentious cases until the second half of the eighteenth century, when parliamentary enclosure provided a sharper legal instrument, offering a real, if, for many, uncomfortable prospect of surgery.

The second question directs our attention to an issue which we have so far barely mentioned, that of agricultural productivity and output. Clearly, whether land is farmed communally or independently is only one of several possible influences on its productivity. The level of technological innovation in agriculture is obviously influential, and this will be surveyed in the next section. Other variables include overall levels of market demand and the composition of that demand, the degree of articulation of the market (i.e., the extent to which producers were supplying non-local needs), the size of the subsistence sector, the amount of land available for agricultural colonisation, farm size and the social distribution of land ownership and farm occupancy, the productivity of labour, and the state of the weather.

Kerridge's (1967) agricultural revolution (so-called) of the sixteenth and seventeenth centuries placed most significant improvement in practice and productivity before 1670. It recognised that enclosure was taking place on a large scale, but did not see it as critical to the process of improvement. Other researchers have been as little inclined as Kerridge to establish links between productivity change and enclosing activity. But at least some offer clearer indications of underlying trends in output during the period concerned. Ever since Jones (1965; 1967) and John (1960; 1965) argued that the stimulus of shifts in market demand and relative prices for cereals and livestock had encouraged innovation and higher output during the second half of the seventeenth and the early part of the eighteenth century, scholars have made sporadic attempts to clothe these suggestions in decent statistical garb. The figures currently available comprise geographically limited, but detailed estimates for East Anglian cereals yields, ingeniously calculated from probate inventories (Overton, 1979), as well as global but somewhat less dependable statistics comprising reworkings of earlier estimates by Deane and Cole (1969) (Crafts, 1976; Jackson, 1985); estimates which reflect supposed changes in per capita food consumption during the eighteenth century (Jones, 1981c); and estimates which give due weight to the quantities of yield data collected by panicking authorities during the Napoleonic Wars (Turner, 1982). Large margins of error are necessarily a feature of all of these estimates.

Setting aside inevitable differences about the magnitude and meaning of such sources of error (Overton, 1984b; Turner, 1984b), the message appears to be that the rate of

increase of agricultural output after about 1760 or 1770 failed
to keep pace with the rate of population growth. Overton's
(1979; 1984a) inventory-based statistics suggest a late seventeenth-century East Anglian farming sector sufficiently
responsive to market pressures as to increase its acreage and
output of barley, and therefore able to benefit from good
livestock prices (barley being used as fodder) and increasing
commercial traffic in that crop. On the other hand, in the
second half of the eighteenth century the global figures
suggest decreasing rates of growth in cereals yields per acre,
and declining output per capita (Jackson, 1985; Turner,
1984b). Clearly, like is not invariably being compared with
like in this exercise. But equally clearly, there is no warrant
for regarding seventeenth century enclosure as an aberrant
feature of an essentially static or backward age, or for
associating parliamentary enclosure with the eighteenth-
century publicist's image of an era of unparalleled achieve-
ment.

INNOVATION

It will be clear from much that has already been said that
agricultural change was not instantaneous in either time or
space. New ways of doing things, new crops, new strains of
livestock, and new items of equipment were each associated
with their own distinctive chronologies and geographies. Just
how distinctive, it is not easy to say. The sources (probate
inventories for the period 1560 to 1730; a variety of sources,
but most usefully farm sales notices for the late eighteenth
and nineteenth centuries) do not provide fully specified
chronological or geographical data, and the information which
they do present is laborious to extract. What, at best, we are
likely to be offered is a partial insight into the pattern of
diffusion of a limited number of innovations in some relatively
small study area, accompanied by speculative reflections on
the probable processes involved. Only the importance of these
processes (in a sense they lie at the very core of agricultural
geography and rural society) can justify the exercise.
    These reservations apply with less force to the chrono-
logical curves than to the spatial patterns and the forces
which have shaped them. The recurrent dilemma of the
historical geographer is that it is easier to reconstruct past
chronologies than to reconstruct past geographies. Historical
geography thus all too often becomes an activity where
history (in the sense of general chronology) is reasonably
clear, but geography (in the sense of spatial variation in the
expression of that chronology) is elusive. If, say, 20% of
farmers' inventories for a specified time period contain
turnips, then one can say that that, or something resembling
it, was the level of adoption among the inventoried farming

population. If, moreover, in the next time period, 25% of farmers' inventories contain turnips, then it is a reasonable proposition that the level of adoption was increasing at approximately that rate. A chronology is established: the fact that the sample represents only a small proportion of the farming population at any one time is not a severe problem. The difficulty that turnips may not have been recorded in some of the inventories of those who actually grew them can be discounted to the extent that the level of omission is likely to have been consistent through time. On the other hand, if we try to establish where, exactly, turnips were being adopted, the problem of missing information becomes more severe. The gaps on the maps have to be interpreted as sensitively as their positive content.

Innovation studies reveal an implicit awareness of these underlying difficulties. Extended discussion of the sources, their limitations, and the methods of analysis employed (Overton, 1977; 1980) are a necessary preliminary to studies which present chronology less tentatively than spatial variation (Overton, 1985; Walton, 1979: 1983), and spatial variation less tentatively than discussions of the factors which underpin it. These studies have been reasonably successful in showing that open field did retard the adoption of some innovations, like turnips (Overton, 1985; Walton, 1978). They are also united in their reluctance to follow Hagerstrand in believing that patterns of adoption may be explained by reference to information flows, and information flows by patterns of interpersonal contact between farmers 'at risk'. The model presupposes a certain type of farming society, apparently different from that of capitalist agrarian Britain.

Somehow, explanations of the diffusion process have to allow for the fact that large numbers of landowners saw their contribution to 'improvement' as a means of enhancing personal prestige and status. The information flows between the elite - the materials they read (Thirsk, 1985; Horn, 1982; Goddard, 1983), societies they joined (Goddard, 1981; Fox, 1979), the letters they wrote to each other (Macdonald, 1979) - become important in determining how certain innovations first appeared in particular localities. Whether the innovations then spread any further depended largely on their appropriateness to the working farmer. There is some evidence that in Wales the deep cultural gulf separating the tenants from their landlords was sufficient to condemn most landlord-inspired innovations to instant oblivion (Colyer, 1978). The same was true of other pieces of pure landlordism elsewhere. At the less elevated level of the farmer, rational judgement of the economics of risk was probably more important than the factors which have traditionally interested geographers. Admittedly, adoption patterns often assumed spatially coherent forms consistent with the hypothesis that interpersonal communication mattered (Emery, 1976; Walton, 1984). But for many innovations, it may be that the patterns reflect communication in the social space of elites, not the geographical space of the farming population as a whole.

THE FRENCH CONNECTION

It will be recalled that, according to Brenner, Britain's conversion to capitalist agriculture contrasted with the experience of France, where the peasantry continued to rule supreme. One theme (the final theme of my survey) which has interested several Anglo-Saxon observers is the circumstances surrounding France's emergence from that condition, and its admission to the fold of 'modernised' rural society. Weber's (1976) magisterial study informs us that the process of modernisation, the emergence of France as an entity which was fully recognised by all of its inhabitants, took place during the years between the Franco-Prussian and Great Wars. Agriculture could only become fully responsive to the challenge of its potential markets once a fully developed railway network had provided access to them. Price (1983) sees that process occurring unevenly and haltingly from the 1850s and 1860s onwards.

The notion that little of any consequence had taken place before the late nineteenth century to disturb the essential tranquillity of an immobile peasant society has not gone unchallenged (Tilly, 1979). Attempts to measure, for example, agricultural output and innovation during the eighteenth and early nineteenth centuries have revealed improvement, to the extent that there has been some necessarily loose talk of the possible existence of an agricultural revolution (Morineau, 1970; Newell, 1973; Grantham, 1978). Yet it is the enduring underlying reality of a predominantly peasant society, responding to change in its own way, according to its own notions of collectivism and individualism which makes the history of rural life in nineteenth-century France quite different from its British counterpart. Dallas's (1982) analysis of the survival strategies of the peasantry of the Loire, and the studies of agricultural syndicalism by Baker (1980; 1984; 1986) and Cleary (1982), all discuss features of the agrarian society and economy which have no exact parallel in the British context.

CONCLUSION

Difference, divergence and contrast are the abiding impression not only of the content of the historical geography of the agrarian sector, but also of its treatment. Greater awareness of broader conceptual and theoretical issues has not trussed the sub-discipline in a conformist straightjacket.

AGRICULTURE AND RURAL SOCIETY

Thankfully, we may still revel in the almost anarchic quality of its diversity.

Diversity reflects not only differences in the sympathies and interests of historical geographers, but also differences in the problems which are being examined. If the cultural landscape has a stronger position in North American than in British historical geography, then that owes something to differences in the historical experience of the two areas. Likewise, when students of Britain emphasise change and those of France the forces which conspired to slow down, prevent or subvert it, then that too reflects important contrasts in the British and French agrarian past.

However, one might wonder whether such contrasts, and especially the last, have been overdrawn. Even in Britain, few contemporaries other than the purveyors of self congratulation and mutual admiration whose writings filled the agricultural press regarded agriculture as dynamic or even adaptable. The stereotype of the country bumpkin, common enough in all other literary productions, was doubtless overstated, but surely not unknown in rural reality. After all, agriculture is a sort of residual, a way of life as much as an economic activity. Unlike industry, its creation and survival is not dependent on the existence and sustenance of a powerful internal dynamic. One might therefore say that, having looked at change for so long, it may be time for historical geographers to view agriculture from the standpoint of its more normal condition - that of immobility or inactivity. How the old survived and how much of it survived are both as interesting and important as questions connected with innovation and change.

REFERENCES

Adams, I.H. (1978) 'The Agricultural Revolution in Scotland: Contributions to the Debate', Area, 10, 198-203

Albert, W. (1976) An Essay on the Peruvian Sugar Industry, 1880-1920, and the Letters of Robert Gordon, Administrator of the British Sugar Company in Canete, 1914-20, School of Social Studies, University of East Anglia

Appleby, A.B. (1980) 'Epidemics and Famine in the Little Ice Age', Journal of Interdisciplinary History, 10, 643-663

Austin, D. (1985) 'Doubts About Morphogenesis', Journal of Historical Geography, 11, 201-9

Axtell, J. (1981) The European and the Indian Essays in the Ethnohistory of Colonial North America, Oxford University Press, Oxford

Baker, A.R.H. (1972) 'Historical Geography in Britain', in A.R.H. Baker (ed.), Progress in Historical Geography, David and Charles, Newton Abbot, 90-110

Baker, A.R.H. (1980) 'Ideological Change and Settlement

Continuity in the French Countryside: The Development of Agricultural Syndicalism in Loir-et-Cher During the Late-nineteenth Century', Journal of Historical Geography, 6, 163-77

Baker, A.R.H. (1983) 'Discourses on British Field Systems', Agricultural History Review, 31, 149-55

Baker, A.R.H. (1984) 'Fraternity in the Forest: The Creation, Control and Collapse of Woodcutters' Unions in Loir-et-Cher 1852-1914', Journal of Historical Geography, 10, 157-73

Baker, A.R.H. (1986) 'The Infancy of France's First Agricultural Syndicate: The Syndicat des Agriculteurs de Loir-et-Cher 1881-1914', Agricultural History Review, 34, 44-59

Baker, A.R.H. and Gregory, D. (1984) 'Some Terrae Incognitae in Historical Geography: An Exploratory Discussion', in A.R.H. Baker and D. Gregory (eds.), Explorations in Historical Geography: Interpretative Essays, Cambridge University Press, Cambridge, 180-94

Baker, A.R.H. and Butlin, R.A. (eds.) (1973) Studies of Field Systems in the British Isles, Cambridge University Press, Cambridge

Barley, M.W. (1961) The English Farmhouse and Cottage, Routledge and Kegan Paul, London

Barley, M.W. (1985) 'Rural Building in England', in J. Thirsk (ed.), The Agrarian History of England and Wales. V: 1640-1750. II: Agrarian Change, Cambridge University Press, Cambridge, 590-685

Beckett, J.V. (1982) 'Regional Variations and the Agricultural Depression, 1730-50', Economic History Review Second series 35, 35-51

Beckett, J.V. (1984) 'The Peasant in England: A Case of Terminological Confusion?' Agricultural History Review, 32, 113-23

Beresford, M.W. (1979) 'The Decree Rolls of Chancery as a Source for Economic History', Economic History Review Second series 32, 1-10

Bettey, J.H. (1982) 'Land Tenure and Manorial Custom in Dorset, 1570-1670', Southern History, 4, 33-54

Bogue, A.G. (1982) 'Farming in the North American Grasslands: A Survey of Publications, 1947-1980', Agricultural History Review, 30, 49-67

Brenner, R. (1976) 'Agrarian Class Structure and Economic Development in Pre-industrial Europe', Past and Present, 70, 30-75

Brenner, R. (1977) 'The Origins of Capitalist Development: A Critique of Neo-Smithian Marxism, New Left Review, 104, 25-93

Brunger, A.G. (1982) 'Geographical Propinquity Among Pre-famine Catholic Irish Settlers in Upper Canada', Journal of Historical Geography, 8, 265-82

Brunskill, R.W. (1971) Illustrated Handbook of Vernacular Architecture, Faber and Faber, London

Bushaway, B. (1982) By Rite Custom, Ceremony and Community in England 1700-1880, Junction Books, London

Butlin, R.A. (1979) 'The Enclosure of Open Fields and Extinction of Common Rights in England c.1600-1750; A Review', in H.S.A. Fox and R.A. Butlin (eds.), Change in the Countryside, Institute of British Geographers, Special Publication, 10, pp. 65-82

Butlin, R.A. (1982) The Transformation of Rural England c.1580-1800: A Study in Historical Geography, Oxford University Press, Oxford

Cameron, R.A.D. (1984) 'The Biology and History of Hedges: Exploring the Connection', Biologist, 31, 203-8

Carson, C. (1976) 'Segregation in Vernacular Building,', Vernacular Architecture, 7, 24-9

Chambers, J.D. and Mingay, G.E. (1966) The Agricultural Revolution 1750-1880, Batsford, London

Chapman, J. (1978) 'Some Problems in the Interpretation of Enclosure Awards', Agricultural History, 26, 108-14

Chapman, J. (1982) 'The Unofficial Enclosure Proceedings: A Study of Horsham Enclosure 1812-1813', Sussex Archaeological Collections, 120, 185-91

Chapman, J. and Harris, T.M. (1982) 'The Accuracy of Enclosure Estimates: Some Evidence from Northern England', Journal of Historical Geography, 8, 261-4

Christopher, A.J. (1976) Southern Africa, Dawson, Folkestone

Christopher, A.J. (1984) Colonial Africa, Croom Helm, London

Clark, T.D. (1984) The Greening of the South: The Recovery of Land and Forest, University Press of Kentucky, Lexington

Clarkson, L.A. (1985) Proto-Industrialization: The First Phase of Industrialization?, Macmillan, London

Cleary, M.C. (1982) 'The Plough and the Cross: Peasant Unions in South-Western France', Agricultural History Review, 30, 127-36

Coleman, D.C. (1983) 'Proto-Industrialization: A Concept Too Many', Economic History Review, Second series 36, 435-48

Colyer, R.J. (1978) 'Limitations to Agrarian Development in Nineteenth-Century Wales', Bulletin of the Board of Celtic Studies, 27, 602-17

Cooper, J.P. (1978) 'In Search of Agrarian Capitalism', Past and Present, 80, 20-65

Crafts, N.F.R. (1976) 'English Economic Growth in the Eighteenth Century: A Re-examination of Deane and Cole's Estimates', Economic History Review, Second series 29, 226-35

Croot, P. and Parker, D. (1978) 'Agrarian Class Structure and Economic Development', Past and Present, 78, 37-47

Crush, J. (1986) 'Towards a People's Historical Geography for South Africa', Journal of Historical Geography, 12, 2-3

Curtin, P.D. (1984) Cross-Cultural Trade in World History, Cambridge University Press, Cambridge

Dallas, G. (1982) The Imperfect Peasant Economy: The Loire Country, 1800-1914, Cambridge University Press, Cambridge

Darby, H.C. (1983) The Changing Fenland, Cambridge University Press, Cambridge

Deane, P. and Cole, W.A. (1969) British Economic Growth 1688-1959 (Second edition), Cambridge University Press, Cambridge

Dodgshon, R.A. (1977) 'A Spatial Perspective', Journal of Peasant Studies, 6, 8-19

Dodgshon, R.A. (1980) The Origin of British Field Systems: An Interpretation, Academic Press, London

Dodgshon, R.A. (1981) Land and Society in Early Scotland, Clarendon Press, Oxford

Donkin, R.A. (1977) Spanish Red: An Ethnogeographical Study of Cochineal and the Opuntia Cactus, Transactions Vol 2 No 5, American Philosophical Society, Philadelphia

Donkin, R.A. (1979) Agricultural Terracing in the Aboriginal New World, University of Arizona Press, Tucson

Dunford, M. and Perrons, D. (1983) The Arena of Capital, Macmillan, London

Emery, F. (1976 'The Mechanics of Innovation: Clover Cultivation in Wales before 1750', Journal of Historical Geography, 2, 35-48

Ernle, Lord (1912, sixth edition 1961) English Farming Past and Present, Heinemann/Cass, London

Fenton, A. and Walker, B. (1981) The Rural Architecture of Scotland, John Donald, Edinburgh

Fogel, R.W. and Elton, G.R. (1983) Which Road to the Past? Two Views of History, Yale University Press, New Haven and London

Fox, H.S.A. (1979) 'Local Farmers' Associations and the Circulation of Agricultural Information in Nineteenth-Century England', in H.S.A. Fox and R.A. Butlin (eds.), Change in the Countryside, Institute of British Geographers, Special Publication 10, 43-63

Gade, D.W. (1979) 'Inca and Colonial Settlement, Coca Cultivation and Endemic Disease in the Tropical Forest', Journal of Historical Geography, 5, 263-79

Galloway, J.H. (1982) 'Agricultural Improvement in Late-Colonial Tropical America: Sources and Issues', in A.R.H. Baker and M. Billinge (eds.), Period and Place: Research Methods in Historical Geography, Cambridge University Press, Cambridge, 79-86

Gauldie, E. (1981) 'Country Homes', in G.E. Mingay (ed.), The Victorian Countryside, Routledge and Kegan Paul, London, 531-41

AGRICULTURE AND RURAL SOCIETY

Glassie, H. (1975) Folk Housing in Middle Virginia: A Struc-
    tural Analysis of Historic Artifacts, University of
    Tennessee Press, Knoxville
Goddard, N. (1981) 'Agricultural Societies', in G.E. Mingay
    (ed.), The Victorian Countryside, pp. 245-59
Goddard, N. (1983) 'The Development and Influence of
    Agricultural Periodicals and Newspapers, 1780-1880',
    Agricultural History Review, 31, 116-31
Grantham, G.W. (1978) 'The Diffusion of the New Husbandry
    in Northern France, 1815-1840', Journal of Economic
    History, 38, 311-37
Gregory, D. (1981) 'Historical Geography', in R.J. Johnston
    (ed.), The Dictionary of Human Geography, Blackwell,
    Oxford, 146-50
Gregson, N. (1984) Context and Structure: Towards Agrarian
    Capitalism in North-West England, Seminar Paper 39,
    Department of Geography, University of Newcastle upon
    Tyne
Gresham, C.A. (1971) 'Gavelkind and the Unit System',
    Archaeological Journal, 128, 174-75
Grigg, D. (1982) The Dynamics of Agricultural Change,
    Hutchinson, London
Hall, C. (1982) 'Private Archives as Sources for Historical
    Geography', in A.R.H. Baker and M. Billinge (eds.),
    Period and Place: Research Methods in Historical Geo-
    graphy, Cambridge University Press, Cambridge, 274-80
Harris, R.C. (1977) 'The Simplification of Europe Overseas',
    Annals of the Association of American Geographers, 67,
    469-83
Harrison, B. and Hutton, B. (1984) Vernacular Houses in
    North Yorkshire and Cleveland, John Donald, Edinburgh
Harrison, C.J. (1971) 'Grain Price Analysis and Harvest
    Qualities, 1465-1634', Agricultural History Review, 19,
    135-55
Harvey, D. (1983) The Limits to Capital, Blackwell, Oxford
Harvey, M. (1978) The Morphological and Tenurial Structure
    of a Yorkshire Township: Preston-in-Holderness 1066-
    1750, Occasional Paper 13, Department of Geography,
    Queen Mary College, London
Hay, D. (1975) Property, Authority and the Criminal Law', in
    D. Hay, P. Linebaugh, J.G. Rule, E.P. Thompson and
    C. Winslow (eds.), Albion's Fatal Tree Crime and Society
    in Eighteenth-Century England, Penguin, Harmonds-
    worth, 17-63
Hennessy, A. (1978) The Frontier in Latin American History,
    Edward Arnold, London
Henretta, J. (1978) 'Families and Farms: Mentalité in Pre-
    industrial America', William and Mary Quarterly, Third
    series 35, 3-32
Hewes, L. (1981) 'Early Fencing on the Western Margin of the
    Prairie', Annals of the Association of American Geo-

graphers, 71, 499-526

Hewes, L. and Jung, C.L. (1981) 'Early Fencing on the Middle Western Prairie', Annals of the Association of American Geographers, 71, 177-201

Hodgshon, R.I. (1979) 'The Progress of Enclosure in County Durham, 1550-1870', in H.S.A. Fox and R.A. Butlin (eds.), Change in the Countryside, Institute of British Geographers, Special Publication 10, 83-102

Hopkins, H. (1985) The Long Affray: The Poaching Wars in Britain, Secker and Warburg, London

Horn, P. (1982) 'The Contribution of the Propagandist in Eighteenth-Century Agricultural Improvement', Historical Journal, 25, 313-29

Hoskins, W.G. (1953) 'The Rebuilding of Rural England 1570-1640', Past and Present, 4, 44-59

Hoskins, W.G. (1964) 'Harvest Fluctuations and Economic History 1480-1619', Agricultural History Review, 12, 28-46

Hoskins, W.G. (1968) 'Harvest Fluctuations and Economic History 1620-1759', Agricultural History Review, 16, 15-31

Houston, R. and Snell, K.D. (1984) 'Proto-industrialization? Cottage Industry, Social Change, and Industrial Revolution', Historical Journal, 27, 473-92

Hoyle, R.W. (1984) 'Lords, Tenants, and Tenant Right in the Sixteenth Century: Four Studies', Northern History, 20, 38-63

Hudson, P. (1981) 'Proto-industrialization: The Case of the West Riding Wool Textile Industry in the Eighteenth and Early Nineteenth Centuries', History Workshop Journal, 12, 34-61

Hudson, P. (1983) 'From Manor to Mill: The West Riding in Transition', in M. Berg, P. Hudson and M. Sonenscher (eds.), Manufacture in Town and Country Before the Factory, Cambridge University Press, Cambridge, 124-44

Jackson, R.V. (1985) 'Growth and Deceleration in English Agriculture, 1660-1790', Economic History Review, Second series 38, 333-51

Jennings, J.N. and Linge, G.J.R. (eds.) (1980) Of Time and Place. Essays in Honour of O.H.K. Spate, Australian National University Press, Canberra

John, A.H. (1960) 'The Course of Agricultural Change, 1660-1760', in L.S. Presnell (ed.), Studies in the Industrial Revolution, Reprinted in W.E. Minchinton (ed.) (1968) Essays in Agrarian History I, David and Charles, Newton Abbot, 221-53

John, A.H. (1965) 'Agricultural Productivity and Economic Growth in England, 1700-1760', Journal of Economic History, 25, 19-34

Johnston, R.J. (1984) 'A Foundling Floundering in World Three', in M. Billinge, D. Gregory and R. Martin (eds.), Recollections of a Revolution: Geography as Spatial Science, Macmillan, London, pp. 39-56

Jones, D.J.V. (1979) 'The Poacher: A Study in Victorian Crime and Protest', Historical Journal, 22, 825-60

Jones, E.L. (1965) 'Agriculture and Economic Growth in England, 1660-1750: Agricultural Change', Journal of Economic History, 25, 1-18

Jones, E.L. (1967) 'Introduction', in E.L. Jones (ed.), Agriculture and Economic Growth in England, 1650-1815, Methuen, London, pp. 1-48

Jones, E.L. (1972) 'The Bird Pests of British Agriculture in Recent Centuries', Agricultural History Review, 20, 107-25

Jones, E.L. (1981a) The European Miracle, Cambridge University Press, Cambridge

Jones, E.L. (1981b) 'Reconstructing Former Bird Communities', Forth Naturalist and Historian, 6, 101-6

Jones, E.L. (1981c) 'Agriculture 1700-1780,', in R. Floud and D. McCloskey (eds.), The Economic History of Britain Since 1700. Vol I. 1700-1860, Cambridge University Press, Cambridge, pp. 66-86

Jones, G.P. (1962) 'The Decline of the Yeomanry in the Lake Counties', Transactions of the Cumberland and Westmorland Antiquarian and Archaeological Society, New series 62, 198-223

Jordan, T.G. (1980) 'Alpine, Alemannic and American Log Architecture', Annals of the Association of American Geographers, 70, 154-80

Kerridge, E. (1967) The Agricultural Revolution, George Allen and Unwin, London

Kerridge, E. (1969a) 'The Agricultural Revolution Reconsidered', Agricultural History, 43, 463-76

Kerridge, E. (1969b) Agrarian Problems in the Sixteenth Century and After, George Allen and Unwin, London

Kniffen, F.B. (1979) 'The Geographer's Craft, I Why Folk Housing?', Annals of the Association of American Geographers, 69, 59-63

Kupperman, K.D. (1980) Settling with the Indians. The Meeting of English and American Cultures in America, 1580-1640, Dent, London

Leff, N.H. (1982) Underdevelopment and Development in Brazil (Two volumes), George Allen and Unwin, London

Lemon, J.T. (1980) 'Early Americans and Their Social Environment', Journal of Historical Geography, 6, 115-31

Levine. D. (1977) Family Formation in an Age of Nascent Capitalism, Academic Press, London

Macdonald, S. (1979) 'The Diffusion of Knowledge Among Northumberland Farmers, 1780-1815', Agricultural History Review, 27, 30-9

Macfarlane, A. (1978) The Origins of English Individualism, Blackwell, Oxford

Macfarlane, A. (1984) 'The Myth of the Peasantry: Family and Economy in a Northern Parish', in R.M. Smith (ed.), Land, Kinship and Life-Cycle, Cambridge University Press, Cambridge, pp. 333-49

Machin, R. (1975) 'The Unit System: Some Historical Explanations', Archaeological Journal, 132, 187-94

Machin, R. (1977a) 'The Great Rebuilding: An Assessment', Past and Present, 77, 35-56

Machin, R. (1977b) 'The Mechanism of the Pre-industrial Building Cycle', Vernacular Architecture, 8, 815-19

McQuillan, D.A. (1978) 'Territory and Ethnic Identity: Some New Measures of an Old Theme in the Cultural Geography of the United States', in J.R. Gibson (ed.), European Settlement and Development in North America: Essays on Geographical Change in Honour and Memory of Andrew Hill Clark, University of Toronto Press, Toronto, pp. 136-69

McQuillan, D.A. (1979) 'The Mobility of Immigrants and Americans: A Comparison of Farmers on the Kansas Frontier', Agricultural History, 53, 576-96

Malcolmson, R.W. (1981) Life and Labour in England 1700-1780, Hutchinson, London

Marshall, H.W. (1981) Folk Architecture in Little Dixie: A Regional Culture in Missouri, University of Missouri Press, Columbia

Marshall, J.D. (1972) '"Statesmen" in Cumbria: The Vicissitudes of an Expression', Transactions of the Cumberland and Westmorland Antiquarian and Archaeological Society, New series 72, 248-73

Marshall, J.D. (1973) 'The Domestic Economy of the Lakeland Yeoman, 1660-1749', Transactions of the Cumberland and Westmorland Antiquarian and Archaeological Society, New series 73, 190-219

Marshall, J.D. (1980) 'Agrarian Wealth and Social Structure in Pre-industrial Cumbria', Economic History Review, Second series 33, 503-21

Martin, J.M. (1977) 'Marriage and Economic Stress in the Felden of Warwickshire in the Eighteenth Century', Population Studies, 31, 519-35

Martin, J.M. (1979a) 'The Small Landowner and Parliamentary Enclosure in Warwickshire', Economic History Review, Second series 32, 328-43

Martin, J.M. (1979b) 'Members of Parliament and Enclosure: A Reconsideration', Agricultural History Review, 27, 101-09

Martin, J.M. (1984) 'Village Traders and the Emergence of a Proletariat in South Warwickshire, 1750-1851', Agricultural History Review, 32, 179-88

Meirion-Jones, G. (1982) The Vernacular Architecture of Brittany, John Donald, Edinburgh

Mercer, E. (1975) English Vernacular Houses, HMSO, London

Mills, D.R. (1976) 'A Scottish Agricultural Revolution?',

Area, 8, 237

Mills, D.R. (1982) 'Rural Industries and Social Structure: Framework Knitters in Leicestershire, 1670-1851', Textile History, 13, 183-203

Mingay, G.E. (1963) 'The Agricultural Revolution in English History: A Reconsideration', Agricultural History, 26, 123-33

Mingay, G.E. (1968) Enclosure and the Small Farmer in the Age of the Industrial Revolution, Macmillan, London

Morineau, M. (1970) 'Was there an Agricultural Revolution in Eighteenth-Century France?', in R.E. Cameron (ed.), Essays in French Economic History, American Economic Association, Holmewood Illinois, 170-82

Munsche, P.B. (1981) Gentlemen and Poachers: The English Game Laws 1671-1831, Cambridge University Press, Cambridge

Nash, R. (1973) Wilderness and the American Mind, (First edition 1967), Yale University Press, New Haven

Neeson, J.M. (1984) 'The Opponents of Enclosure in Eighteenth-Century Northamptonshire', Past and Present, 105, 114-39

Newell, W.H. (1973) 'The Agricultural Revolution in Nineteenth-Century France', Journal of Economic History, 33, 697-730

Newson, L.A. (1976) Aboriginal and Spanish Colonial Trinidad: A Study in Culture Contact, Academic Press, London

Ostergren, R. (1979) 'A Community Transplanted: The Formative Experience of a Swedish Immigrant Community in the Upper Middle West', Journal of Historical Geography, 5, 189-212

Outhwaite, R.B. (1986) 'Progress and Backwardness in English Agriculture, 1500-1650', Economic History Review, Second series 39, 1-18

Overton, M. (1977) 'Computer Analysis of an Inconsistent Data Source: The Case of Probate Inventories', Journal of Historical Geography, 4, 317-26

Overton, M. (1979) 'Estimating Crop Yields from Probate Inventories: An Example from East Anglia, 1585-1735', Journal of Economic History, 39, 363-78

Overton, M. (1980) 'English Probate Inventories and the Measurement of Agricultural Change', A.A.G. Bijdragen, 23, 205-15

Overton, M. (1984a) 'Agricultural Revolution? Development of the Agrarian Economy in Early Modern England', in A.R.H. Baker and D. Gregory (eds.), Explorations in Historical Geography, Cambridge University Press, Cambridge, 118-39

Overton, M. (1984b) 'Agricultural Productivity in Eighteenth-Century England: Some Further Speculations, Economic History Review, Second series 37, 244-51

Overton, M. (1985) 'The Diffusion of Agricultural Innovations in Early Modern England: Turnips and Clover in Norfolk and Suffolk, 1580-1740', Transactions, Institute of British Geographers, New series 10, 205-11

Palacios, M. (1980) Coffee in Colombia, 1850-1970: An Economic, Social and Political History, Cambridge University Press, Cambridge

Parry, M.L. (1975) 'Secular Climatic Change and Marginal Agriculture, Transactions, Institute of British Geographers, 64, 1-13

Parry, M.L. (1976a) 'The Abandonment of Upland Settlement in Southern Scotland, Scottish Geographical Magazine, 92, 50-60

Parry, M.L. (1976b) 'A Scottish Agricultural Revolution?', Area, 8, 238-39

Parry, M.L. (1978) Climatic Change, Agriculture and Settlement, Dawson, Folkestone

Perkins, J.A. (1975) 'Working-class Housing in Lindsey, 1780-1870', Lincolnshire History and Archaeology, 10, 49-55

Peters, J.E.C. (1969) The Development of Farm Buildings in Western Lowland Staffordshire up to 1880, Manchester University Press, Manchester

Pollard, E., Hooper, M.D., and Moore, N.W. (1974) Hedges, Collins, London

Porter, J. (1975) 'A Forest in Transition: Bowland, 1500-1650', Transactions of the Historic Society of Lancashire and Cheshire, 125, 40-60

Porter, J. (1978) 'Wasteland Reclamation in the Sixteenth and Seventeenth Centuries: The Case of South-East Bowland', Transactions of the Historic Society of Lancashire and Cheshire, 127, 1-23

Portman, D. (1974) 'Vernacular Building in the Oxford Region in the Sixteenth and Seventeenth Centuries', in C.W. Chalklin and M.A. Havinden (eds.), Rural Change and Urban Growth 1500-1800, Longman, London, pp. 135-68

Powell, J.M. (1981) 'Wide Angles and Convergences: Recent Historical-Geographical Interaction in Australasia', Journal of Historical Geography, 7, 407-13

Price, R. (1983) The Modernization of Rural France: Communications Networks and Agricultural Market Structures in Nineteenth-Century France, Hutchinson, London

Rackman, D. (1980) Ancient Woodland: Its History, Vegetation and Uses in England, Edward Arnold, London

Reed, M. (1984) 'The Peasantry of Nineteenth-Century England: A Neglected Class?', History Workshop Journal, 18, 53-76

Rice, J.G. (1977) 'The Role of Culture and Community in Frontier Prairie Farming', Journal of Historical Geography, 3, 155-75

Rogers, A. (1981) 'Rural Industries and Social Structure:

The Framework Knitting Industry of South Nottingham-
shire, 1670-1840', Textile History, 12, 7-36
Rotberg, R.I. and Rabb, T.K. (1981) Climate and History,
Princeton University Press, Princeton
Sandall, K.L. (1975) 'The Unit System in Essex', Archae-
ological Journal, 132, 195-201
Searle, C.E. (1986) 'Custom, Class Conflict and Agrarian
Capitalism: The Cumbrian Customary Economy in the
Eighteenth Century', Past and Present, 110, 106-33
Sharp, B. (1980) In Contempt of All Authority: Rural
Artisans and Riot in the West of England, 1586-1660,
University of California Press, Berkeley
Sheail, J. (1978) 'Rabbits and Agriculture in Post-medieval
England', Journal of Historical Geography, 4, 343-55
Sheail, J. and Wells, T.C.E. (1983) 'The Fenlands of
Huntingdonshire, England: A Case Study in Catastrophic
Change', in A.J.P. Core (ed.), Swamp, Bog, Fen and
Moor. B. Regional Studies, Amsterdam, 375-93
Sheehan, B. (1980) Savagism and Civility, Indians and
Englishmen in Colonial Virginia, Cambridge University
Press, Cambridge
Sheppard, J.A. (1979) The Origins and Evolution of Field and
Settlement Patterns in the Herefordshire Manor of
Marden, Occasional Paper 15, Department of Geography,
Queen Mary College, London
Skipp, V.H.T. (1970) 'Economic and Social Change in the
Forest of Arden', Agricultural History Review, 18,
Supplement, 84-111
Skipp, V.H.T. (1978) Crisis and Development: An Ecological
Case Study of the Forest of Arden, 1570-1674,
Cambridge University Press, Cambridge
Smith, P. (1975) Houses of the Welsh Countryside, HMSO,
London
Snell, K.D.M. (1985) Annals of the Labouring Poor, Social
Change and Agrarian England 1660-1900, Cambridge
University Press, Cambridge
Spence, R.T. (1984) 'The Backward North Modernised? The
Cliffords, Earls of Cumberland and the Socage Manor of
Carlisle', Northern History, 20, 64-87
Spufford, M. (1974) Contrasting Communities: English
Villagers in the Sixteenth and Seventeenth Centuries,
Cambridge University Press, Cambridge
Stilgoe, J.R. (1982) Common Landscape of America, 1580-
1845, Yale University Press, New Haven
Stoddart, D.R. (1981) 'The Paradigm Concept and the History
of Geography', in D.R. Stoddart (ed.), Geography,
Ideology and Social Concern, Blackwell, Oxford, 70-80
Storch, R.D. (1982) 'Popular Festivity and Consumer Protest:
Food Price Disturbances in the Southwest and Oxford-
shire in 1867', Albion, 14, 463-89
Sturgess, R.W. (1966) 'The Agricultural Revolution on the

English Claylands', Agricultural History Review, 14, 104-21

Thirsk, J. (1961) 'Industries in the Countryside', in F.J. Fisher (ed.), Essays in the Economic and Social History of Tudor and Stuart England, Cambridge University Press, Cambridge, 70-88

Thirsk, J. (1985) 'Agricultural Innovations and Their Diffusion', in J. Thirsk (ed.), Agrarian History of England and Wales V.II 1640-1750, Cambridge University Press, Cambridge, 533-89

Thompson, E.P. (1971) 'The Moral Economy of the English Crowd in the Eighteenth Century', Past and Present, 50, 72-136

Thompson, E.P. (1976) 'The Grid of Inheritance: A Comment', in J. Goody, J. Thirsk and E.P. Thompson (eds.), Family and Inheritance Rural Society in Western Europe 1200-1800, 328-60

Thompson, F.M.L. (1968) 'The Second Agricultural Revolution, 1815-1880', Economic History Review, Second series 21, 62-77

Thwaites, W. (1985) 'Dearth and the Marketing of Agricultural Produce: Oxfordshire c.1750-1800', Agricultural History Review, 33, 119-31

Tilly, C. (1979) 'Did the Cake of Custom Break?' in J.M. Merriman (ed.), Consciousness and Class Experience in Nineteenth-Century Europe, Holmes and Meier, New York, 17-44

Tribe, K. (1978) Land, Labour and Economic Discourse, Routledge and Kegan Paul, London

Turner, M. (1975) 'Parliamentary Enclosure and Landownership Change in Buckinghamshire', Economic History Review, Second series 27, 565-81

Turner, M. (1980) English Parliamentary Enclosure, Dawson, Folkestone

Turner, M. (1982) 'Agricultural Productivity in England in the Eighteenth Century: Evidence from Crop Yields', Economic History Review, Second series 35, 489-510

Turner, M. (1984a) Enclosure in Britain 1750-1830, Macmillan, London

Turner, M. (1984b) 'Agricultural Productivity in Eighteenth-Century England: Further Strains of Speculation', Economic History Review, Second series 37, 252-57

Turner, M. (ed.) (1978) 'W.E. Tate', A Domesday of English Enclosure Acts and Awards, University of Reading Library, Reading

Turnock, D. (1982) The Historical Geography of Scotland Since 1707, Cambridge University Press, Cambridge

Wallerstein, I. (1974) The Modern World-System. Capitalist Agriculture and the Origins of the European World-Economy in the Sixteenth Century, Academic Press, London

Wallerstein, I. (1980) The Modern World-System. Mercantilism and the Consolidation of the European World Economy 1600-1750, Academic Press, London

Walsh, M. (1981) The American Frontier Revisited, Macmillan, London

Walton, J.R. (1975) 'The Residential Mobility of Farmers and its Relationship to the Pariamentary Enclosure Movement in Oxfordshire', in A.D.M. Phillips and B.J. Turton (eds.), Environment, Man and Economic Change, Longman, London, pp. 238-52

Walton, J.R. (1978) 'Agriculture 1730-1900', in R.A. Dodgshon and R.A. Butlin (eds.), An Historical Geography of England and Wales, Academic Press, London, pp. 239-65

Walton, J.R. (1979) 'Mechanization in Agriculture: A Study of the Adoption Process', in H.S.A. Fox and R.A. Butlin (eds.), Change in the Countryside, Institute of British Geographers, Special Publication 10, pp. 23-42

Walton, J.R. (1983) 'The Diffusion of Improved Sheep Breeds in Eighteenth- and Nineteenth-Century Oxfordshire', Journal of Historical Geography, 9, 175-95

Walton, J.R. (1984) 'The Diffusion of the Improved Shorthorn Breed of Cattle in Britain During the Eighteenth and Nineteenth Centuries', Transactions, Institute of British Geographers, New series 9, 22-36

Weber, E. (1976) Peasants into Frenchmen. The Modernization of Rural France 1870-1914, Chatto and Windus (paperback edition 1979), London

Whittington, G. (1975) 'Was There a Scottish Agricultural Revolution?' Area, 7, 204-6

Whyte, I.D. (1975) 'Rural Housing in Lowland Scotland in the Seventeenth Century: The Evidence of Estate Papers', Scottish Studies, 19, 55-68

Whyte, I.D. (1978) 'The Agricultural Revolution in Scotland: Contributions to the Debate', Area, 10, 203-5

Whyte, I.D. (1979) Agriculture and Society in Seventeenth-Century Scotland, John Donald, Edinburgh

Whyte, K.A. (1984) Register of Research in Historical Geography, Historical Geography Research Series 14, Geo Books, Norwich

Wiliam, E. (1982) Traditional Farm Buildings in North-East Wales, 1550-1900, National Museum of Wales (Welsh Folk Museum), Cardiff

Williams, M. (1970) 'The Enclosure and Reclamation of Waste Land in England and Wales in the Eighteenth and Nineteenth Centuries', Transactions, Institute of British Geographers, 51, 55-69

Williams, M. (1982) 'Clearing the United States Forests: Pivotal Years 1810-1860', Journal of Historical Geography, 8, 12-28

Wolf, E.R. (1982) Europe and the People Without History,

University of California Press, Berkeley

Wonders, W.C. (1979) 'Log Dwellings in Canadian Folk Architecture', Annals of the Association of American Geographers, 69, 187-207

Wood-Jones, R.B. (1963) Traditional Domestic Architecture in the Banbury Region, Manchester University Press, Manchester

Wordie, J.R. (1983) 'The Chronology of English Enclosure, 1500-1914', Economic History Review, Second series 36, 483-505

Wrigley, E.A. and Schofield, R.S. (1981) The Population History of England, 1541-1871, Edward Arnold, London

Yelling, J.A. (1977) Common Field and Enclosure in England 1450-1850, Macmillan, London

Yelling, J.A. (1978) 'Agriculture 1500-1730', in R.A. Dodgshon and R.A. Butlin (eds.), An Historical Geography of England and Wales, Academic Press, London, pp. 151-72

Zell, M.L. (1984) 'Fixing the Custom of the Manor: Slindon, West Sussex, 1568', Sussex Archaeological Collections, 122, 101-6

Chapter Six

THE HISTORICAL GEOGRAPHY OF INDUSTRIAL CHANGE

C.G. Pooley

INTRODUCTION

Although recent reviews have proclaimed the health and
vigour of British historical geography (Baker and Gregory,
1984, pp. 180-81), study of the industrial revolution and of
the process and impact of industrial change is something of a
lacunae in geographical research. The main emphasis of
geographical research on the period from the late-eighteenth
century is on urban and social themes. Although such studies
tackle issues which frequently arose from industrial change,
the pattern. and process of industrial development is mostly
implicit. With a few notable exceptions the study of industry
and industrial change in the nineteenth century has been
strangely neglected by historical geographers.
    This fact is illustrated by the most recent register of
research in historical geography (Whyte, 1984). Of 495
entries listed in the register, only 2 are classified under the
heading, 'industrial revolution' and a further 16 under
'industry and industrialisation'; although there are undoubtedly other studies which impinge on industrial change
but are classified under a different heading. The same trend
is demonstrated by an analysis of articles published in the
Journal of Historical Geography since its inception in 1975. Of
some 174 articles published in the journal only about 8 appear
to be directly concerned with industry and industrial change
from the late-eighteenth century. Moreover, at least half of
these are from authors working in departments of economics
or economic history. The fact that the majority of contributions on industrial themes have appeared since 1980
suggests that there may be the beginnings of an upsurge of
interest in industrial change, but it also highlights the dearth
in the 1970s. From another perspective, the almost complete
lack of contributions by geographers to the Economic History
Review, and indeed the paucity of references to geographical
material in articles published in this and other major journals
of economic history, confirms that geographers are making

little contribution to debates on industrialisation and economic change.

Assessment of the impact of geographical research on the study of industrial change is made difficult by the indeterminate nature of the concepts concerned. The classic period of industrial revolution' in Britain is usually located between about 1780 and 1840 (Mathias, 1969, Deane, 1969), but concentration on this period produces a restricted view of industrial change. The concept of 'proto-industrialisation', introduced by Mendels (1972) and developed in many subsequent papers (Mendels, 1976 Medick, 1976: Kriedte, Medick and Schlumbohm, 1981; Hudson 1981; Coleman, 1983; Langton and Hoppe, 1983) has forced scholars of industrial change to delve into the early-eighteenth century and earlier; other studies of national economic trends emphasise the importance of economic changes in the second half of the nineteenth century (Lee 1971; McCloskey, 1981; Harley, 1982; Crafts, 1983, Lee 1984). As with studies of agrarian change (Overton, 1984) the term 'revolution' is an inappropriate description of the pace and nature of industrial change and, indeed, it may be positively misleading.

Not only is the timing of industrial change difficult to pin down, but also the definition of what such change embraces is elusive. As Mathias (1969) stated in his classic book, the problem of writing about the industrial revolution is to 'limit its definition' (p.1); the definition which he uses would "involve changes sooner or later, in every aspect of a country's history and its institutions" (p.3). At one level it can be argued that any study of nineteenth- or twentieth-century Britain must take account of the rapid industrial change that occurred between 1780 and 1840 and the adjustments that have continued up to the present. Thus all modern historical geography and social and economic history is concerned with the pattern, process and impact of industrial change. However, for most geographers these themes have remained implicit, and have appeared at best as a static backdrop to a stage on which other changes in urban and rural society were acted out. It has been left largely to economic historians to examine the processes and mechanisms of industrial change.

One of the more significant recent contributions to the study of industrial development in Britain is the Atlas of the Industrial Revolution (Langton and Morris, 1986), produced as a multidisciplinary project by geographers and economic historians. In defining their terms of reference the editors of the atlas have taken a broad definition of the industrial revolution. Although the bulk of maps (41 per cent) cover the period 1850-1899, 9 per cent refer to the eighteenth century and 23 per cent relate to the twentieth century (mostly before 1914). The classic period of industrial revolution in the first half of the nineteenth century is allocated only 27 per cent of contributions, reflecting recent historical emphasis on industrial development in the second half of the nineteenth century (Samuel, 1977). The contents of the atlas also range widely over topics as disparate as population, resources, agriculture, manufacturing industry, personal wealth, disease, sport, language, education, religion and politics, with 50 per cent of maps relating to 'economic' themes and 47 per cent focusing on 'social' issues (Langton and Morris 1985; Withers, 1985). The editors thus dodge the problem of definition and range widely over the many aspects of economy and society that were affected by industrial change during the period circa 1780-1914.

The remainder of this paper takes a similarly catholic approach to the concept of industrial revolution and the study of industrial change. Although 'proto-industrialisation' is beyond the scope of this essay, discussion will range over the whole of the period 1780 to 1940, on the assumption that significant structural changes in the economy occurred throughout this period, and that readjustments of the late-nineteenth century and economic restructuring during the inter-war years were just as important as the accelerated growth and structural change which occurred during the classic period of industrial revolution. Moreover, examples will not be narrowly confined to economic issues, but will range widely over the whole spectrum of economy and society that was touched by industrial change. Although examples are drawn mainly from Britain, comparisons with North America are made where appropriate. European and other literature on industrial change is beyond the scope of this essay.

THE GEOGRAPHICAL STUDY OF INDUSTRIAL CHANGE

The study of industrial change since 1780 is necessarily interdisciplinary in nature and should not be constricted by artificial boundaries. Geographical study thus interacts with the work of social, economic and political historians, and no areas of research are the exclusive preserve of particular disciplines. However, it is possible to identify four research areas where geographers have and should have contributed; where geographers have engaged in lively debate with other scholars; or where work - which may originate from many disciplines - clearly falls within a geographical tradition.

First, there are studies of specific industries or groups of industries, which focus on such themes as the organisational and business history of the firm, technological developments affecting operations, changes in working conditions and in employment practices affecting the labour force, and the origins and impact of changing locational decisions. Second, a group of studies focuses on the regional structure of industrial change, ranging over such issues as the extent to which industrialisation destroyed traditional regional economies, the degree to which regional diversity existed within a mature industrial economy, and the emergence of the 'regional problem' of depressed industrial areas. Third, there are studies which focus on various mechanisms of industrial change and particularly on the significance for industrial development of interaction and exchange over space. The role of transport and communications in industrial change, the development of telecommunications, demographic change and labour migration, the flow of capital and ideas between enterprises and areas and changing patterns of demand are all fruitful areas in which geographers may work. Fourth, a wide range of studies examine the impact of industrial change on the wider economy and society, focussing on such themes such as the development of urban structures, changing class relations, employment and unemployment, and the effects of industrial change on the social, political, and cultural experiences of the labour force. Although not exhaustive, this classification covers the main traditions of geographical research on industrial change. The significance of selected research in each field will be discussed below.

There are many other areas of debate where a geographical input has been minimal. Research on the causes of industrial change, the nature of capital formation and investment, the supply of raw materials, the development and impact of new technology, and the economic consequences of industrial change at the regional and national level has been scarcely tackled by geographers despite a wealth of research by economic historians and others (Crafts, 1977, 1983; Harley, 1982; McCloskey, 1981; Payne, 1974; Kenwood, 1978; Lee, 1981). Not surprisingly, historical geographies of the causes of industrial change are heavily dependent on the work of non-geographers (Pawson, 1978, 1979), and it is really only the early work of Wrigley (1962, 1972) which has made a major impact in this field. Likewise, debates over the effects of industrialisation on the standard of living in Britain seem to have passed geographers by, despite a continued airing by other scholars (Barnsby, 1971; Cage, 1983; Lindert and Williamson, 1983; Taylor, 1975, Neale, 1985). The reasons for these gaps in the geographical literature are not immediately obvious as they are all topics where a geographical focus on place and spatial analysis could be rewarding.

At a wider level, it can be argued that geographers have a contribution to make to the study of industrial change through the development of theory and method, and the application of geographical ideas and techniques to a range of topics. However, the extent to which this has been achieved is debatable. The majority of research by historical geographers remains empirical, and, apart from the development of mapping skills and the representation of distributional data

HISTORICAL GEOGRAPHY OF INDUSTRIAL CHANGE

(Darby, 1973, Langton and Morris, 1986), geographical contributions at a methodological and theoretical level have been minimal. There are some notable exceptions: the work of Massey (1979, 1983, 1984) on the regional effects of economic restructuring and the spatial division of the labour market has begun to influence economic historians (Heim, 1983); Langton (1979) made explicit use of a systems approach and economic location theory in his study of the Lancashire coal industry; and Gregory and others (Gregory, 1982, 1984; Billinge, 1982, 1984) have attempted to apply social theory mainly derived from Giddens (1979) to the regional and cultural impacts of industrial change. Although the use of structuration theory to aid understanding of the complex processes that operated in a period of rapid industrial change is challenging and, as one reviewer puts it, 'raises questions of long-standing importance' (Hudson, 1983), the theoretical formulations propounded by Gregory have not yet been successfully linked to empirical examination of the evidence, and the usefulness of the theory itself seems to have gained little acceptance outside a small group at Cambridge (Pratts and Pringle, 1985).

THE GEOGRAPHICAL STUDY OF SPECIFIC INDUSTRIES

There is still a great deal to learn about the spatial structure of British industry in the nineteenth century and the regional impact of industrial change. Although unfashionable, the only way in which these topics can be tackled is through the careful analysis of the development, location and organisation of specific industries. This also necessitates the analysis of distributional data at a spatial scale much smaller than the county or region, and the examination of a far wider range of industries than the traditional heavy manufacturing trades of textiles, chemicals and iron or steel which are usually seen as the lynchpin of the industrial revolution.

The Atlas of the Industrial Revolution (Langton and Morris, 1986) goes some way towards providing this crucial empirical data for the study of industrial change. The twin aims of the editors in compiling the atlas are to provide a resource of illustrative material for teaching and research, and to begin to utilise these maps to examine the internal geography of the processes and patterns which made up the industrial revolution (Withers, 1985). Inevitably, constraints imposed by data sources, time, and publisher's requirements have meant that there are omissions, and by no means all industries are represented in comparable detail. However, the atlas does cover seven branches of manufacturing industry, together with the service sector, banking and finance, retailing and public utilities; it provides not only a valuable repository of information, but also a series of maps and

commmentaries that are themselves provocative, and likely to
lead to further research into the reasons for the spatial
patterns which emerge.

Although more than 80 per cent of maps in the Atlas of
the Industrial Revolution are drawn at the national scale, it is
well-recognised that most industries also have a significant
internal geography at the micro-scale. Hudson (1981, 1983)
has recently re-emphasised this with respect to the Yorkshire
woollen and worsted industry: 'The economic, technological
and spatial changes of the early-nineteenth century differed
enormously between the two major branches (as well as within
them)' (Hudson, 1983, p.315). From the perspective of those
who organised and worked in the industry, small-scale variations from place to place and from firm to firm were far more
significant than national distributions. Although more difficult
to map and analyse, the true local and regional importance of
industrial change can only be recognised through the micro-
scale analysis of the internal organisation and spatial location
of specific industries.

Recent work on the internal structure of the cotton
industry has developed from the research of Rodgers (1960)
which clearly demonstrated the extent of local specialisation.
Gatrell (1977) examined the size structure of the Lancashire
cotton industry in 1841, and concluded that small and single-
process firms remained numerous and competitive well into
mid-century. He suggests that the size distribution of firms
remained fairly static, with managerial and other constraints
preventing the further growth of large firms, and with few
incentives for small firms to grow. Thus there existed a
diversity of structure within the cotton industry, with
different types of firm surviving because of their different
labour and capital requirements. Gatrell does not specifically
explore spatial variations in this structure, but some insights
are provided by Gregory (1984), who examines the sub-
regional impact of economic crises in the cotton industry in
1841. Power loom weavers, and the coarse spinners who
supplied them, were most affected by economic crisis; consequently those towns and parts of towns which specialised in
these processes bore the brunt of the crisis in trade, while
fine spinning areas were scarcely affected. The distinctive
sub-regional structure of the industry meant that every-day
experiences of industrial change within the cotton industry
varied greatly from place to place and person to person.
These are subtleties which cannot be captured in national-
level studies of economic change. Kenny (1982) has also
extended analysis of sub-regional specialisation in the cotton
industry to the late-nineteenth century, and concludes that
the internal geography of the industry remained distinctive
until 1914. He argues that the varied capital and labour
requirements of different sectors of the industry (particularly
the labour-intensive nature of weaving compared with the

163

capital-intensive nature of spinning) perpetuated differences between the sectors and led to significant sub-regional variations within a small geographical area.

The diversity of experience which can exist within an industry over a small geographical area has also been highlighted in studies of the coal industry. Cromar (1977) analyses the internal dynamics of the coal industry on Tyneside in the late-eighteenth century, emphasising the varied experiences within the Tyneside coalfield, while Langton (1979) reaches similar conclusions in a detailed study of the south-west Lancashire coalfield. He emphasises the "spatial complexity of the economic structure of the coalfield, the cleavages between different parts of it, the variety of measure systems, markets, economic growth rates and prices ...' in the seventeenth and early eighteenth centuries. Although this was an area which measured only 12 square miles in extent, it was the late-eighteenth century before the majority of the area was 'articulated into a single economic system' (Langton, 1979 p.240). Such small-scale variations were unlikely to be confined to a few industries, but would be the norm for most industries at certain stages of their development. Undoubtedly, more geographical studies of micro-scale variations in industrial organisation and location are required.

Other recent studies may not have stressed internal variability, but have ranged over a wide spectrum of topics and trades. The cotton and coal industries continue to attract research (Lee, 1972; Cuca, 1977; Farnie, 1979; Lazonick 1981, 1983; Hirsch and Hausman, 1983; Walters, 1980; Saxonhouse and Wright, 1984), but less thoroughly researched trades have also received attention, including the pin industry (Dutton and Jones, 1983) soap and chemical manufacture (Gittins, 1982; Warren, 1980), gas and electricity supply (Falkus, 1967, 1982; Morgan, 1983) and car manufacture (Foreman-Peck, 1983). Although little of this work is explicitly concerned with spatial analysis, these single-industry studies pave the way for more detailed geographical research. The service sector has received less exposure than most in studies of the development of the British economy although the service economy of seaside resorts and the development of retailing have received attention (Walton, 1981, 1983; Johnson and Pooley, 1982, Winstanley, 1983; Crossick, 1984). Lee (1984) also provides a welcome corrective in his detailed statistical analysis of regional variations in the growth of the service sector. Lee emphasises the importance of service industries, accounting for 59 per cent of national income by 1907, and refutes the suggestion that the service sector was dependent on manufacturing industry, growing only in response to industrial demand. He demonstrates that the service sector was concentrated in London and the South East, and that this growth was neither led by manufacturing industry, nor dependent on the transfer of wealth from manufacturing to services. Rather he argues that the growth of the service sector was quite independent of the manufacturing economy of northern England, and was a self-perpetuating process generated by high overseas investment, London-based commerce, and high incomes in both urban and rural parts of the South East. Thus the service sector was an independent and significant part of the Victorian economy, which itself demonstrated a high degree of regional specialisation. Here lies yet another opportunity for detailed geographical study which could begin to unravel the organisational complexities and regional impacts of the service industries.

INDUSTRIAL CHANGE AND GEOGRAPHICAL REGIONS

The regional impact of specific industries has been stressed in the previous section, but this leads to a much broader consideration of the impact of industrial change on the regional geography of Britain. Everitt characterises England before the railway age as a country of great local and regional diversity where one should not "expect to find a homogeneous or coherent pattern of evolution, but a piecemeal, localised and fragmented one: a pattern of regional paradoxes and survivals" in which regional development was "elusive and kaleidoscopic" (Everitt, 1979, pp. 106-7). One view of the effect of the industrial revolution was that it removed this regional distinctiveness, and united the people of Britain for the first time with a common culture and identity centred around an emerging but unified working-class consciousness. This position is challenged by Langton (1984) who argues that the process of industrialisation did not destroy the distinctiveness of regional culture and identity in England, and that in some instances industrial change even served to highlight regional differences.

Langton develops his analysis of the regional geography of England during the industrial revolution by demonstrating the continuing regional fragmentation of social and political movements, the extent to which regional identity and regional consciousness flourished in the newly industrialising areas, and the degree to which economic activity retained a strong regional base. He argues that pressure groups usually had a narrow regional base, that even Luddism was "never .. more than a conglomeration of disjointed regionally based disturbances", (p.152), and that trade unions continued to be concerned mainly with regional issues. As the new industrial areas developed so they took on a new identity and, Langton argues, regional differences accentuated by industrialisation began to enter 'popular consciousness' (p.156). This revived regional identity was itself strengthened through dialect, literature and entertainment. Transport was an essential factor influencing the development of regional economies, and Langton argues that the dependence on a discontinuous canal network for much industrial traffic heightened the importance of regional economies before the railway age. The railway itself eventually broke down this regional distinctiveness, but initially it accentuated contrasts as easier travel emphasised the differences between regions and cultures.

This assessment of the impact of industrial change on the regional geography of England gains general support from Freeman (1984), who qualifies Langton's conclusions on the importance of transport. Freeman suggests that Langton overstates economic regionalism in the early phases of industrialisation, but understates it in the railway age. He argues that Langton places too much emphasis on the role of canal transport, and not enough on roads and coastal shipping which provided a much more efficient network of national linkages in the pre-railway period, and that he overstates the impact of rail transport on regional economies by ignoring the multiplicity of railway companies and pricing mechanisms which meant that rail freight in particular continued to have a strong regional structure. Gregory also lends support to the argument that a strong regional geography survived the industrial revolution, when he concludes a survey of the process of industrial change by suggesting that 'at the close of the nineteenth century the space-economy was still made up of regional mosaics. Many of the traditional parochialisms had been dissolved by the processes of industrial change, but they were yet to be reconstituted at the national level' (Gregory, 1978, p.307). His own study of the Yorkshire woollen industry also emphasises that both economic crisis and popular reaction were articulated at a regional level (Gregory, 1982), but he tempers this with caution by also arguing that regional experience should be set within a national and international context (Gregory, 1984).

A number of other recent articles impinge on the debate over the impact of industrialisation on regional structures. Meyer (1983) has examined the emergence of regional manufacturing systems in the United States, showing the gradual development of a national system during the nineteenth century, while Persky and Moses (1984) assess the economic performance of specialised industrial cities within the context of American regional economic systems. Although differences in scale invalidate close comparison between Britain and America, such studies continue to emphasise the significance of regional studies of industrial change. Such research also interacts with the interests of local and regional historians whose work has emphasised the continuing importance of regional economies and cultures despite the process of industrialisation and the penetration of the railway to remote corners of Britain (Marshall, 1983). As Langton (1979) argues, it is travel time and opportunity which crucially affect the level of interaction between regions. Even in the early railway age swift and cheap travel was not open to all, and it is this "expansion of space as one regressess backwards in time that makes the analysis of regional patterns and processes vitally necessary in studies of the economies and societies of the past" (Langton, 1979, p.242).

The emergence of 'the regional problem' in Britain in the inter-war period is a further area of study which should attract historical geographers interested in the impact of economic change on regional structures, but it is a theme that has remained strangely neglected. The inter-war years have been characterised as the 'Dark Ages' of historical geography (Baker and Gregory, 1984, p.187), and Jones (1984) has recently commented on the way in which both economic historians and historical geographers have neglected to study the regional impact of economic change between the wars. Booth and Glynn (1975) and Jones (1984) have both emphasised the regional impact of unemployment and the need for regional analysis to provide a proper understanding of economic trends during the period, but historical geographers have been slow to turn their attention to this period. Heim (1983) has examined the impact of industrial organisations on regional development between the wars. Drawing on the work of Massey (1979) on the present regional problem, he argues that depressed regions suffered particularly in the inter-war period because they not only missed out on the development of new industry that occurred in the south, but also that the new industries were not yet in a position to expand from the South-East and establish a branch-plant economy in the regions where they could exploit low labour costs. Study of the regional problem in the inter-war period is surely one area which lends itself to further interdisciplinary research by economic historians, regional economists, and historical and economic geographers.

GEOGRAPHICAL ANALYSIS OF THE MECHANISM OF INDUSTRIAL CHANGE

One essential component of the mechanism of industrial change is the development of interaction within and between regions and industrialising areas. This may take the form of flows of commodities, capital, ideas and information, or labour. All of these form fruitful areas for geographical study and lend themselves to spatial analysis, but geographical analyses of the mechanism of industrial change are scarce. Conzen (1975) has examined the role of transport in the development of urban regions in the American Mid-West, arguing that improvements in transport and access to large metropolitan centres had more direct effects upon the primary sector than on manufacturing and service industries, which depended much more on the growth of local urban centres for their distribution. In England, Freeman (1982) has explored internal trade within the industrial region of the West Riding of Yorkshire during the industrial revolution, emphasising the impact which different transport modes had on the trade in wool and cloth, while several other studies have highlighted different aspects of transport during the industrial revolution (Freeman, 1980; Hawke, 1970; Davis, 1979; Kenwood, 1981; Le Guillan, 1975; Aldcroft and Freeman, 1983).

The significance of transport in the industrial revolution is discussed from a geographical perspective in Aldcroft and Freeman (1983). Freeman argues that transport developments were far from revolutionary in pace and kind, with gradual change in a multiplicity of traditional modes of transport typical of the pre-railway age. Following Wrigley (1962) he examines the 'geography' of transport demand in the industrial revolution, accepting that this was related to transition from local and regional markets to a system of national exchange, but arguing that in the pre-railway age the road-carrying business was far better fitted to fulfil this demand than the canal network. The relationships between transport and economic development are complex, and Freeman rejects suggestions of simple relationships between transport and industrialisation. Rather, he argues that there can be no 'universal rules concerning the relationship between transport and economic growth' but that it varies 'over space and through time' (Aldcroft and Freeman, 1983, p.18). He also echoes Langton (1979) in arguing for more explicit analyses of spatial processes in studies of the role of transport in economic growth.

It is not only the movement of goods and passengers that is important for industrial development, but also the communication of information and ideas. The historical geography of communication systems remains relatively under-researched in Britain, despite Robson's (1973) brief but pioneering study of the diffusion of telephone exchanges and the development of the telephone trunk network in Britain; but two recent studies have examined the significance of telecommunications in the United States (Langdale, 1978; Brooker-Gross, 1981). In concluding her study of the news wire services in nineteenth-century America, Brooker-Gross (p.178) postulates three main ways in which the emergence of a national network of wire service news facilitated urban-isation and industrialisation and led to the break-up of regional economic structures. First, the transfer of economic and financial news allowed the economic system to function more efficiently at a national level, with the dual effect of strengthening the position of New York by tying all markets into it, and of promoting economic development in regional centres. Second, the wire service promoted a greater national coherence and consciousness, as readers throughout the country were faced with similar news items and were more closely linked in with national events. Third, the wire services were themselves part of a transition from regional to national economic systems and promoted debate on regionalism and the dominance of New York. The development of telecommunications, allowing the flow of information and ideas, is thus a significant factor in industrial change, and one which itself links back to debates over the regional geography of industrialisation.

In addition to the movement of commodities and information, industrial development in any period also depends on the flow of capital between investors, entrepreneurs and areas. Although there are a number of studies of capital formation in specific areas and industries (Kenwood, 1978, 1981; Walters, 1980; Gatrell, 1977), few geographers have attempted to provide empirical evidence of the spatial flow of capital in the industrial revolution to back up the theoretical formulations of Harvey (1975a, 1982). Attracting slightly more attention are the institutional structures which allow the accumulation and movement of capital. Conzen (1977) has investigated the development of the American banking system from an initial primate city pattern to one of increasing complexity, leading to an integrated network of financial interdependence within the maturing urban system; and Robson (1973) has provided a preliminary analysis of the development of building societies within the British urban system. There is much scope for further geographical study of both capital accumulation and the development of other factors of production, especially power, technology and entrepreneurship, within the industrial revolution (Musson, 1976; Von Tunzleman, 1978; Payne, 1974). The significance of regional analysis of these factors is clearly demonstrated by Rubenstein (1977a, 1977b), who analyses the distribution of the wealthy from tax assessments, and emphasises the importance of London and the South East for wealth creation and capital formation in Victorian Britain. London was already the centre of investment in the second half of the nineteenth century and was developing at the expense of traditional industrial areas of the North and West.

One process which geographers have studied is the movement of labour. Labour migration, bringing workers to the centres of production and also altering the pattern of demand for goods within the country, was a further essential factor in the mechanism of industrialisation. Unfortunately, most studies of migration rely on published or unpublished census evidence (Lawton and Pooley, 1980), which restricts analysis to the second half of the nineteenth century. There have been few attempts to develop Redford's (1926) pioneering study of labour migration in the period before 1850, although Southall's (1982) research using trade union records has shed important light on the movement of certain skilled workers within the engineering trades. Other research on labour migration has focussed on movement to particular towns or within specific industries in the second half of the nineteenth century. Jackson (1981) has demonstrated the way in which skilled workers were recruited to the St. Helens glass industry from other established glass-making centres, especially in the North East, and Cromar (1977) studied the relationship between labour migration and suburban development. Other research has examined the extent to which long-distance migration streams could be established where new industries were expanding rapidly and labour with specific skills was required. Hence Middlesbrough and Barrow both recruited skilled workers over quite long distances, including movement from the West Midlands and South Wales (Gwynne and Sill, 1976; Pooley, 1983; Saunders, 1984). However, most studies of the migration process remain largely descriptive and are conceived at the aggregate scale. Examination of the complex links between migration and industrialisation, including study of the contribution of labour migration to industrial change and the effect of migration on individual and group employment opportunities, requires the detailed longitudinal analysis of individual data and the careful integration of information from a range of quantitative and qualitative sources (Pooley and Doherty, 1985; Doherty, 1986).

GEOGRAPHICAL STUDIES OF THE IMPACT OF INDUSTRIAL CHANGE

The impact of industrial change can be studied at many levels, ranging from direct assessment of the effect of changing work practices and industrial technology on individual working lives to the assessment of much more wide-ranging effects such as the impact of industrialisation on urban growth, housing and residential differentiation, social structure, health and environment. It is in the latter area, dealing with wide-ranging effects of industrialisation, that geographers have made the most prolific contributions, but these studies are rarely concerned with the interaction between industrialisation and social change. Most often the process of industrialisation is a static and taken-for-granted framework within which the development of society is set.

One of the most fundamental questions about the impact of industrial change relates to the effect of changing work practices on individual patterns of behaviour, on family relations and on the social structure of small-scale communities. Following Thompson (1967, 1968) it is usually assumed that the change from small-scale and dispersed domestic and workshop production to large-scale and centralised factory production created a fundamental fission in traditional life-styles, work-practices and inter-personal relations, and that it was this process which led to a new form of industrial society within which a coherent and unified working-class consciousness could emerge. This view has been challenged and refined in a number of recent works. Hopkins (1982) demonstrates that a model of fundamental disruption to traditional working practices and conditions does not apply equally to all industries and areas, showing that there was little change in working hours and conditions in Birmingham prior to 1850. Similarly, Joyce (1980) argues that even in the textile industry, most affected by rapid industrial change, the consequences of factory industry for communities and social life were not necessarily dramatic. Joyce argues that throughout the nineteenth century workers quickly adapted to the routine of large factory work, and most potential sources of disruption within the workforce were avoided because of the paternalist nature of factory employment which led to a stable factory community with which individuals could identify. Despite fundamental changes, the status quo of factory communities was maintained and workers could thus successfully adapt to a new working regime.

Geographical contributions to the debate have largely taken the work of E.P. Thompson as their starting point and have tended to concentrate on the impact of changing work practices on the social and spatial relations of every day life. Both Gregory (1982) and Pred (1981) also draw on structuration theory (Giddens, 1979) to place their studies within a theoretical context, whilst Pred uses the concept of 'time geography' (Hagerstrand, 1975; Thrift, 1977; Carlstein, Parkes and Thrift, 1978) for his study of the impact of the change from home and workshop production to factory-based production in the United States of America. Using a time-geographic framework, he attempts to identify individual life 'paths' which are linked to 'projects' that embrace groups and institutions, in an attempt to uncover the 'social practices' which form a link between individual actions and societal constraints. The study focuses on the time and space constraints imposed by factory work, and particularly its impact on the division of labour within the family, the level of contact and interaction between different family members, and the availability of leisure time. As with some other work of this genre, the argument is often clouded by dense prose and there is a distinct lack of empirical evidence to back up theoretical observations; but the study does highlight an important theme which is basic to research on the impact of industrial change on society. There is need for more studies which appraise the impact of industrial change on spatial and temporal patterns of everyday interaction, and which explain more fully the interactions between individual behaviour and societal constraints. Some other geographical work has begun to explore the nature of community formation and spatial interaction in the nineteenth century (Dennis, 1984, Bramwell, 1984), but these studies rarely explicitly examine the links between changing industrial practices and community development. On a broader scale, the links between popular disturbances and industrial change is a further important theme (Charlesworth 1982, Bohstedt, 1983).

The impact of industrial and economic change on employment practices has also been explored by Green (1985), who tackles the daunting task of studying the complex links between labour market change and poverty in London in the first half of the nineteenth century. In effect, the research is an extension of the pioneering work of Stedman-Jones (1976); it is set within a theoretical framework drawn from a Marxist analysis of the transition from handicraft manufacture to factory industry, but also clearly located within a geographical tradition of spatial analysis. The thesis demonstrates that economic pressures within the metropolis led to a continual process of reduced labour costs and consequent de-skilling of the workforce. Attention is focussed on the way in which the artisan workforce attempted to resist the process of economic restructuring through labour disputes and protest, and the impact of industrial change on the nature and distribution of poverty within London. The spatial analysis of manufacturing employment and poverty effectively amplifies the analysis of economic restructuring, and the distinctive nature of the manufacturing sectors of East and West London is clearly demonstrated. The study is also marked by its effective use of a range of sources from conventional census statistics to newspaper reports of strikes and protests, and clearly indicates the value of a geographical approach to analysis of the impact of industrial change on employment. Other studies of employment change following industrial restructuring are sadly lacking, although Southall (1984) provides a valuable analysis of regional trends in unemployment in the second half of the nineteenth century, and Glynn and Booth (1983) have recently reappraised the regional impact of unemployment in the inter-war depression. Although data sources are often difficult to handle, and long time-series are hard to interpret (Lee, 1979), there is much scope for further geographical analysis of this topic.

The societal impact of industrial change has attracted a considerable amount of attention although the links between economic and social change are often implicit. Most central to the discussion is the work of those geographers who, drawing particularly on the work of Thompson (1968, 1974), Foster (1974) and Stedman-Jones (1975; 1976), seek to study the impact of industrial change on class relations. Billinge (1984) examines the hegemonic control exercised by certain groups within the city of Manchester and the impact of this on cultural change and class relations, while Gregory (1984)

more specifically seeks to examine the effects of industrialisation on class struggle. He critically examines the work of
Foster and Thompson in an attempt to establish the relationship between different phases of industrialisation and the
responses of the population as seen through changing class
relations. However, with these exceptions, there have been
few attempts to develop a geographical dimension to the
debate over changing relationships between classes during the
industrial revolution, despite the continued attention which
social historians give to this topic (Gray, 1977; 1981; Morris,
1979; Joyce, 1980; Calhoun, 1982; Glen, 1984).

Although changing class relations were undoubtedly
manifest in changes in the residential structure and patterns
of social interaction within the city (Harvey, 1975b), the
large number of studies of residential differentiation in
Victorian cities rarely give more than passing mention to the
links between residential space, social relations and industrial
change (Cannadine, 1982; Ward, 1980; Pooley, 1984; Dennis
1984). This is partly because reliance on census sources for
the study of residential separation has led to a concentration
on the second half of the nineteenth century (rather than on
the period of most rapid industrial change before 1840), and
has led to a relative neglect of less easily quantifiable sources
that would illuminate changing class relations more clearly
than can be done through census studies. There is need for
further study of the impact of industrial change on class
relations and social interaction, and of the relationship
between these changes and the residential structure of nineteenth-century cities.

The impact of industrial change was also felt in other
parts of nineteenth-century society. The movement of labour
created a more heterogeneous population in centres of production where problems of segregation and assimilation of
those from distinctive cultural backgrounds inevitably arose.
Consequently, studies of the impact of migration on urban
growth, and especially of the effect of large-scale movement
of unskilled labour from Ireland to England, are common
(Pooley, 1977; O'Tuathaigh, 1981; Swift and Gilley, 1985).
Industrialisation and urbanisation also led to a deterioration in
the urban environment, with massive overcrowding in insani-
tary housing conditions in most large cities. Not surprisingly,
rates of morbidity and mortality were high and several geographical studies have recently examined spatial variations in
living conditions and disease mortality in a number of British
towns (Woods and Woodward, 1984). The repercussions of
industrial change were also felt across a whole spectrum of
other areas of social life, including leisure pursuits,
education and religion, but few studies of these and other
aspects of the social history of Britain directly relate the
changing experiences within society to the restructuring of
economy and work practices which was simultaneously occur-

ring. Unusual in this respect is the recent work of Oddy (1983), which examines the impact of the crisis in employment in the Lancashire cotton industry caused by the Cotton Famine of 1861-63 on the diet and health of the working population. It is particularly difficult to correlate specific periods of economic or industrial change with their social effects, not least because the effects may only become obvious some years after industrial change has occurred. Although the results of Oddy's study are inconclusive, he does demonstrate the significance of links between industrial change and social conditions in a much more effective way than most studies of nineteenth-century society and environment.

CONCLUSIONS

The conclusions to be drawn from this review of recent work on the historical geography of industrial change are threefold. First, it has been demonstrated that the pattern and process of industrial change in the past is an area of study which lends itself to geographic analysis, focussing on such traditional areas of geographical research as the spatial analysis of industrial production, the impact of change upon regional structures, the development of flows of goods, ideas and labour between centres of production and the complex interactions between economic, social and environmental change that occurred in the past. Second, with only a handful of notable exceptions, geographers have made little contribution to current debates on the processes that caused industrial change, despite their obvious geographical relevance. It has too often been left to economic and social historians to explore themes such as the regional impact of industrial change, while many other important research fields remain almost completely untilled. Third, it is hoped that this necessarily incomplete review of the literature will act as a stimulus both to further research by geographers and to research by historians and others on themes of geographical relevance, thereby leading to a greater and more productive interchange between scholars from different disciplinary backgrounds.

There seem to be three main priorities for future research on the historical geography of industrial change. First, there is a pressing need for more empirical studies of the process and impact of change, which span a range of different industrial sectors, focus on a variety of regional scales and extend across a wide time period. Studies of industrial change should not only focus on the leading sectors of manufacturing industry during the classic period of 'industrial revolution' in the first half of the nineteenth century, but should research the gradual longitudinal changes which occurred from the period of proto-industrialisation through to the economic restructuring which occurred in the interwar years. Second, the development of appropriate theory to interpret and explain the processes of industrial change should go hand in hand with the careful examination of empirical evidence. Although theory is important, there is a danger that slavish adherence to particular theoretical formulations will obscure more than it reveals, and will deflect research from greater historical understanding. Third, studies of the social geography of the nineteenth century must make more explicit the links between economic and industrial change on the one hand, and the changing structure of society on the other. Too many social geographers neglect economic issues: it is time that social geographers of the past emerged from their comfortable niche in census-based studies of urban society and confronted the reality of the complex interactions between society and economic change. Clearly the historical geography of industrial change is a very catholic area of study in which definitions are nebulous and traditional disciplinary boundaries have little meaning. Within this broadly-based research field there is room for a great diversity of approach, where theoretical and empirical studies can go forward side by side, where social and economic geographers can both contribute to research, and in which geographers can bring their particular skills to bear on important interdisciplinary debates on the history of industrial change.

Acknowledgements
Grateful thanks to John Walton and Marilyn Pooley for their helpful comments on an earlier draft of this essay.

REFERENCES

Aldcroft, D.H., and Freeman, M.J. (eds.) (1983) Transport in the Industrial Revolution, Manchester University Press, Manchester

Baker, A.R.H., and Gregory, D. (eds.) (1984) Explorations in Historical Geography, Cambridge University Press, Cambridge

Barnsby, G.J. (1971) 'The Standard of Living in the Black Country during the Nineteenth Century', Economic History Review, 2nd Series, XXIV, 220-39

Billinge, M. (1982) 'Reconstructing Societies in the Past: The Collective Biography of Local Communities' in A.R.H. Baker and M. Billinge (eds.), Period and Place: Research Methods in Historical Geography, Cambridge University Press, Cambridge, pp. 19-32

Billinge, M. (1984) 'Hegemony, Class and Power in late Georgian and Early Victorian England: Towards a Cultural Geography' in A.R.H. Baker and D. Gregory (eds.), Explorations in Historical Geography, Cambridge University Press, Cambridge, pp. 28-67

Bohstedt, J. (1983) Riots and Community Politics in England and Wales 1790-1810, Harvard University Press, Cambridge, Mass. and London

Booth, A.E., and Glynn, S. (1975) 'Unemployment in the Interwar Period: A Multiple Problem', Journal of Contemporary History, 10, 611-36

Bramwell, W. (1984) 'Pubs and Localised Communities in Mid-Victorian Birmingham', Occational Paper, Queen Mary College, London, 22

Brooker-Gross, S. (1981) 'News Wire Services in the Nineteenth-Century United States', Journal of Historical Geography, 7, 167-80

Cage, R.A. (1983) 'The Standard of Living Debate: Glasgow 1800-1850', Journal of Economic History, XLIII, 175-82

Calhoun, C. (1982) The Question of Class Struggle: Social Foundations of Popular Radicalism During the Industrial Revolution, Blackwell, Oxford

Cannadine, D. (1982) 'Residential Differentiation in Nineteenth-Century Towns: From Shapes on the Ground to Shapes in Society' in J.H. Johnson and C.G. Pooley (eds.), The Structure of Nineteenth-Century Cities, Croom Helm, London, pp. 235-52

Carlstein, T., Parkes, D., and Thrift, N. (1978) (eds.), Timing Space and Spacing Time, Arnold, London

Charlesworth, A. (1982) (ed.), An Atlas of Rural Protest in Britain 1549-1900, Croom Helm, London

Coleman, D.C. (1983) 'Proto-Industrialization: A Concept Too Many', Economic History Review, 2nd Series, XXXVI, 435-448

Conzen, M.P. (1975) 'A Transport Interpretation of the Growth of Urban Regions: An American Example', Journal of Historical Geography, 1, 361-82

Conzen, M.P. (1977) 'The Maturing Urban System in the United States 1840-1910', Annals of the Association of American Geographers, 67, 88-108

Crafts, N.F.R. (1977) 'Industrial Revolution in England and France: Some Thoughts on the Question "Why was England First?"', Economic History Review, 2nd Series, 30, 429-41

Crafts, N.F.R. (1983) 'British Economic Growth 1700-1831: a Review of Evidence', Economic History Review, Second Series, 36, 177-99

Cromar, P. (1977) 'The Coal Industry on Tyneside 1771-1800: Oligopoly and Spatial Change', Economic Geography, 53, 79-94

Cromar, P. (1980) 'Labour Migration and Suburban Expansion in the North of England: Sheffield in the 1860s and 1870s' in P. White and R. Woods (eds.), The Geographical Impact of Migration, Longman, London, pp. 129-51

Crossick, G. (eds.) (1984) Shopkeepers and Master Artisans in Nineteenth-Century Europe, Methuen, London

Cuca, J.R. (1977) 'Industrial Change and the Progress of Labour in the English Cotton Industry', International Review of Social History, XXII, 211-55

Darby, H.C. (eds.) (1973) A New Historical Geography of England, Cambridge University Press, Cambridge

Davis, R. (1979) The Industrial Revolution and British Overseas Trade, Leicester University Press, Leicester

Deane, P. (1969) The First Industrial Revolution, Cambridge University Press, Cambridge

Dennis, R. (1984) English Industrial Towns in the Nineteenth Century, Cambridge University Press, Cambridge

Doherty, J. (1986) 'Short-Distance Migration in Nineteenth-Century Lancashire', unpublished Ph.D. thesis, University of Lancaster

Dutton, H.I., and Jones, S.R.H. (1983) 'Invention and Innovations in the British Pin Industry 1790-1850', Business History Review, 57, 175-93

Everitt, A. (1979) 'Country, County and Town: Patterns of Regional Evolution in England', Transactions of the Royal Historical Society, 5th Series, 29, 79-108

Falkus, M.E. (1967) 'The British Gas Industry before 1850', Economic History Review, 2nd Series, XX, 494-508

Falkus, M.E. (1982) 'The Early Development of the British Gas Industry 1790-1815', Economic History Review, 2nd Series, XXXV, 217-234

Farnie, D.A. (1979) The English Cotton Industry and the World Market 1815-1896, Clarendon Press, Oxford

Foreman-Peck, J. (1983) 'Diversification and the Growth of the Firm: The Rover Company to 1914', Business History, XXV, 179-92

Foster, J. (1974) Class Struggle and the Industrial Revolution. Early Industrial Capitalism in Three English Towns, Weidenfeld and Nicolson, London

Freeman, M.J. (1980) 'Road Transport in the English industrial Revolution: An Interim Assessment', Journal of Historical Geography, 6, 17-28

Freeman, M.J. (1982) A Perspective on the Geography of English Internal Trade During the Industrial Revolution: The Trading Economy of the Textile District of the Yorkshire West Riding circa 1800, University of Oxford, School of Geography, Research Paper No. 29

Freeman, M. (1984) 'The Industrial Revolution and the Regional Geography of England: A Comment', Transactions of the Institute of British Geographers N.S. 9, 507-12

Gatrell, V.A.C. (1977) 'Labour Power and the Size of Firms in Lancashire Cotton in the Second Quarter of the Nineteenth Century', Economic History Review, 2nd Series, XXX, 95-139

Giddens, A. (1979) Central Problems in Social Theory: Action, Structure and Contradiction in Social Analysis, Macmillan, London

Gittins, L. (1982) 'Soap Making in Britain 1824-1851: A Study in Industrial Location', Journal of Historical Geography, 8, 12-28

Glen, R. (1984) Urban Workers in the Early Industrial Revolution, Croom Helm, London

Glynn, S., and Booth, A. (1983) 'Unemployment in Inter-War Britain: A Case for Relearning the Lessons of the 1930s', Economic History Review, 2nd Series, XXXVI, 329-348

Gray, R.Q. (1977) 'Bourgeois Hegemony in Victorian Britain' in J. Bloomfield (ed.), Class, Hegemony and Party, Lawrence and Wishart, London, pp. 73-94

Gray, R.Q. (1981) The Aristocracy of Labour in Nineteenth-Century Britain, Macmillan, London

Green, D.R. (1985) 'From Artisans to Paupers. The Manufacture of Poverty in Mid-Nineteenth Century London', unpublished Ph.D. thesis, University of Cambridge

Gregory, D. (1978) 'The Process of Industrial Change 1730-1900' in R.A. Dodgshon and R.A. Butlin (eds.), An Historical Geography of England and Wales, Academic Press, London, pp. 291-312

Gregory, D. (1982) Regional Transformation and Industrial Revolution: A Geography of the Yorkshire Woollen Industry, Macmillan, London

Gregory, D. (1984) 'Contours of Crisis? Sketches for a Geography of Class Struggle in the Early Industrial Revolution' in D. Gregory and A.R.H. Baker (eds.), Explorations in Historical Geography, Cambridge University Press, Cambridge, pp. 68-117

Gwynne, T., and Sill, M. (1976) 'Census Enumerators' Books: A Study of Mid-Nineteenth Century Migration', Local Historian, 12, 74-79

Hagerstrand, T. (1975) 'Space, Time and Human Conditions' in A. Karlqvist, L. Lundquist and F. Snickars (eds.), Dynamic Allocation of Urban Space, Saxon House, Farnborough, pp. 3-14

Harley, C.K. (1982) 'British Industrialization before 1841: Evidence of Slower Growth during the Industrial Revolution', Journal of Economic History, XLII, 267-89

Harvey, D. (1975a) 'The Geography of Capitalist Accumulation: A Reconstruction of the Marxian Theory', Antipode, 7, 9-21

Harvey, D. (1975b) 'Class Structure in a Capitalist Society and the Theory of Residential Differentiation' in R. Peel, M. Chisholm and P. Haggett (eds.), Processes in Physical and Human Geography: Bristol Essays,

Heinemann, London, pp. 354-69
Harvey, D. (1982) The Limits to Capital, Basil Blackwell, Oxford
Hawke, G.R. (1970) Railways and Economic Growth in England and Wales 1840-1870, Clarendon, Oxford
Heim, C. (1983) 'Industrial Organisation and Regional Development in Inter-War Britain', Journal of Economic History, XLII, 931-52
Hirsch, B.T., and Hausman, W.J. (1983) 'Labour Productivity in the British and S. Wales Coal Industry 1874-1914', Economica, 50, 145-57
Hopkins, E. (1982) 'Working Hours and Conditions during the Industrial Revolution: A Reappraisal', Economic History Review, 2nd Series, XXXV, 52-66
Hudson, P. (1981) 'Proto-Industrialization: The Case of the West Riding Wool Textile Industry in the Eighteenth and Early Nineteenth Centuries', History Workshop, 12, 34-61
Hudson, P. (1983) 'Review of Gregory, D. (1982) Regional Transformation and Industrial Revolution', Journal of Historical Geography, 9, 313-15
Jackson, J.T. (1981) 'Long-Distance Migrant Workers in Nineteenth-Century Britain: A Case Study of the St. Helens Glassmakers', Transactions of the Historic Society of Lancashire and Cheshire, 131, 113-38
Johnson, J.H., and Pooley, C.G. (1982) The Structure of Nineteenth Century Cities, Croom Helm, London
Jones, M.E.F. (1984) 'The Economic History of the Regional Problem in Britain 1920-38', Journal of Historical Geography, 10, 385-95
Joyce, P. (1980) Work Society and Politics: The Culture of Factory in Later Victorian England, Harvester Press, Brighton
Kenny, S. (1982) 'Sub-Regional Specialization in the Lancashire Cotton Industry 1884-1919: A Study in Organizations and Locational Change', Journal of Historical Geography, 8, 41-67
Kenwood, A.G. (1978) 'Fixed Capital Formation on Merseyside, 1800-1913', Economic History Review, 2nd Series, XXXI, 214-37
Kenwood, A.G. (1981) 'Transport Capital Formation and Economic Growth on Teeside 1820-50', Journal of Transport History, 2, 53-72
Kriedte, P., Medick, H., and Schlumbohm, J. (1981) Industrialization Before Industrialization, Cambridge University Press, Cambridge
Langdale, J.V. (1978) 'The Growth of Long-Distance Telephony in the Bell System: 1875-1907', Journal of Historical Geography, 4, 145-60
Langton, J. (1979) Geographical Change and Industrial Revolution: Coal Mining in South-West Lancashire 1590-1799, Cambridge University Press, Cambridge

HISTORICAL GEOGRAPHY OF INDUSTRIAL CHANGE

Langton, J. (1984) 'The Industrial Revolution and the Regional Geography of England', Transactions of the Institute of British Geographers N.S., 9, 145-67
Langton, J., and Hoppe, G. (1983) 'Town and Country in the Development of Early Modern Western Europe', Historical Geography Research Series, Geo Books, Norwich, 11
Langton, J., and Morris, R. (1985) 'Compiling an Atlas of the Industrial Revolution', unpublished paper presented to the I.B.G. annual conference, Leeds
Langton, J., and Morris, R. (1986) (eds.), An Atlas of the Industrial Revolution, Methuen, London
Lawton, R., and Pooley, C.G. (1980) 'Problems and Potentialities for the Study of Internal Population Mobility in Nineteenth-Century England', Canadian Studies in Population, 5 (special issue), 69-84
Lazonick, W.H. (1981) 'Production Relations, Labour Productivity and Choice of Technique: British and U.S. Cotton Spinning', Journal of Economic History, XLI, 491-516
Lazonick, W.H. (1983) 'Industrial Organization and Technological Change: The Decline of the British Cotton Industry', Business History Review, LVII, 195-236
Lee, C.H. (1971) Regional Economic Growth in the United Kingdom Since the 1880s, McGraw, London
Lee, C.H. (1972) A Cotton Enterprise 1795-1840: A History of M'Connel Kennedy, Fine Cotton Spinners, Manchester University Press, Manchester
Lee, C.H. (1979) British Regional Employment Statistics 1841-1971, Cambridge University Press, Cambridge
Lee, C.H. (1981) 'Regional Growth and Structural Change in Victorian Britain', Economic History Review, 2nd Series, 34, 438-52
Lee, C.H. (1984) 'The Service Sector, Regional Specialization and Economic Growth in the Victorian Economy', Journal of Historical Geography, 10, 139-56
Lees, L.H. (1979) Exiles of Erin, Manchester University Press, Manchester
Le Guillon, M. (1975) 'Freight Rates and Their Influence on the Black Country Iron Trade in a Period of Growing Domestic and Foreign Competition 1850-1914', Journal of Transport History N.S., III, 108-18
Lindert, P.H., and Williamson, J.G. (1983) 'English Workers' Living Standards during the Industrial Revolution: A New Look', Economic History Review, 2nd Series, XXXVI, 1-25
McCloskey, D. (1981) 'The Industrial Revolution 1780-1860: A Survey' in R. Floud and D. McCloskey (eds.), The Economic History of Britain Since 1700, Cambridge University Press, Cambridge
Marshall, J.D. (1983) 'The Rise and Transformation of the Cumbrian Market Towns 1660-1900', Northern History,

<u>19</u>, 128-209

Massey, D. (1983) 'In What Sense a Regional Problem?', Regional Studies, 13, 233-43

Massey, D. (1984) Spatial Divisions of Labour - Social Structures and the Geography of Production, Macmillan, London

Mathias, P. (1969) The First Industrial Nation, Methuen, London

Medick, H. (1976) 'The Proto-Industrial Family Economy: The Structural Function of Household and Family during the Transition from Peasant Society to Industrial Capitalism', Social History, 3, 291-315

Mendels, F. (1972) 'Proto-Industrialization: The First Phase of the Process of Industrialization', Journal of Economic History, 32, 241-61

Mendels, F. (1976) 'Social Mobility and Phases of Industrialization', Journal of Interdisciplinary History, 7, 193-216

Meyer, D.R. (1983) 'Emergence of the American Manufacturing Belt: An Interpretation', Journal of Historical Geography, 9, 145-74

Morgan, R.H. (1983) 'The Development of the Electricity Supply Industry in Wales to 1919', Welsh History Review, 11, 317-37

Morris, R.J. (1979) Class and Class-Consciousness in the Industrial Revolution, Macmillan, London

Musson, A.E. (1976) 'Industrial Motive Power in the United Kingdom 1800-70', Economic History Review, 2nd Series, XXIX, 415-39

Neale, R.S. (1985) Writing Marxist History, Blackwell, Oxford

Oddy, D.J. (1983) 'Urban Famine in Nineteenth-Century Britain: The Effect of the Lancashire Cotton Famine on Working-Class Diet and Health', Economic History Review, 37, 68-86

O'Tuathaigh, M.A.G. (1981) 'The Irish in Nineteenth-Century Britain: Problems of Integration', Transactions of the Royal Historical Society, 31, 149-73

Overton, M. (1984) 'Agricultural Revolution? Development of the Agrarian Economy in Early Modern England' in A.R.H. Baker and D. Gregory (eds.), Explorations in Historical Geography, Cambridge University Press, Cambridge, pp. 118-39

Pawson, E. (1978) 'The Framework of Industrial Change 1730-1900' in R.A. Dodgshon and R.A. Butlin (eds.), An Historical Geography of England and Wales, Academic Press, London, pp. 267-290

Pawson, E. (1979) The Early Industrial Revolution: Britain in the Eighteenth Century, Batsford, London

Payne, P.L. (1974) British Entrepreneurship in the Nineteenth Century, Macmillan, London

Persky, J., and Moses, R. (1984) 'Specialised Industrial Cities in the United States 1860-1930', Journal of

Historical Geography, 10, 37-51

Pooley, C.G. (1977) 'The Residential Segregation of Migrant Communities in Mid-Victorian Liverpool', Transactions of the Institute of British Geographers N.S., 2, 363-82

Pooley, C.G. (1983) 'Welsh Migration to England in the Mid-Nineteenth Century', Journal of Historical Geography, 9, 287-306

Pooley, C.G. (1984) 'Residential Differentiation in Victorian Cities: A Reassessment', Transactions of the Institute of British Geographers N.S., 9, 131-44

Pooley, C.G., and Doherty, J. (1985) 'The Longitudinal Study of Population Migration in Nineteenth-Century Britain', unpublished paper presented to I.B.G. conference on population geography, Liverpool

Pratts, J.D., and Pringle, T.R. (1985) 'Rethinking "Studies in Historical Geography": Towards a New Hegemony', Area, 17, 50-51

Pred, A. (1981) 'Production, Family and Free-Time Projects: A Time Geographic Perspective on the Individual and Societal Change in Nineteenth-Century U.S. Cities', Journal of Historical Geography, 7, 3-36

Redford, A. (1926) Labour Migration in England 1800-1850, Manchester University Press, Manchester.

Robson, B.T. (1973) Urban Growth: An Approach, Methuen, London

Rodgers, H.B. (1960) 'The Lancashire Cotton Industry in 1840', Transactions of the Institute of British Geographers, 28, 135-53

Rubenstein, W.D. (1977a) 'The Victorian Middle Classes: Wealth, Occupation and Geography', Economic History Review, 2nd Series, XXX, 602-23

Rubenstein, W.D. (1977b) 'Wealth, Elites and the Class Structure of Modern Britain', Past and Present, 76, 99-126

Samuel, R. (1977) 'The Workshop of the World: Steampower and Hand Technology in Mid-Victorian Britain', History Workshop, 3, 6-72

Saunders, M. (1984) 'Migration to 19th Century Barrow-in-Furness: An Examination of the Census Enumerators' Books 1841-1871', Transactions of the Cumberland and Westmorland Antiquarian and Archaeological Society, LXXXIV, 215-25

Saxonhouse, G., and Wright, G. (1984) 'New Evidence on the Stubborn English Mule and the Cotton Industry 1878-1920', Economic History Review, 37, 507-19

Southall, H. (1982) 'Migration as a way of Life: The Social Meaning of Mobility Among Nineteenth-Century English Artisans', unpublished paper presented to I.B.G. conference on historical demography, Cambridge

Southall, H. (1984) 'Antecedents of the Great Depression: Unemployment, Growth and Regional Economic Structure

in Britain 1851-1914', unpublished paper presented at
I.B.G. annual conference, Durham
Stedman-Jones, G. (1975) 'Class Struggle and the Industrial
Revolution', New Left Review, 90, 35-69
Stedman-Jones, G. (1976) Outcast London: A Study in the
Relationship between Classes in Victorian London,
Penguin edition, Harmondsworth
Swift, R., and Gilley, W. (1985) (eds.), The Irish in the
Victorian City, Croom Helm, Beckenham
Taylor, A.J. (1975) The Standard of Living in Britain in the
Industrial Revolution, Methuen, London
Thompson, E.P. (1967) 'Time, Work-Discipline and Industrial
Capitalism', Past and Present, 37, 56-97
Thompson, E.P. (1968) The Making of the English Working
Class, Penguin edition, Harmondsworth
Thompson, E.P. (1974) 'Patrician Society, Plebian Culture',
Journal of Social History, 7, 382-405
Thrift, N. (1977) 'Time and Theory in Human Geography',
Progress in Human Geography, 1, 413-57
Von Tunzleman, G.N. (1978) Steam Power and British Indus-
trialization to 1860, Clarendon Press, Oxford
Walters, R. (1980) 'Capital Formation in the South Wales Coal
Industry 1840-1914', Welsh History Review, 10, 68-91
Walton, J.K. (1981) 'The Demand for Working-Class Seaside
Holidays in Victorian England', Economic History Review,
34, 249-65
Walton, J.K. (1983) The English Seaside Resort: A Social
History, Leicester University Press, Leicester
Ward, D. (1980) 'Environs and Neighbours in the "Two
Nations": Residential Differentiation in Mid-Nineteenth
Century Leeds', Journal of Historical Geography, 6,
133-62
Warren, K. (1980) Chemical Foundations: The Alkali Industry
in Britain to 1926, Clarendon Press, Oxford
Whyte, K. (1984) Register of Research in Historical Geo-
graphy, Historical Geography Research Series, 14,
Geobooks, Norwich
Winstanley, M. (1983) The Shopkeepers World 1830-1914,
Manchester University Press, Manchester
Withers, C.W.J. (1985) 'Mapping the Industrial Revolution',
Journal of Historical Geography, 11, 196-7
Woods, R., and Woodward, J. (1984) (eds.), Urban Disease
and Mortality in Nineteenth-Century England, Batsford,
London
Wrigley, E.A. (1962) 'The Supply of Raw Materials in the
Industrial Revolution', Economic History Review, 2nd
Series, 15, 1-16
Wrigley, E.A. (1972) 'The Process of Modernization and the
Industrial Revolution in England', Journal of Inter-
disciplinary History, 3, 225-59

Chapter Seven

PEOPLE AND HOUSING IN INDUSTRIAL SOCIETY

R. Dennis

As the focus of human geography has shifted from pattern to process, and from landscape to society, so in the field of urban historical geography, the last twenty years have witnessed a blossoming of interest in the structure and organisation of urban industrial societies. This interest originated partly in the quantitative revolution, which stimulated the analysis of large data sets and a search for sources, such as the nineteenth-century census enumerators' books, amenable to statistical analysis, and partly in the rise of contemporary urban social geography, in which ecological models and social area theory were tested through multivariate analysis of small-area statistics in modern censuses (Robson, 1969; Timms, 1971; Herbert and Johnston, 1976).

More recently, research has become less technical and quantitative, reflecting geographers' growing awareness of debates in social history, and also the emergence of human-istic geography, emphasising the lived experience of individuals (Ley and Samuels, 1978), and marxist analysis, stressing the need to explore the past roots of current economic crises, inner-city problems and de-industrialisation (Dear and Scott, 1981; Dunford and Perrons, 1983). The conjuncture of these two strands in marxian humanism, or in Giddens' theory of structuration, focusing on the interface between structure and human agency (Gregory, 1982), has generated some distinctive - but less obviously 'urban' - contributions to social historical geography; for example, examining the role of institutions, whether organisations (School Boards, Poor Law Guardians, Philanthropic Trusts) or the buildings they erected, in reflecting and promoting ideology and shaping behaviour (Cosgrove, 1984; Driver, 1985; King, 1980; 1984), or researching the relationships between culture, class and community and the connections between geographical, historical and sociological interpret-ations of those terms (Billinge, 1982; 1984). What, to date, has been lacking, is much empirical investigation of the links between economy and society: how did changes in the labour process - specialisation, mechanisation, deskilling, the emergence of a labour aristocracy and a non-manual clerical lower middle class, and changes in the sexual division of labour - feed through into changes in urban spatial structure, and in the nature as well as the extent of residential differentiation in cities?

This chapter will first review progress in positivist, quantitative analyses of nineteenth-century cities, mostly structured around analysis of census books for the period 1841-1881, but exploring patterns and processes of community and social interaction as well as residential differentiation (Dennis, 1984). Then the links between this kind of social geography and the built environment will be discussed, particularly by considering how housing was managed and allocated in Victorian and Edwardian cities, and how housing management impinged on residential mobility. The chapter is necessarily selective. For example, I pay almost no attention to recent research on the medical geography of nineteenth-century cities (see Woods and Woodward, 1984; Kearns, 1985) or on urban retailing (see Shaw and Wild, 1979; Shaw, 1982a, 1985).

Most of the chapter will focus on English towns and cities, especially between 1850 and 1914, but at the outset it is important to recognise differences between England, Scotland, North America and continental Europe, particularly with regard to housing systems and consequent social patterns (Cannadine, 1980a; Daunton, 1983; Radford, 1981; Sutcliffe, 1974; Ward, 1984; Ward and Radford, 1983). In England most working-class families rented accommodation from private landlords on weekly terms: their homes were most likely to be all or parts of two- or three-storey terraced houses, whether 'through houses' (with a back door and yard), 'back-to-back', or built around courtyards, perhaps sharing water supply and toilet facilities. In Scotland and continental Europe the urban working classes were more likely to live in tenements, managed by a 'factor' on behalf of an absentee landlord and let on yearly terms. In North America rates of home ownership were higher (Barrows, 1983; Kirk and Kirk, 1981), although most poorer families still rented from private landlords, and there was little philanthropic provision to compare with agencies like the Peabody Trust in London, and less state intervention, in the form of either local government bye-laws and building regulations, or council housing (Birch and Gardner, 1981). It remains to be demonstrated whether this diversity of housing provision led to an equal diversity of residential patterns and rates of mobility.

SOURCES FOR RESEARCH ON THE SOCIAL GEOGRAPHY OF NINETEENTH- AND EARLY TWENTIETH-CENTURY CITIES

The principal source for social research has been the nineteenth-century census, which assumed its modern form in

PEOPLE AND HOUSING IN INDUSTRIAL SOCIETY

Britain with effect from the 1841 census. From 1801 to 1831 the published census records the numbers of males and females and the numbers of families engaged in agriculture, trade and manufacturing in each parish, but few individual returns survive. Laxton (1981) employed a manuscript enumeration of Liverpool in 1801, and there are private censuses of particular localities in the late eighteenth century (Corfield, 1982); but from 1841 onwards the country was divided into enumeration districts, enumerators delivered schedules to be completed by each household head, collected them on the day after census night, and transcribed the replies into enumerators' books which are preserved in the Public Record Office, and are open to inspection once they are more than one hundred years old. At the time of writing, therefore, returns are available for censuses up to 1881. The range of information included in nineteenth-century censuses, sources of error and unreliability, problems of interpretation and methods of categorising and analysing data have been discussed in a succession of guides (Armstrong, 1966; Wrigley, 1972; Lawton, 1978). Suffice to note here that while the census provides abundant information on family and household structure, on social status - measured not only in terms of occupation, but also by noting the presence of domestic servants and lodgers, teenage children still recorded as 'scholars', and women and children in paid employment (Cowlard, 1979) - and on birthplace, it tells us nothing about housing conditions, tenure or recent migration, except insofar as migration can be inferred from the birthplaces of successive children in the same family. One unfortunate consequence of these omissions is that studies of the socio-spatial structure of cities have been conducted independently of research on the built environment. Several researchers have discussed the need to relate 'social space' and 'physical space' (e.g. Pooley, 1982), but most attempts to do so have been stronger on the former than the latter (Shaw, 1979; Jackson, 1981).

Some information on housing is available on a systematic basis in ratebooks, which have been used to plot rateable values, assumed to indicate housing quality (Gordon, 1979; Lewis, 1985), and to calculate rates of owner-occupation and scales of landlordism (Daunton, 1976; Pritchard, 1976). Most local authorities have preserved only a few sample ratebooks, but where there are continuous series it would be possible to trace the frequency with which property was bought and sold, and the way in which tenure changed over time, within settlements as a whole, and as individual properties aged. It seems probable that while new dwellings were increasingly likely to start life owner-occupied in late Victorian and Edwardian England, older properties were acquired by absentee landlords as part of a filtering process in which they came to be occupied by successively lower income households. At times of low rates of new building, levels of owner-

occupation consequently fell (Dennis, 1984, pp. 142-3). There
should also have been changes in the spatial pattern of
tenure as owner-occupiers concentrated in suburbia and
inner-city areas converted to private renting. There is more
potential for this kind of research in North America, where
assessment records include much more information on both
owners and occupiers, and where continuous series of data
appear to have survived more frequently (Harris, Levine and
Osborne, 1981; Harris, 1986).

Directories are another valuable source (Shaw, 1982b;
1984). Most city directories included an alphabetical list of
businesses and inhabitants who were sufficiently respectable
to warrant inclusion - either because they had something to
sell, or because they were potential clients. Residents were
also listed in a street-by-street gazetteer, and businesses
were recorded in an A-Z directory of trades. Assuming that
entry in a directory was a measure of status for private
inhabitants, especially where there was a separate listing of
the 'court' or 'gentry' of a town, we can easily identify
high-status areas and their changing location over time
(Tunbridge, 1977). But directories can also be used more
imaginatively, for example, to reconstruct journey-to-work
patterns among businessmen, whose home and work addresses
were both normally recorded. Census books rarely distinguished between master craftsmen, journeymen and unskilled
operatives, all of whom might be assigned the same
occupation, such as 'weaver', or between workers in the
factory and domestic systems. But if a census householder
was also listed in a directory, we may assume that he was an
employer or at least self-employed. Whereas censuses were
produced only every ten years, directories often appeared
annually or at least every two years, and so are valuable for
measuring the persistence of households at the same address
or their pattern of residential mobility (Pooley, 1979).

Other sources reveal behaviour, attitudes and patterns
of social interaction (Dennis and Daniels, 1981). Marriage
registers indicate the premarital addresses, occupations and
parental occupations of brides and grooms. They have been
used by historians to measure inter-generational social
mobility, by comparing the occupations of fathers and sons,
and to calculate rates of inter-occupational social interaction
as a guide to class consciousness: did labourers' sons marry
craftsmen's daughters or were marriages concentrated within
narrowly defined status groups? (Crossick, 1978; Foster,
1974). Geographers have used the same sources to measure
'marriage distances', in rural as well as urban areas, showing
how the breakdown of rural isolation was reflected in the
selection of more distant marriage partners, where courtship
depended upon use of the bicycle, the attainment of a
sufficiently advanced level of literacy for the exchange of
letters by penny post, and the availability of enough leisure

time to travel between villages several miles apart (Perry, 1969). In cities we can identify 'urban villages', including relatively self-contained inner-city slums or suburban industrial colonies where most brides and grooms drew spouses from the same neighbourhood (Dennis, 1977a; 1984).

Church and society membership registers offer another insight into the community structure of past societies. Many nonconformist churches kept extraordinarily detailed records, noting the regularity with which members attended communion, expelling some members for inexcusable absence or moral impropriety, and commissioning others to found new daughter congregations in suburbia (Dennis, 1984; McLeod, 1974; 1977). So it is feasible to reconstruct the catchment areas from which organisations drew their members and, by linking membership records to other nominal lists such as census books and directories, to establish their social composition.

Prior to the Secret Ballot Act of 1872 pollbooks were published after many parliamentary elections, recording how electors had used their votes (Vincent, 1967). Most pollbooks listed only the names and addresses of each candidate's supporters, but when they are linked to other nominal records we can see how different occupational groups, or members of different churches, or the tenants of different landlords, cast their votes (Morris, 1983a; Dennis, 1986a). So we can build up evidence not only of the existence of territorially defined communities or urban villages, but also of the foundations of community structure: were communities based on deference, fear, mutuality of interest, class consciousness? (Joyce, 1980).

In addition to these quantifiable records, urban historical geographers have also made increasing use of literary and artistic sources (Pocock, 1981). Novels and paintings may fall foul of the accusation that they are elitist art, telling us little of how ordinary people thought about or experienced urbanisation or industrialisation (Burgess and Gold, 1985); but even if they were not representative or typical of popular sentiment, they were powerful shapers of public opinion. As yet, geographers are still less than expert in their handling of literature, too often seeking a factual content that treats novels as no different from documentary sources. They have been more adept at using diaries, autobiographies and oral histories, which reveal the thoughts and processes behind social patterns; for example, the way in which families sought to move house (Lawton and Pooley, 1975; Roberts, 1984). Of course, anybody who kept a diary over a long period, especially anybody from the working classes, may be considered atypical; and autobiographies and oral histories record the experiences of survivors, usually looking back to their childhood in a nostalgic or sentimental fashion, or perhaps exaggerating the hardships and poverty of their youth. Their formative years may be interpreted through the filter of subsequent experience. Robert Roberts, the author of two acclaimed accounts of Edwardian childhood, The Classic Slum (1971) and A Ragged Schooling (1976), describing life in Salford before World War I, became a prison visitor and educational reformer, widely read in sociological literature which must have influenced his views of class and class consciousness in ways unknown to his schoolmates who failed to climb out of the slum. Nonetheless, biographical sources are an important complement to the lists of individuals that we find in censuses, directories and other nominal records (Burnett, 1974; 1983; Thompson, 1973).

Between the extremes of census enumerators' returns and novels, a variety of records focus on social problems of special interest to contemporary observers. Victorian cities acted as honeypots attracting foreign visitors or roving newspaper correspondents (e.g. Tocqueville, 1958; Engels, 1969; Taine, 1957) whose reports offer a potent mixture of ideology, observation and presupposition (Coleman, 1973; Lees, 1983; 1985; Marcus, 1974; Ward, 1978). The members of newly founded statistical societies undertook surveys of housing and employment for publication in periodicals like the Journal of the Statistical Society of London, while concern for textile workers in the 1830s, sanitation in the 1840s and working-class housing in the 1880s, among countless other social issues, was expressed in the repeated attention they received from parliamentary Select Committees, Royal Commissions and other public inquiries (e.g. Flinn, 1965). Each report included a general statement of findings and resolutions, the cross-examination of witnesses and the submission of detailed evidence on particular places. The appointment of local Medical Officers of Health triggered the collection of statistics on different causes of mortality and the incidence of infectious diseases (Woods and Woodward, 1984); and the suspicion that improved living standards still left a substantial minority in dire poverty, was confirmed by both detailed poverty surveys (e.g. Booth, 1902; Rowntree, 1901) and sensational tracts (Keating, 1976).

COMMUNITY AND SEGREGATION IN NINETEENTH-CENTURY CITIES

Much recent research has focused on residential segregation and the spatial structure of social and ethnic groups. Its original objective was summarised in the title of an important paper by Ward (1975): Victorian cities - how modern? The aim was to compare cities past and present in the context of ecological and social area theory. Did nineteenth- (or even eighteenth-) century cities exhibit the sectoral and zonal patterns of economic and family status characteristic of Hoyt's and Burgess' theories of North American urban structure? Were levels of residential segregation increasing? Where could Victorian cities be located along a continuum from Sjoberg's pre-industrial city, where the rich lived in the centre in contrast to poor and alien outcasts on the urban fringe, to Burgess' industrial city, in which the relative locations of rich and poor were reversed? Or along a continuum from a one-dimensional feudal city, in which socio-economic, family and ethnic status were interrelated, to a multi-dimensional modern city, in which each kind of status varied independently?

It soon became clear that answers to these questions depended on not only the size and type of city being investigated, but also the techniques used to assess modernity (Pooley, 1984). Researchers using factor analysis invariably found evidence of a trend towards a multi-dimensional urban structure, where separate factors could be associated with socio-economic status, family status (or stage in the life cycle) and ethnic or migrant status. But this was not surprising, given that the census data employed in most analyses related to occupation, servant-keeping, household structure, age-sex ratios and birthplace. There was also a danger of overinterpreting what were often quite low correlations between variables loading on the same factor and equally low levels of 'variance explained'. Nonetheless, Lawton and Pooley (1976), working on Liverpool in 1871, Carter and Wheatley (1982) on Merthyr Tydfil in 1851, Shaw (1977) on Wolverhampton in 1851 and 1871, Dennis (1984) on Huddersfield in 1851 and 1861, and Warnes (1973) on the small Lancashire town of Chorley in 1851, all found evidence for the emerging modernity of mid-Victorian cities. All followed in the footsteps of Goheen (1970) who argued that Toronto rapidly acquired the characteristics of a modern city during the 1870s. Applying techniques of factor analysis, trend surface analysis and multiple regression to Canadian assessment records, and also plotting journey-to-work patterns for selected occupational groups, Goheen concluded that 'By 1880 Toronto was caught up in the process of transition from the old city, represented by 1860 and 1870, to something quite different' (Goheen, 1970, p. 172). But given the rate at which Toronto grew in the late nineteenth century (from 56,000 inhabitants in 1870 to 377,000 in 1911), structural change was inevitable and we may question whether Goheen was observing the contrast between a small city and a metropolis, rather than that between an early and a late Victorian city.

Indeed, other researchers such as Cannadine (1977) using less sophisticated statistical techniques, have observed that large cities were more likely to be 'modern', perhaps because they were provided with public transport systems which facilitated the separation of different grades of worker, as well as of rich and poor; also because the scale of housing construction was larger in big cities, leading to the development of extensive one-class areas compared to much smaller units of social homogeneity in lesser towns; and perhaps reflecting the impersonality and anonymity of big-city life, such that the new middle classes and labour aristocracy were obliged to separate themselves geographically in order to protect and proclaim their superior status. Yet, if modernity requires the complete segregation of different social classes, even large mid-Victorian cities could not be called modern. Ward (1980), studying Leeds through all four censuses from 1841 to 1871, found few enumeration districts that were exclusively middle-class or working-class. Even in the city's north-west suburbs, where the middle classes were relatively concentrated, they were substantially outnumbered by poorer households.

This inequality of class sizes complicates interpretation of simple measures of concentration and segregation, such as Location Quotients and the Index of Dissimilarity (Peach, 1975). The latter was originally commended for being an easily understood, symmetrical index whose value was independent of the size of groups being compared. For example, Lawton and Pooley (1976) calculated an index of 77 for the residential dissimilarity of professional and unskilled workers in Liverpool in 1871, indicating that 77 per cent of either group would have had to move between enumeration districts for their distributions to have coincided at that scale. But since Lawton and Pooley's city-wide sample included only 181 professional workers but 7,480 unskilled workers, we may suspect that each group experienced segregation differently. The minority middle class might have thought of their cities as undifferentiated because even in districts where they were proportionally over-represented (as indicated by a high location quotient), there were plenty of working-class inhabitants. But most unskilled workers lived in districts with few professional residents. They would have been much more conscious of the geographical separations of 'two nations'.

To counter this problem of differential experience Robinson (1980) proposed the use of Lieberson's Isolation Index (P*), which measures the probability of contact between different groups, based on the assumption that social isolation is correlated with spatial separation. In the example cited above, the probability of an unskilled worker encountering somebody of professional status is less than the probability of a professional meeting an unskilled worker. Clearly, there is considerable potential for the use of P* as an additional index of residential differentiation in Victorian cities.

Whichever index is employed, its numerical value will depend upon the scale of analysis (Carter and Wheatley,

1980). Few enumeration districts and even fewer wards were dominated by only one class or ethnic group, but particular streets or courts were sometimes the preserve of a single group. So location quotients and indices of dissimilarity or isolation will increase, the finer the scale of analysis.

What was the significance to contemporaries of different scales of segregation? Married women and young children spent most time among their immediate neighbours; youths and working men had experience of much wider areas (Meacham, 1977). In small places everybody would have known something about the character of almost every street; in larger cities individuals made do with more generalised mental maps. Liverpudlians would have known the contrasting reputations of Toxteth, Everton and Scotland Road, but they might not have been familiar with individual streets in each of those areas. So the critical scale at which segregation affected behaviour varied according to age, sex, class, ability to travel and locale.

In general, in early Victorian cities, in smaller cities and in older established towns, segregation operated at the scale of individual streets, often between front streets which were quite prosperous and back streets and courts occupied by the poor (Figure 7.1), or where an eighteenth-century middle-class terrace was subsequently swamped by streets of jerry-built back-to-backs. In larger places, in the later nineteenth century, and in newer industrial towns, where all the buildings dated from the same period, and where there had not been time for neighbouring streets to acquire distinctive reputations, or for buildings to change their use, one-class areas were more extensive. For example, a variegated and small-scale pattern of differentiation in the old industrial town of Wigan contrasted with larger, more homogeneous social areas in the nearby but newer glass-making town of St Helens (Jackson, 1981).

More attention has recently been paid to mechanisms of development and social processes underlying these patterns. Pooley (1982) distinguished elements of choice and constraint, examining how housing was built and managed, the locations of workplaces and the needs of different groups to live close to their work, the nature of migration and the tendency for migrants to cluster, for reasons of choice, discrimination or economic circumstances, and the personal attributes and aspirations of individual households. Pooley (1984) has also focused on the significance and meaning, as well as the causes of segregation, concentrating on its implications for the workings of the urban economy; for an urban environment in which residents of different areas experienced different levels of pollution, sanitation and water supply, and where ill-health and mortality rates also varied through space; and for social structure and its active expression in class consciousness. A similar review of the political and social

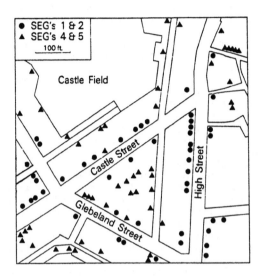

Figure 7.1: Front-street/back-street segregation in the centre of Merthyr Tydfil, 1851 (from Carter and Wheatley, 1980).

implications of segregation has been undertaken in a Canadian context by Harris (1984a,b).

Few geographers have examined what contemporary observers thought about the socio-spatial structure of their cities. What emerges from the writings of authors as different in time and ideology as De Tocqueville (writing in 1835), Engels (1844), Reach and other reporters writing for the Morning Chronicle (1849-51), Taine (1860s) and Rowntree (1901), is an emphasis on the separation of extremes of wealth and poverty, but much less concern for differences within the working classes. Elements of a more modern pattern of residential differentiation in Manchester and London were attributable as much to ethnic diversity (Irish rookeries and Jewish East End) as to distinctions within the English working classes. The contrast was also made between cities based on factory production, of which Manchester was the prime example, and those developed around workshops or a surviving domestic system, such as Birmingham and Sheffield (Briggs, 1968). In the former, the enormous gulf between employers and employed was mirrored by their geographical separation; in the latter, differences between masters, journeymen and apprentices were less marked, the mutuality of their interests more apparent, and social mobility more common, all reflected in a cellular, almost pre-industrial form of spatial structure, where workplace and industry mattered more than status.

Contemporaries also indicate that segregation occurred in time as well as space. Different classes pursued very different daily timetables, such that they used the same streets or crossed one another's territories without meeting face-to-face. Again, however, the relationship was asymmetrical. Many of the poor depended for their livelihood on providing personal services to rich clients - as domestic servants, street traders, messengers, washerwomen, dressmakers. But, aside from philanthropic and evangelistic activities, the middle classes shunned all unnecessary contact with the poor, for fear of contagious disease and, perhaps, preferring the comfort of ignorance to the unease of too intimate a knowledge of the conditions of poverty. Occasional recent excursions in time geography (e.g. Pred, 1984) will perhaps prompt more research on this important dimension of urban life.

Contemporary observers rarely wrote as geographers. Each had an ideological axe to grind, using accounts of urban structure to highlight hopes for reform or revolution or fears about socialism, atheism or the polarisation of 'two nations'. In a series of papers Ward has examined the political and ideological basis to writings by American and British commentators. Dealing initially with the first half of the nineteenth century, he concluded that contemporary descriptions could be viewed as 'geopolitical images designed to justify reform' (Ward, 1978, p. 189). Subsequently, he compared British and American attitudes to inner-city problems, contrasting the perspectives of 'progressives', who promoted state intervention to rescue the poor from perpetual poverty, and the Chicago School, who optimistically envisaged each generation of slum dwellers escaping to suburban affluence and respectability. Even progressives did not attempt to remove the underlying structural causes of poverty, but merely used the power of the state 'to remove obstacles to individual advancement' (Ward, 1984, p. 308). All believed in what Pooley and Irish (1984, p. 112) summarise as 'the principle of the market economy and the ideology of self-help and [in the field of housing] eventual owner-occupancy'.

Most nineteenth-century authors regarded 'segregation' as the undesirable antithesis of 'community', even if it developed through the workings of the market economy, free from discriminatory practices of social or ethnic apartheid. 'Community', meaning a balanced local social structure, organised on hierarchical and paternalistic or exemplary principles, where the elite were responsible for the moral and physical well-being of the poor, was to be engineered at the expense of segregation. Since many Victorian intellectuals were also anti-urban in outlook, their ideal communities were limited in size and surrounded by countryside, or at least aimed to resurrect a feudal social order within cities, as in the establishment of settlement houses and public school missions in inner-city slums in the late nineteenth century (Coleman, 1973; Glass, 1955).

Among historical geographers, 'community' has attracted less research than 'segregation'. Dennis (1984) has used some of the sources outlined earlier in this chapter to reconstruct patterns of interaction and links between the residents of different social areas in Huddersfield. Billinge (1982) has disputed the relevance of territorial definitions of community, instead emphasising its political and hegemonic dimensions and the institutions that sustain communities of influence and authority. Contrary to most studies of class consciousness, which have focused on the working classes, Billinge examined divisions among the middle classes, 'the important and hitherto much neglected class struggle waged between capital and land', as an urban bourgeoisie assumed power in place of landed gentry at the beginning of the nineteenth century (Billinge, 1984, p. 31). The struggle was observed through the operations of the Manchester Literary and Philosophical Society, the premier institutional embodiment of nonconformist Liberalism. Billinge was obliged to assume that members accepted the creed of the organisation to which they belonged, so that the attitudes of the silent majority of members could be inferred from the pronouncements of a few spokesmen. This may be valid for a select organisation like the Manchester Literary and Philosophical Society, but it cannot be assumed for members of more modest associations, like local churches. In these cases, membership may indicate the probability of social interaction, but not the class consciousness of members.

Billinge was not particularly concerned with the spatial dimensions of his elite community, which presumably reflected the location of the society's rooms, the nonconformist congregations to which members belonged, and the areas of bourgeois residence in suburban Manchester. Lower middle-class institutions, like suburban churches, sports clubs and friendly societies had more restricted catchment areas. But all these communities were class-selective so that, given the propensity of different classes to reside in different localities, community and segregation were not opposites, as contemporaries believed, but opposite sides of the same coin. Close-knit community life was most likely to develop in one-class neighbourhoods. Territorially restricted 'urban villages' were confined to working-class districts, as evidenced by contemporary accounts in diaries and autobiographies (Roberts, 1971; Jasper, 1969), and by analyses of marriage patterns, journeys to church, work or pub (Bramwell, 1984), and residential mobility.

PEOPLE AND HOUSING IN INDUSTRIAL SOCIETY

CLASS AND ETHNICITY

Studies of residential differentiation presuppose that the categories to which we allocate individuals or households are useful analytically and meaningful in terms of contemporary experience. In practice, geographers have generally used a version of the five-class model devised by Armstrong (1966; 1972) and based on the Registrar General's socio-economic classification of occupations in the 1951 census. Households were categorised as Professional (I), Intermediate (II), Skilled (III), Semi-Skilled (IV) or Unskilled (V). This classification usually produced a barrel-shaped social structure, often with more than half of the population concentrated in Class III. Attention then focused on the geographical distributions of the extremes - merchants, manufacturers and members of the professions in a few remaining central high-status streets but mostly located in new suburbs, and unskilled and casual labourers concentrated in inner-city slums. Much less attention was paid to patterns of differentiation within Class III. Yet social historians have been more concerned to disentangle complexities in the middle of the hierarchy - between an emergent labour aristocracy of skilled mechanics and engineers who built and maintained the new technology that served to de-skill less fortunate craftsmen, a petty bourgeoisie of shopkeepers and surviving independent master craftsmen whose trades had not yet been mechanised, and a lower middle class of clerks, agents and schoolteachers (Crossick, 1977; 1983; Crossick and Haupt, 1984; Gray, 1981; Morris, 1983b).

The five-category classification also lends a false impression of stability to a dynamic system. Not only did occupations change in status and relative income, as a result of mechanisation, de-skilling or professionalisation, but individuals could pursue social and economic advancement through class-based organisations like trade unions or by self-help leading to personal upward mobility. Neale (1981) distinguished between three kinds of working class, one deferential and socially immobile, another proletarian, class-conscious and militant, and thirdly an individually ambitious middling class. His model was devised with early nineteenth-century society in mind, but it is also applicable to later periods, despite marked differences in class relations and class consciousness between towns and over time. For example, Foster (1974) distinguished between Oldham, a cotton textiles town, South Shields, a seaport and ship-building centre, and Northampton, based on the shoemaking industry. According to Foster, Oldham comprised Marx's two classes: a small but powerful bourgeoisie and a vast proletariat, the latter encompassing both skilled manual workers and unskilled factory operatives and casual labourers. In South Shields and Northampton, trade was more important than class as the basis for social interaction, residence and political behaviour. But this contrast applied only until the 1840s. Thereafter, even Oldham's labour aristocracy succumbed to embourgoisement, and working-class consciousness was replaced by a system in which skill, status and income differentials were jealously defended. Other researchers have questioned, confirmed or re-evaluated Foster's thesis for other places and periods (e.g. Gadian, 1978; Glen, 1984; Joyce, 1980; Smith, 1982).

In essence, there was a trend from three 'interest groups' in the eighteenth century - land, capital and labour - where only the former was sufficiently well-integrated to function at a national scale (Laslett's (1983) idea of a one-class society before the industrial revolution), to two 'classes' by the 1830s as land and capital joined forces, to a multitude of status groups, each selfishly defending its corner while comprising individuals earnestly striving to attain promotion to the group above. But how was this changing social structure related to changes in spatial structure? In Cannadine's (1982) terms, what were the connections between 'shapes on the ground' and 'shapes in society'? To answer such questions geographers must adopt more active definitions of class and status, derived from information on voting behaviour, property ownership, the membership of political and religious organisations, and intermarriage, focusing especially on social patterns within areas inhabited by the skilled working class and lower middle class: streets of bye-law terraced housing, estates developed by building clubs and freehold land societies, and leasehold houses, where restrictive covenants imposed by ground landlords were intended to ensure at least a minimum level of respectability.

If class is more than occupation, so ethnicity is more than birthplace. Numerous studies of the residential patterns of different migrant groups have been undertaken (e.g. Pooley, 1983; Jackson, 1982; Lewis, 1980), but few have examined the experience of second and subsequent generations of migrants, who may have regarded themselves as just as Irish or Welsh as their parents. More wide ranging accounts by social historians include Lees' (1979) and Swift and Gilley's (1985) studies of Irish migrants and Buckman's (1983) and Williams' (1985) analyses of provincial Jewry. Pooley (1977) attempted to distinguish between 'ethnic communities' and 'ghettoes', classifying migrants to Liverpool according to their socio-economic and cultural similarities to the local-born host society, and the degree to which their geographical concentration was attributable to elements of choice or constraint.

In-migrants exhibited high levels of residential segregation in most cities, partly because of their dependence on ethnic institutions such as Catholic schools and churches, Jewish shops and synagogues, and Welsh-language Baptist chapels, and partly because word-of-mouth information about accommodation and job vacancies inevitably led them to housing areas and workplaces where fellow migrants were already established. But there were also elements of discrimination, especially where Irish factory operatives or dock labourers were used as strike-breakers, or where Jewish sweated labour was regarded as undermining indigenous workshop production. Most segregation, however, reflected the concentration of immigrants in low-paid, casual employment - as unskilled labourers, newspaper vendors, street-sellers, etc. They could not afford the rent for a whole house, so were forced into overcrowding and multi-occupancy; nor could they provide the employers' references or guarantee the regular payment of rent demanded by philanthropic and local authority landlords. In effect, their geographical distribution reflected their housing class.

HOUSING AND MOBILITY

During the 1970s contemporary human geographers became increasingly interested in the political economy of housing, initially through the wide publicity given to Rex and Moore's (1967) theory of housing classes. Rex and Moore argued that individuals' positions in the housing market reflected their ability to satisfy the criteria used by different 'gatekeepers', such as estate agents, building society managers and council officials, to allocate the relatively scarce resources at their disposal. Homeownership was assumed preferable to renting, a suburban semi preferable to an inner-city terrace. Renting from the council was better than renting from a private landlord, since it was usually cheaper and, in practice, more secure, and the quality of council accommodation was generally higher. So competition for housing, as a vital aspect of consumption, was considered more important than the class struggle for control of production. Moreover, housing class would also affect patterns of social interaction and political behaviour. Owner-occupiers were more likely to vote Conservative, oppose rate increases and contest any action likely to reduce property values. Tenants, especially those who received rate rebates, or whose rate contribution was subsumed in the rent they paid their landlord, would be less sensitive to changes in the rates, less concerned about long-term trends in property prices and more enthusiastic about local government expenditure on council housing and central government intervention to protect tenants' rights.

Subsequently, housing class theory attracted a more critical press (Bassett and Short, 1980; Saunders, 1981), but it has spawned an enduring debate on the political economy of housing which is now being replicated in the context of Victorian and Edwardian cities.

PEOPLE AND HOUSING IN INDUSTRIAL SOCIETY

Production of the built environment has been thoroughly researched, with countless monographs on the roles of pre-urban landowners (e.g. Cannadine, 1980b; Olsen, 1982), speculative builders and developers (Beresford, 1974; Dyos, 1968), and specialist agencies such as building clubs (Rimmer, 1963), freehold land societies (Gaskell, 1971), philanthropic and company landlords (Tarn, 1973), and local authorities acting as both legislators and builders (Wohl, 1977). Different classes and styles of nineteenth-century dwelling have each attracted their historian (e.g. Chapman, 1971; Daunton, 1983; Muthesius, 1982; Simpson and Lloyd, 1977; Sutcliffe, 1974). But the management of dwellings once they had been built is less well documented and correspondingly less well researched. Most studies have examined minority, institutional tenures, like council housing (Pooley, 1985), or management crises, such as rent strikes (Melling, 1983) and the continuing class struggle between landlords and tenants (Englander, 1983).

Prior to World War I more than 90 per cent of households rented their accommodation from private landlords. But this proportion varied both regionally and between areas within cities. Swenarton and Taylor (1985) note that before 1914 owner-occupation was relatively common in regions with well-paid and secure adult male employment, especially if men's wages could be supplemented by women's or children's earnings (e.g. in textile towns); in remote areas, like South Wales mining valleys, which lacked the resident professional and commercial middle classes who were important landlords in big cities; and where there was a tradition of working-class self-help, often nonconformist in origin, and reflected in the existence of local friendly societies and building clubs. In all these areas, owner-occupation was predominantly working-class. But in large cities, where land values and building costs were higher, and home-ownership correspondingly more expensive, and where in the nineteenth century there was no shortage of commercial landlords, there were few owner-occupiers outside of new middle-class suburbs. In Cardiff and Leicester, some suburban wards had rates of owner-occupancy of between 10 and 20 per cent; but fewer than 5 per cent of householders were owner-occupiers in inner-city wards (Pritchard, 1976; Daunton, 1977). In some planned suburbs, like West Hill Park in Halifax, where mortgages were provided by the newly established Halifax Building Society, more than half of the first occupiers owned their home, but the home-ownership rate soon declined. Swenarton and Taylor (1985, p. 380) summarise the findings of the Select Committee on Town Holdings (1889):

> over time, as the social character of a district changed, property tended to move from possession by the occupier to possession by investors and middlemen ... owner-

occupation declined in the older areas at the same time as it increased in the newer suburbs.

In both Huddersfield and Cardiff, two towns for which rate-books have been analysed (Springett, 1979; Daunton, 1977), levels of owner-occupation declined over time (Table 7.1). Evidently, home-ownership was not regarded as preciously as it is today. Many who could afford to buy still preferred the flexibility of renting.

In the largest cities, including London, Liverpool, Glasgow and Dublin, some dwellings were provided by philanthropic agencies and, by the early twentieth century, by local councils (Tarn, 1973; Wohl, 1977; Aalen, 1984). Although, in total, housing agencies such as the Peabody and Guinness Trusts, five per cent companies such as the Improved Industrial Dwellings Company, and local authorities like the London County Council owned only about 2 per cent of the nation's housing stock in 1914, they were locally more significant. For example, in 1891, more than 8 per cent of residents in Westminster lived in philanthropic block dwellings (Arkell, 1891). By 1914, the LCC had provided 9822 and other local authorities in and around London 5500 dwellings (Wohl, 1977); Liverpool Corporation had built 2895 dwellings, almost all central-area flats, and between 1899 and 1914 provided 7.6 per cent of all new dwellings built in the city (Pooley, 1985).

Most pre-1914 council flats were erected on slum clearance sites, usually in inner-city areas. Under the terms of the Cross Act (1875), local authorities could compulsorily purchase and demolish slum property, but had to offer cleared sites for sale to private, usually philanthropic, agencies. Whether rebuilding was undertaken by councils or private enterprise, rents for the new 'model dwellings' were set sufficiently high to repay the loans that had been raised to finance each scheme, overcrowding was strictly controlled, and subletting rarely permitted. One consequence was that the tenants of the new buildings were very different from the slum dwellers who had previously occupied the same sites (Table 7.2); another was that councils anticipated the viability of rehousing when they chose which slums to demolish. In London, philanthropic agencies happily acquired Cross Act sites in the West End and around the fringes of the City, but they declined to purchase East End sites close to the Thames, where the population comprised mainly casually employed dock labourers (Yelling, 1981; 1982). When the LCC began building and managing its own dwellings, from the 1890s onwards, it rarely encountered problems letting flats in central London, but Dockland estates proved hard to let: dwellings were often empty and many tenants were evicted for non-payment of rent (Dennis, 1986b). In Liverpool too, the corporation failed to house 'the poorest poor' (Pooley, 1985).

PEOPLE AND HOUSING IN INDUSTRIAL SOCIETY

Table 7.1:   Owner-occupation in England and Wales.

Place	Pre-1914 Year	%	1938 Place	%
Leeds	1839	3.7	Nottingham	14.3
Leicester	1855	4.0	Manchester	15.6
Durham	1850	17.0	Sheffield	18.0
	1880	17.5	Merthyr Tydfil	22.5
Huddersfield	1847	10.7	Birmingham	23.5
	1896	9.3	Stoke-on-Trent	25.9
Cardiff	1884	9.6	Burnley	35.1
	1914	7.2	Oxford	44.3
Oldham	1906	8.3	Bristol	51.1
			Plymouth	68.5
			England & Wales*	35.0
			Built pre-WWI	27.1
			Built post-WWI	49.1
			Gross value < £20.10s.	19.0
			Gross value > £20.20s.	55.6

* excluding separately rated flats and houses let wholly in tenements.

Sources: Dennis 1984, Swenarton and Taylor 1985.

     Councils and philanthropists trusted to a process of 'levelling up', whereby the very poor would benefit by moving into dwellings vacated by the better-off working class, who could afford model dwellings or suburban cottages. In practice, 'levelling up' rarely worked, partly because urban populations increased too quickly for housebuilders to keep up, and also because the expansion of commercial and business districts resulted in large-scale demolitions of working-class property, in excess of those for which replacement housing was constructed (Stedman Jones, 1971). Rather than the tenants levelling up, it was the property which levelled down, as in the industrial suburb of West Ham, where terraced houses intended for individual families were subdivided, sublet or let off in rooms to households squeezed out of inner London (Howarth and Wilson, 1907).
     Despite the growing involvement of local authorities, most families in all social classes rented their homes privately. Most landlords owned only a few houses, perhaps occupying one house in a terrace and letting the rest. Although some landlords managed their own properties, increasing numbers employed agents and many agents managed houses owned by

PEOPLE AND HOUSING IN INDUSTRIAL SOCIETY

Table 7.2: Occupational Structure on the London County Council's Boundary Street Estate.

Selected Occupations	Before Clearance Adult Workers		After Rebuilding (1899) Household Heads	
	No.	%	No.	%
Policemen	5	0.5	40	5.5
Labourers	149	14.1	14	1.9
Cabinetmakers	120	11.4	44	6.0
Charwomen	11	1.0	1	0.1
Costermongers	23	2.2	-	-
Hawkers	126	11.9	-	-
Carmen	29	2.7	26	3.6
Carpenters	10	0.9	27	3.7
Clerks	-	-	21	2.9
Boot/Shoemakers	74	7.0	18	2.5
Porters (market)	31	2.9	10	1.4
Postmen	-	-	30	4.1
Total	1057		728	

Source: London County Council (1900) The Housing Question in London between the Years 1855 to 1900.

several landlords (Daunton, 1983). In West Ham, for example, thirteen agents were collectively responsible for more than 5000 dwellings in 1905, about one-eighth of all the cottage property in the borough (Howarth and Wilson, 1907). The implication is that effective control was much more concentrated than figures on ownership might suggest.

Very few landlords left records detailing their day-to-day activities, but from the evidence of novels (e.g. Bennett, 1911a; 1911b), local and central government inquiries (see the references in Daunton, 1983, and Offer, 1981), a few personal papers and the occasional social survey (like Howarth and Wilson's study of West Ham), historians are beginning to reconstruct how houses were managed. In England, working-class property was usually let on weekly terms, allowing maximum flexibility for both landlords and tenants. Although tenants were legally liable to pay rates, in practice most landlords paid rates direct to the council, recouping the payment by adding a few pence to the weekly rent demand. It was only middle-class tenants occupying houses on yearly or at least monthly terms who expected to pay property rates themselves. There were also differences in tenants' responsibilities for repairs and decoration. Working-class dwellings would usually be whitewashed or repapered whenever a new tenant moved in, and some families reputedly moved house (or threatened to do so) simply to obtain newly decorated accommodation; as now, many landlords had to be persuaded into meeting their contractual obligations!

Most landlords owned their properties on mortgage; their profit comprised the difference between the rents they received and the mortgage payments, rates and maintenance costs they paid out. When rates and interest rates rose, about the turn of the century, landlords had to increase rents to maintain profit margins. But this strategy could only work while demand was buoyant. If there were lots of empty houses, tenants could move to cheaper dwellings rather than pay increased rents. Yet landlords repaying mortgages could ill afford to own empty dwellings. So the situation in Edwardian England was not always, or often, one of ruthless landlords exploiting helpless tenants. Landlord-tenant relations were much more complex, and it could be argued that private landlordism was already facing a terminal crisis before 1914.

From 1915 onwards, rent control held working-class rents at pre-war levels, but costs of materials and labour rapidly increased, reducing the incentive for people to become landlords. Instead, the inter-war years witnessed the construction of subsidised council housing and the growth of owner-occupation (Daunton, 1984). Before 1914, owner-occupancy had been largely working-class, concentrated in industrial areas, but between the wars it became more middle-class and more concentrated in the south of England (Table 7.1) (Swenarton and Taylor, 1985).

Why should these changes in tenure interest historical geographers? Firstly because, as Rex and Moore (1967) argued, different social and ethnic groups had different housing opportunities. Housing classes existed in the nineteenth century just as they do today; in particular, different groups within the working and lower middle classes had access to different kinds of renting, according to their income, regularity of employment, family circumstances and the region in which they lived. So the geography of housing ownership and management helped to determine patterns of residential differentiation; although, of course, the relationship was two-way: builders and landlords could not defy popular taste indefinitely.

Secondly, housing class was related to behaviour. We currently know very little about the links between housing class and voting behaviour or religious affiliation, but it is perfectly feasible to link information on named individuals in ratebooks, pollbooks and church registers. As a foretaste, it is worth noting that in Huddersfield, usually a Liberal stronghold, the only districts in which the majority of electors voted Conservative in 1868 were also areas with the highest rates of owner-occupation (Dennis, 1986a). Since even in these areas homeowners were in the minority, we may be in danger of the ecological fallacy: the correlation may not work at the level of individuals. But at least it deserves further investigation.

Much more clearly, housing class was related to residential mobility, certainly to <u>rates</u> of population turnover and possibly to <u>patterns</u> of movement. Several Canadian studies have shown that homeowners moved less frequently than tenants (Katz, 1975; Katz, Doucet and Stern, 1982). Among British studies, Pritchard (1976, p. 179) observed that turnover rates declined dramatically after World War I: 'Whereas a century ago, 1 in every 4½ households could be expected to change their address in any year, today that has been reduced to about 1 in 11'. He attributed the decline to several factors:

1. The ageing of the population. Several studies have shown that the elderly move less frequently than young adults (Dennis, 1977b; Pooley, 1979). So twentieth-century increases in life expectancy and in the number of elderly couples or widows constituting separate one- or two-person households inevitably produce reductions in overall mobility.

2. A decline in the number of empty dwellings. When there is a housing shortage, as occurred after World War I, it is obviously more difficult to move than when empties abound and landlords and estate agents vie for custom by reducing rents and house prices.

3. The switch in tenure from insecure private renting before 1915 to a mixture of controlled private renting, owner-occupation where legal complexities and costs discouraged the kind of move, common among private tenants, that produced only marginal changes in housing, and council housing, where transfers between dwellings depended on the efficiency and sympathy of local bureaucrats.

Although mobility rates were high before 1914, most people moved very short distances, reflecting their dependence on friends and neighbours for help in times of crisis and local sources of information on suitable vacancies. Working-class families were reluctant to leave areas in which they were known and where they could obtain credit at local shops. Even if the household head obtained employment at a distance, there was no guarantee that the job would last. If the family moved nearer the new workplace they would have to begin again to establish their credentials among new neighbours and shopkeepers (Pember Reeves, 1914). We may also envisage information on housing being transmitted by neighbours, workmates and rent collectors. It would be worth investigating how often households moved between dwellings owned by the same landlord or managed by the same agent.

Patterns of residential mobility provide important evidence of community structure in nineteenth-century cities. Persistence rates indicate the proportion of households who remained at the same address, within the same city or - most usefully - within the same district, over a given period, usually between successive directories or censuses. In most cities fewer than 20 per cent stayed at the same address for as long as a decade, but 30-40 per cent remained in the same area. Movements between contiguous enumeration districts also indicate the existence of relatively closed migration systems within cities; and studies of individual mobility, based on diaries or biographies, tell the same story (Lawton and Pooley, 1975; Dennis, 1984). As Balfour (1891, p. 412) observed in south London:

> '"Four houses in four months", "Five houses in eighteen months", so run my notes. But these moves are seldom further than three streets away, and a year or two will very probably witness the return of the exiles to within a few doors of one of their many forsaken homes.'

Charles Booth, to whose mammoth survey Balfour was contributing, himself noted of East Enders that they 'often cling from generation to generation to one vicinity, almost as if the set of streets which lie there were an isolated country village' (quoted in Pritchard, 1976, p. 64).

Mobility patterns also provide the context for residential differentiation. We may doubt the value of segregation studies at the microscale, comparing the class or birthplace of neighbours who were unlikely to stay neighbours for long. Residential patterns assume more significance if mobility rates were low, or if those moving were replaced by households with identical characteristics. Hence the necessity of integrating studies of segregation, community and mobility.

NORTH AMERICAN COMPARISONS

American social historians have produced numerous studies of persistence, usually complementing analyses of occupational mobility: it is possible to measure the extent of social mobility only among people who remain in one place long enough to appear in successive censuses, directories or tax assessments. Average within-city persistence rates of 40-50 per cent per decade conceal a range from about 20 per cent in some southern and western cities in the 1850s to 70 per cent in late nineteenth-century Indianapolis (Barrows, 1981). Persistence was much higher among property-owners: in

PEOPLE AND HOUSING IN INDUSTRIAL SOCIETY

Newburyport, Mass., 80 per cent of those who owned property remained in the town between 1850 and 1860, compared to only 31 per cent of those who did not (Thernstrom, 1964), and in Hamilton, Ontario, 43 per cent of homeowners in 1861 had been resident in the city in 1851 compared to 12 per cent of renters (Katz, 1975). However, the direction of causality is open to debate: did householders become homeowners because they had decided to stay put, or did they remain at the same address because, having become homeowners, it was easier to improve their existing home than to move elsewhere? Did renters move frequently because it was cheap and easy to do so, or did people whose circumstances forced them to move frequently choose renting?

What were the consequences of high mobility rates? Radford (1981, p. 264) commented that in Buffalo, frequent moves 'did not undermine the strength of the Italian community, which was bound by emotional ties transcending place and distance'. Responding to both American and British persistence studies, Anderson (1982) questioned the links between mobility and social stability, noting that a sense of community may develop among a highly transient population, especially if it can be sustained by the continued presence of just a few key individuals. Certainly, the scale at which persistence occurs is critical.

Compared to Britain, homeownership was common in North America. The US census recorded owner-occupancy from 1890 onwards, when rates varied from 6 per cent of dwellings in New York City to over 40 per cent in cities bordering the Great Lakes (Barrows, 1983). In Canada, assessment records indicate rates of 30-35 per cent in late nineteenth-century Hamilton and Kingston, slightly lower in Toronto, and much lower in Francophone Montreal, where there was also a tradition of tenement living (Harris, 1986). In general, as in Britain, the larger the city, the lower the homeownership rate. But there were few differences between social classes and, despite the overall higher level of owner-occupation, the majority of households in all classes were still more likely to be private tenants. Yet, apart from work on Kingston (Harris, Levine and Osborne, 1981), there has been even less interest by historical geographers in the structure of renting in North America than in Britain.

Reviewing research on North American cities, Radford (1981, p. 260) noted that:

> The "social geography" of the nineteenth century US city is largely written in passing references to spatial or environmental changes by historians in the pursuit of other, more pressing, topics.

He also lamented the lack of an adequate theoretical framework, questioning - as I have done - whether industrial-

isation really prompted the disintegration of community, as
Warner (1968) implied. I have argued that while the socially
balanced community did disappear, if indeed it had ever
existed, a new form of working-class community life emerged
within the segregated environment of industrial urbanism.
Radford (1981, p. 281) was equally critical of models that
envisaged the city in transition between Sjoberg's and
Burgess' stereotypes, commenting that:

> although attractive as a pedagogical device for assessing
> the impact of industrialisation on the city, such a con-
> struct is seriously deficient as a research hypothesis.

One of Radford's promising alternatives - Walker's
marxist perspective on the city as the spatial expression of
successive stages of industrial accumulation - has attained
maturity in recent years (Walker, 1978; 1981), especially in
Harvey's attempts to link the economic structure of urbanisation to issues of experience and consciousness (Harvey,
1985a; 1985b). Politically neutral ecological theory, linked to a
benign faith in 'progress' or 'modernisation', seems less
appropriate to our own times than perspectives which situate
social order - of which urban structure is a component - in
the realities of political and economic change. Just as
historians constantly reinterpret and reinterrogate the past in
the light of their current beliefs and concerns, so historical
geographers cannot avoid using the present to understand the
past. The impact of privatisation on the distribution of wealth
and equality of access to services in the 1980s should prompt
research on the gradual erosion of private property rights
and the social consequences of municipalisation in the nineteenth century. The socio-geographic implications of industrial
restructuring were as critical then as they are today
(Stedman Jones, 1971; Green, 1985).

I have interpreted social patterns in the context of a
changing class structure and a particular system of housing
provision. Yet both are expressions of changes in ideology,
political philosophy and the needs of capital: what functions
were housing and social areas required to fulfil at different
stages in the evolution of industrial capitalism? These are
larger questions than I have attempted to review here, and
they indicate the impossibility of separating the social
geography of cities from economic and demographic issues
discussed elsewhere in this volume, or from broader themes,
such as secularisation or imperialism.

REFERENCES

Aalen, F.H.A. (1984) 'Approaches to the Working-Class
Housing Problem in Late Victorian Dublin: The Dublin

Artisans Dwellings Company and the Guinness (later Iveagh) Trust', Mannheimer Geographische Arbeiten, 17, 161-90

Anderson, M. (1982) 'Indicators of Population Change and Stability in Nineteenth-Century Cities: Some Sceptical Comments' in J.H. Johnson and C.G. Pooley (eds.), The Structure of Nineteenth Century Cities, Croom Helm, London, pp. 283-98

Arkell, G.E. (1891) 'Blocks of Model Dwellings' in C. Booth (ed.), Labour and Life of the People. Volume II, Williams and Norgate, London, pp. 236-62

Armstrong, W.A. (1966) 'Social Structure from the Early Census Returns', in E.A. Wrigley (ed.), An Introduction to English Historical Demography, Weidenfeld and Nicolson, London, pp. 209-37

Armstrong, W.A. (1972) 'The Use of Information about Occupation' in E.A. Wrigley (ed.), Nineteenth-Century Society, Cambridge University Press, Cambridge, pp. 191-310

Balfour, G. (1891) 'Battersea' in C. Booth (ed.), Labour and Life of the People. Volume II, Williams and Norgate, London, pp. 407-13

Barrows, R. (1981) 'Hurryin' Hoosiers and the American "Pattern": Geographic Mobility in Indianapolis and Urban North America', Social Science History, 5, 197-222

Barrows, R.G. (1983) 'Beyond the Tenement: Patterns of American Urban Housing 1870-1930', Journal of Urban History, 9, 395-420

Bassett, K., and Short, J. (1980) Housing and Residential Structure: Alternative Approaches, Routledge and Kegan Paul, London

Bennett, A. (1911a) The Card, Methuen, London

Bennett, A. (1911b) Hilda Lessways, Methuen, London

Beresford, M.W. (1974) 'The Making of a Townscape: Richard Paley in the East End of Leeds 1771-1803' in C.W. Chalklin and M.A. Havinden (eds.), Rural Change and Urban Growth, Longman, London, pp. 281-320

Billinge, M. (1982) 'Reconstructing Societies in the Past: The Collective Biography of Local Communities' in A.R.H. Baker and M. Billinge (eds.), Period and Place: Research Methods in Historical Geography, Cambridge University Press, Cambridge, pp. 19-32

Billinge, M. (1984) 'Hegemony, Class and Power in Late Georgian and Early Victorian England: Towards a Cultural Geography' in A.R.H. Baker and D. Gregory (eds.), Explorations in Historical Geography: Interpretative Essays, Cambridge University Press, Cambridge, pp. 28-67

Birch, E.L., and Gardner, D.S. (1981) 'The Seven-Percent Solution: A Review of Philanthropic Housing 1870-1910', Journal of Urban History, 7, 403-38

PEOPLE AND HOUSING IN INDUSTRIAL SOCIETY

Booth, C. (eds.) (1902) Life and Labour of the People of London, (17 volumes), Macmillan, London

Bramwell, W. (1984) 'Pubs and Localised Communities in Mid-Victorian Birmingham', Queen Mary College (University of London) Occasional Paper, Department of Geography, 22

Briggs, A. (1968) Victorian Cities, Penguin, Harmondsworth

Buckman, J. (1983) Immigrants and the Class Struggle: The Jewish Immigrant in Leeds 1880-1914, Manchester University Press, Manchester

Burgess, J.A., and Gold, J. (eds.) (1985) Geography, The Media and Popular Culture, Croom Helm, London

Burnett, J. (1974) (ed.), Useful Toil, Allen Lane, London

Burnett, J. (1983) (ed.), Destiny Obscure, Allen Lane, London

Cannadine, D. (1977) 'Victorian Cities: How Different?', Social History, 2, 457-82

Cannadine, D. (1980a) 'Urban Development in England and America in the Nineteenth Century: Some Comparisons and Contrasts', Economic History Review, 33, 309-25

Cannadine, D. (1980b) Lords and Landlords: The Aristocracy and the Towns, 1774-1967, Leicester University Press, Leicester

Cannadine, D. (1982) 'Residential Differentiation in Nineteenth-Century Towns: From Shapes on the Ground to Shapes in Society' in J.H. Johnson and C.G. Pooley (eds.), The Structure of Nineteenth Century Cities, Croom Helm, London, pp. 235-51

Carter, H., and Wheatley, S. (1980) 'Residential Segregation in Nineteenth-Century Cities', Area, 12, 57-62

Carter, H., and Wheatley, S. (1982) Merthyr Tydfil in 1851, University of Wales Press, Cardiff

Chapman, S.D. (ed.) (1971) The History of Working-Class Housing, David and Charles, Newton Abbot

Coleman, B.I. (ed.) (1973) The Idea of the City in Nineteenth-Century Britain, Routledge and Kegan Paul, London

Corfield, P.J. (1982) The Impact of English Towns 1700-1800, Oxford University Press, Oxford

Cosgrove, D. (1984) Social Formation and Symbolic Landscape, Croom Helm, London

Cowlard, K.A. (1979) 'The Identification of Social (Class) Areas and Their Place in Nineteenth-Century Urban Development', Transactions of the Institute of British Geographers N.S., 4, 239-57

Crossick, G. (ed.) (1977) The Lower Middle Class in Britain 1870-1914, Croom Helm, London

Crossick, G. (1978) An Artisan Elite in Victorian Society: Kentish London 1840-1880, Croom Helm, London

Crossick, G. (1983) 'Urban Society and the Petty Bourgeoisie in Nineteenth-Century Britain' in D. Fraser and A.

PEOPLE AND HOUSING IN INDUSTRIAL SOCIETY

Press, Cambridge
Sutcliffe (eds.), The Pursuit of Urban History, Edward
Arnold, London, pp. 307-26
Crossick, G., and Haupt, H-G. (1984) (eds.), Shopkeepers
and Master Artisans in Nineteenth Century Europe,
Methuen, London
Daunton, M.J. (1976) 'House-Ownership from Rate Books',
Urban History Yearbook, 21-7
Daunton, M.J. (1977) Coal Metropolis: Cardiff 1870-1914,
Leicester University Press, Leicester
Daunton, M.J. (1983) House and Home in the Victorian City:
Working-Class Housing 1850-1914, Edward Arnold,
London
Daunton, M.J. (ed.) (1984) Councillors and Tenants: Local
Authority Housing in English Cities 1919-1939, Leicester
University Press, Leicester
Dear, M., and Scott, A.J. (eds.) (1981) Urbanisation and
Urban Planning in Capitalist Society, Methuen, London
Dennis, R. (1977a) 'Distance and Social Interaction in a
Victorian City', Journal of Historical Geography, 3,
237-50
Dennis, R. (1977b) 'Intercensal Mobility in a Victorian City',
Transactions of the Institute of British Geographers
N.S., 2, 349-63
Dennis, R. (1984) English Industrial Cities of the Nineteenth
Century: A Social Geography, Cambridge University
Press, Cambridge
Dennis, R. (1986a) 'Housing, Class and Voting Behaviour in
West Riding Textile Towns: A Geographical Analysis' in
C. Withers (ed.), Geography of Population and Mobility
in Nineteenth-Century Britain, Historical Geography
Research Group (papers to Economic History Society
Conference, Cheltenham, April 1986)
Dennis, R. (1986b) '"Hard to Let" in Edwardian London',
paper to Urban History Group Conference, Cheltenham,
April 1986
Dennis, R., and Daniels, S.J. (1981) '"Community" and the
Social Geography of Victorian Cities', Urban History
Yearbook, 7-23
Driver, F. (1985) 'Power, Space, and the Body: A Critical
Assessment of Foucault's Discipline and Punishment',
Environment and Planning D: Society and Space, 3,
425-46
Dunford, M., and Perrons, D. (1983) The Arena of Capital,
Macmillan, London
Dyos, H.J. (1968) 'The Speculative Builders and Developers
of Victorian London', Victorian Studies, 11, 641-90
Engels, F. (1969) The Condition of the Working Class in
England, Panther, London (originally 1845)
Englander, D. (1983) Landlord and Tenant in Urban Britain
1838-1918, Oxford University Press, Oxford
Flinn, M.W. (ed.) (1965) Chadwick's Report on the Sanitary

Condition of the Labouring Population of Great Britain, Edinburgh University Press, Edinburgh

Foster, J. (1974) Class Struggle and the Industrial Revolution, Weidenfeld and Nicolson, London

Gadian, D.S. (1978) 'Class Consciousness in Oldham and Other North West Industrial Towns 1830-1850', Historical Journal, 21, 161-72

Gaskell, S.M. (1971) 'Yorkshire Estate Development and the Freehold Land Societies in the Nineteenth Century', Yorkshire Archaeological Journal, 43, 158-65

Glass, R. (1955) 'Urban Sociology in Great Britain', Current Sociology, 4(4)

Glen, R. (1984) Urban Workers in the Early Industrial Revolution, Croom Helm, London

Goheen, P. (1970) Victorian Toronto 1850-1900, University of Chicago, Chicago

Gordon, G. (1979) 'The Status Areas of Early to Mid-Victorian Edinburgh', Transactions of the Institute of British Geographers N.S., 4, 168-91

Gray, R. (1981) The Aristocracy of Labour in Nineteenth-Century Britain c. 1850-1914, Macmillan, London

Green, D.R. (1985) 'From Artisans to Paupers: The Manufacture of Poverty in Mid-Nineteenth Century London', unpublished Ph.D. thesis, University of Cambridge

Gregory, D.J. (1982) 'Action and Structure in Historical Geography' in A.R.H. Baker and M. Billinge (eds.), Period and Place: Research Methods in Historical Geography, Cambridge University Press, Cambridge, pp. 244-50

Harris, R. (1984a) 'Residential Segregation and Class Formation in the Capitalist City: A Review and Directions for Research', Progress in Human Geography, 8, 26-49

Harris, R. (1984b) 'Residential Segregation and Class Formation in Canadian Cities: A Critical Review', Canadian Geographer, 28, 186-96

Harris, R. (1986) 'Home Ownership and Class in Modern Canada', International Journal of Urban and Regional Research, 10, 67-86

Harris, R., Levine, G., and Osborne, B.S. (1981) 'Housing Tenure and Social Classes in Kingston Ontario 1881-1901', Journal of Historical Geography, 7, 271-89

Harvey, D. (1985a) Consciousness and the Urban Experience, Blackwell, Oxford

Harvey, D. (1985b) The Urbanisation of Capital, Blackwell, Oxford

Herbert, D.T., and Johnston, R.J. (1976) (eds.), Social Areas in Cities, (2 volumes), Wiley, London

Howarth, E.G., and Wilson, M. (1907) West Ham: A Study in Social and Industrial Problems, Dent, London

Jackson, J.T. (1981) 'Housing Areas in Mid-Victorian Wigan and St Helens', Transactions of the Institute of British

Geographers N.S., 9, 413-32
Jackson, J.T. (1982) 'Long-Distance Migrant Workers in Nineteenth-Century Britain: A Case Study of the St Helens' Glassmakers', Transactions of the Historic Society of Lancashire and Cheshire, 131, 113-37
Jasper, A.S. (1969) A Hoxton Childhood, Barrie and Rockliff, London
Joyce, P. (1980) Work, Society and Politics: The Culture of the Factory in Later Victorian England, Harvester, Brighton
Katz, M. (1975) The People of Hamilton, Canada West, Harvard University Press, Cambridge, Mass.
Katz, M., Doucet, M., and Stern, M. (1982) The Social Organisation of Early Industrial Capitalism, Harvard University Press, Cambridge, Mass.
Kearns, G. (1985) Urban Epidemics and Historical Geography: Cholera in London 1848-9, Geo Books, Norwich
Keating, P. (ed.) (1976) Into Unknown England 1866-1913: Selections from the Social Explorers, Fontana, London
King, A.D. (ed.) (1980) Buildings and Society. Essays on the Social Development of the Built Environment, Routledge and Kegan Paul, London
King, A.D. (1984) 'The Social Production of Building Form: Theory and Research', Environment and Planning D: Society and Space, 2, 429-46
Kirk, C.T., and Kirk, G.W. (1981) 'The Impact of the City on Home Ownership', Journal of Urban History, 7, 471-98
Laslett, P. (1983) The World We Have Lost: Further Explored, Methuen, London
Lawton, R. (ed.) (1978) The Census and Social Structure: An Interpretative Guide to 19th Century Censuses for England and Wales, Cass, London
Lawton, R., and Pooley, C.G. (1975) 'David Brindley's Liverpool', Transactions of the Historic Society of Lancashire and Cheshire, 125, 149-68
Lawton, R., and Pooley, C.G. (1976) The Social Geography of Merseyside in the Nineteenth Century, Report to SSRC, Liverpool
Laxton, P. (1981) 'Liverpool in 1801: A Manuscript Return for the First National Census of Population', Transactions of the Historic Society of Lancashire and Cheshire, 130, 73-113
Lees, A. (1983) 'Perceptions of Cities in Britain and Germany 1820-1914' in D. Fraser and A. Sutcliffe (eds.), The Pursuit of Urban History, Edward Arnold, London, pp. 151-65
Lees, A. (1985) Cities Perceived: Urban Society in European and American Thought 1820-1940, Manchester University Press, Manchester
Lees, L.H. (1979) Exiles of Erin, Manchester University

Press, Manchester

Lewis, C.R. (1980) 'The Irish in Cardiff in the Mid-Nineteenth Century', Cambria, 7, 13-41

Lewis, C.R. (1985) 'Housing Areas in the Industrial Town: A Case Study of Newport, Gwent 1850-1880', The National Library of Wales Journal, 24, 118-46

Ley, D., and Samuels, M. (eds.) (1978) Humanistic Geography, Maaroufa Press, Chicago

McLeod, H. (1974) Class and Religion in the Late Victorian City, Croom Helm, London

McLeod, H. (1977) 'White Collar Values and the Role of Religion' in G. Crossick (ed.), The Lower Middle Class in Britain 1870-1914, Croom Helm, London, pp. 61-88

Marcus, S. (1974) Engels, Manchester and the Working Class, Weidenfeld and Nicolson, London

Meacham, S. (1977) A Life Apart: The English Working Class 1890-1914, Thames and Hudson, London

Melling, J. (1983) Rent Strikes: Peoples' Struggle for Housing in West Scotland 1890-1916, Polygon Books, Edinburgh

Morris, R.J. (1983a) 'Property Titles and the Use of British Urban Poll Books for Social Analysis', Urban History Yearbook, 29-38

Morris, R.J. (1983b) 'The Middle Class and British Towns and Cities of the Industrial Revolution 1780-1870' in D. Fraser and A. Sutcliffe (eds.), The Pursuit of Urban History, Edward Arnold, London, pp. 286-305

Muthesius, S. (1982) The English Terraced House, Yale University Press, New Haven, Conn.

Neale, R.S. (1981) Class in English History 1680-1850, Blackwell, Oxford

Offer, A. (1981) Property and Politics 1870-1914, Cambridge University Press, Cambridge

Olsen, D.J. (1982) Town Planning in London: The Eighteenth and Nineteenth Centuries, Yale University Press, New Haven, Conn.

Peach, C. (ed.) (1975) Urban Social Segregation, Longman, London

Pember Reeves, M.S. (1914) Round About a Pound a Week, Bell, London

Perry, P.J. (1969) 'Working-Class Isolation and Mobility in Rural Dorset 1837-1936', Transactions of the Institute of British Geographers, 46, 121-41

Pocock, D.C.D. (ed.) (1981) Humanistic Geography and Literature, Croom Helm, London

Pooley, C.G. (1977) 'The Residential Segregation of Migrant Communities in Mid-Victorian Liverpool', Transactions of the Institute of British Geographers N.S., 2, 364-82

Pooley, C.G. (1979) 'Residential Mobility in the Victorian City', Transactions of the Institute of British Geographers N.S., 4, 258-77

Pooley, C.G. (1982) 'Choice and Constraint in the Nineteenth-

Century City: A Basis for Residential Differentiation' in J.H. Johnson and C.G. Pooley (eds.), The Structure of Nineteenth Century Cities, Croom Helm, London, pp. 199-233

Pooley, C.G. (1983) 'Welsh Migration to England in the Mid-Nineteenth Century', Journal of Historical Geography, 9, 287-305

Pooley, C.G. (1984) 'Residential Differentiation in Victorian Cities: A Reassessment', Transactions of the Institute of British Geographers N.S., 9, 131-44

Pooley, C.G. (1985) 'Housing for the Poorest Poor: Slum-Clearance and Rehousing in Liverpool 1890-1918', Journal of Historical Geography, 11, 70-88

Pooley, C.G., and Irish, S. (1984) The Development of Corporation Housing in Liverpool 1869-1945, Centre for North West Regional Studies, University of Lancaster

Pred, A. (1984) 'Structuration, Biography Formation, and Knowledge: Observations on Port Growth During the Late Mercantile Period', Environment and Planning D: Society and Space, 2, 251-75

Pritchard, R.M. (1976) Housing and the Spatial Structure of the City, Cambridge University Press, Cambridge

Radford, J. (1981) 'The Social Geography of the Nineteenth-Century U.S. City' in D.T. Herbert and R.J. Johnston (eds.), Geography and the Urban Environment: Progress in Research and Applications, Volume 4, Wiley, Chichester, pp. 257-94

Rex, J., and Moore, R. (1967) Race, Community and Conflict, Oxford University Press, Oxford

Rimmer, W.G. (1963) 'Alfred Place Terminating Building Society 1825-1843', Publications of the Thoresby Society, 46, 303-30

Roberts, E. (1984) A Woman's Place: An Oral History of Working-Class Women, 1890-1940, Blackwell, Oxford

Roberts, R. (1971) The Classic Slum, Manchester University Press, Manchester

Roberts, R. (1976) A Ragged Schooling, Manchester University Press, Manchester

Robinson, V. (1980) 'Lieberson's Isolation Index: A Case Study Evaluation', Area, 12, 307-12

Robson, B.T. (1969) Urban Analysis: A Study of City Structure, Cambridge University Press, Cambridge

Rowntree, B.S. (1901) Poverty: A Study of Town Life, Macmillan, London

Saunders, P. (1981) Social Theory and the Urban Question, Hutchinson, London

Shaw, G. (1982a) 'The Role of Retailing in the Urban Economy' in J.H. Johnson and C.G. Pooley (eds.), The Structure of Nineteenth Century Cities, Croom Helm, London, pp. 171-94

Shaw, G. (1982b) British Directories as Sources in Historical

Geography, Geo Books, Norwich

Shaw, G. (1984) 'Directories as Sources in Urban History: A Review of British and Canadian Material', Urban History Yearbook, 36-44

Shaw, G. (1985) 'Changes in Consumer Demand and Food Supply in Nineteenth-Century British Cities', Journal of Historical Geography, 11, 280-96

Shaw, G., and Wild, M.T. (1979) 'Retail Patterns in the Victorian City', Transactions of the Institute of British Geographers N.S., 4, 278-91

Shaw, M. (1977) 'The Ecology of Social Change: Wolverhampton 1851-71', Transactions of the Institute of British Geographers N.S., 2, 332-48

Shaw, M. (1979) 'Reconciling Social and Physical Space: Wolverhampton 1871', Transactions of the Institute of British Geographers N.S., 4, 192-213

Simpson, M.A., and Lloyd, T.H. (1977) (eds.), Middle-Class Housing in Britain, David and Charles, Newton Abbot

Smith, D. (1982) Conflict and Compromise: Class Formation in English Society 1830-1914, Routledge and Kegan Paul, London

Springett, J. (1979) 'The Mechanics of Urban Land Development in Huddersfield 1770-1911', unpublished Ph.D. thesis, University of Leeds

Stedman Jones, G. (1971) Outcast London: A Study in the Relationship Between Classes in Victorian Society, Oxford University Press, Oxford

Sutcliffe, A. (1974) 'Introduction' in A. Sutcliffe (ed.), Multi-Storey Living, Croom Helm, London, pp. 1-18

Swenarton, M., and Taylor, S. (1985) 'The Scale and Nature of Owner-Occupation in Britain Between the Wars', Economic History Review, 38, 373-92

Swift, R., and Gilley, S. (eds.) (1985) The Irish in the Victorian City, Croom Helm, London

Taine, H. (1957) Notes on England, translated with introduction by E. Hyams, Thames and Hudson, London

Tarn, J.N. (1973) Five Per Cent Philanthropy, Cambridge University Press, Cambridge

Thernstrom, S.A. (1964) Poverty and Progress: Social Mobility in a Nineteenth-Century City, Harvard University Press, Cambridge, Mass.

Thompson, P. (1973) 'Voices from Within' in H.J. Dyos and M. Wolff (eds.), The Victorian City: Images and Realities, Routledge and Kegan Paul, London, pp. 59-80

Timms, D.W.G. (1971) The Urban Mosaic: Towards a Theory of Residential Differentiation, Cambridge University Press, Cambridge

Tocqueville, A. de (1958) Journeys to England and Ireland, (ed. J.P. Mayer), Faber, London

Tunbridge, J.E. (1977) 'Spatial Change in High-Class Residence: The Case of Bristol', Area, 9, 171-4

Vincent, J.R. (1967) Pollbooks: How Victorians Voted, Cambridge University Press, Cambridge

Walker, R. (1978) 'The Transformation of Urban Structure in the Nineteenth Century and the Beginnings of Suburbanisation' in K.R. Cox (ed.), Urbanization and Conflict in Market Societies, Methuen, London

Walker, R. (1981) 'A Theory of Suburbanization: Capitalism and the Construction of Urban Space in the United States' in M. Dear and A.J. Scott (eds.), Urbanization and Urban Planning in Capitalist Society, Methuen, London, pp. 383-429

Ward, D. (1975) 'Victorian Cities: How Modern?', Journal of Historical Geography, 1, 135-51

Ward, D. (1978) 'The Early Victorian City in England and America' in J.R. Gibson (ed.), European Settlement and Development in North America, Toronto University Press, Toronto, pp. 170-89

Ward, D. (1980) 'Environs and Neighbours in the "Two Nations": Residential Differentiation in Mid-Nineteenth Century Leeds', Journal of Historical Geography, 6, 133-62

Ward, D. (1984) 'The Progressives and the Urban Question: British and American Responses to the Inner City Slums 1880-1920', Transactions of the Institute of British Geographers N.S., 9., 299-314

Ward, D., and Radford, J.P. (1983) North American Cities in the Victorian Age, Geo Books, Norwich

Warner, S.B. (1968) The Private City, University of Pennsylvania Press, Philadelphia

Warnes, A.M. (1973) 'Residential Patterns in an Emerging Industrial Town' in B.D. Clark and M.B. Gleave (eds.), Social Patterns in Cities, Institute of British Geographers, London, pp. 169-89

Williams, B. (1985) The Making of Manchester Jewry 1740-1875, Manchester University Press, Manchester

Wohl, A.S. (1977) The Eternal Slum: Housing and Social Policy in Victorian London, Edward Arnold, London

Woods, R., and Woodward, J. (eds.) (1984) Urban Disease and Mortality in Nineteenth-Century England, Batsford, London

Wrigley, E.A. (ed.) (1972) Nineteenth-Century Society: Essays in the Use of Quantitative Methods for the Study of Social Data, Cambridge University Press, Cambridge

Yelling, J. (1981) 'The Selection of Sites for Slum Clearance in London 1875-88', Journal of Historical Geography, 7, 155-65

Yelling, J. (1982) 'LCC Slum Clearance Policies 1889-1907', Transactions of the Institute of British Geographers N.S., 7, 292-303

Chapter Eight

HISTORICAL DEMOGRAPHY

P.E. Ogden

It is always tempting for the writer of a review to assert that the field has shown signs of great vitality, thus rendering the task more worthwhile and the essay of greater interest to the reader. Yet, in the case of historical demography, no invention is required, for the last thirty years have brought about a transformation of the field. Its exponents have been as imaginative in their uses of technique and source as they have been aware of the breadth of interpretation required for their findings. The field has emerged as lively and productive, with the 1970s and 1980s in particular witnessing a rapid increase in research and publication. Despite the occasionally disparaging remarks of more traditionally-minded historians, historical demography has had little difficulty in establishing itself as a distinct field. To its preoccupations with the measurement of fertility and mortality have been added the study of marriage behaviour, of family and household and of migration. To the technical virtuosity of the modern demographer has been added the insight and imagination of the historian. The success of historical demography has been to show how the reconstruction of demographic behaviour of individuals and communities may provide a key to the understanding of past societies more generally.

Developments in historical demography are significant for the geographer too. There is a growing awareness of the relevance of geographical questions and methodologies to the study of population geographies in the past, as historical demography forces itself from the constraints of analysis at either the national or parochial scale which have thus far been its principal concern. Geographers have, indeed, made distinguished contributions to historical population studies and one of the explicit themes of this essay is to illustrate the extent to which we have advanced in the understanding of pressures and patterns of demographic change at a variety of different geographical scales and historical periods. For it has proved a very reasonable supposition that men and women in different continents and countries, and indeed in different regions and parishes, showed marked dissimilarities in their attitude towards birth and child-rearing, in their proneness to disease and death or in their likelihood to move from place to place. The geographical scope of historical demography has widened markedly and, while many of the examples cited in this chapter will be from Britain, a conscious attempt to use material from elsewhere will be made. It is not the purpose, however, to describe British demographic trends in great detail, since this has been done elsewhere, for example by Smith (1978), by the recent publications of Wrigley and Schofield, fully referenced below, or by Tranter (1985).

HISTORY, DEMOGRAPHY AND GEOGRAPHY

The emergence of historical demography as a distinct branch of history or, as some would argue, a discipline in its own right, dates from the years after the Second World War. The invention of new techniques allowed a new mastery over parish records of baptisms, burials and marriages and there-by the application of precise demographic measures to the past. Yet, a concern with past population trends is deep-seated: as Jacques and Michel Dupâquier (1985) have recently reminded us, demography itself has a distinguished history. The nineteenth century saw a great growth in recording by census and vital statistics and in a concern for longer-term population changes; but the work of John Graunt, William Petty, Edmund Halley, Gregory King and others in England in the seventeenth and eighteenth centuries, of Johann Peter Süssmilch in Germany and of many others ensured that the debate on demographic issues was well in train before Malthus's treatise was published in the late eighteenth century. The ideas of Malthus, as we shall see below, have indeed proved enduring and the questions he posed still underpin much of the discussion of the causes of population growth. During the nineteenth and early twentieth centuries, the periodic publication of censuses in the European countries, the concern about the relationship between size of population and national power, and the onset of declining fertility and population growth provoked a large literature.

The sense in which historical demography is a child of the post-war years relates to two factors: changes in tech-nique and methodology, and changes in general approach to the interpretation of findings. First, the move towards longi-tudinal analysis was made possible by the invention of family reconstitution. Fleury and Henry (1956; 1965) set out a clear methodology to turn the crude data on baptisms, burials and marriages into demographic rates:

> following a birth, the subsequent marriage, progeny, and burial of an individual could be extracted from the

registers, thus permitting the precise establishment of
that individual's demographic behaviour - his or her age
at marriage and perhaps at remarriages, age at birth of
children and therefore the intervals between these
births, age at birth of the last child, and age at death
and, therefore the length of life itself (Flinn, 1981, p.
2).

Thus, Dupaquier (1984, p. 20) has suggested that historical
demography was able to establish itself as a new field because
it applied new techniques to sources that were established
originally not for 'scientific' purposes but for a variety of
political, military, religious or administrative reasons. Many of
the early studies were concerned primarily with the technique
itself and with descriptive statistical results. Gradually,
though, historians began to see the significance of the
findings for a wider interpretation of past economies and
societies. By 1981, Flinn was able to trace the diffusion of
family reconstitution from France to many other countries. In
Britain, Wrigley's use of the technique for the parish of
Colyton in Devon (Wrigley, 1966a,b; 1968) gave new focus
and impetus to the study of past populations. He, with his
colleagues Laslett and Schofield, set up the Cambridge Group
for the History of Population and Social Structure from which
has flowed scholarly work distinguished both for its technical
sophistication and the breadth of issues raised by the
results.

The Cambridge Group was particularly influential in
another technical advance. This followed from the limitations
of family reconstitution: first, that the amount of work involved even for a small community is very large and,
secondly, because it is difficult to keep track of migrants,
some have questioned the representativeness of studies based
on reconstitution for the population as a whole. The technique
of back projection has been designed to overcome the problems posed by the absence of censuses before the nineteenth
century. This involves moving backwards in time from a date
at which the size and age structure of the population is
known, and yields estimates of net migration as well as providing quinquennial 'censuses' (Wrigley, 1981b, p. 214). Back
projection is particularly useful for regional and national
analyses, as the major volume on England by Wrigley and
Schofield (1981) showed in providing a convincing outline of
the history of fertility and mortality. Reconstitution and back
projection complement each other, the former providing details
of marital fertility and nuptiality which help to explain the
general trends identified by the latter. The ramifications of
these techniques are discussed in more detail later. Added to
the mastery of parish registers came an increased awareness
of the potential of the census for the nineteenth century
onwards, illustrated for Britain in Wrigley (1972).

In addition to these technical advances, and partly as a consequence of them, we may also note a change in scholarly approach. Historical demography has been created partly by applying the methods of the demographer to the past and, more especially, by historians looking at their sources and periods with fresh vision. Some, for example those investigating the history of the family (Anderson, 1980, p. 27), have questioned the reliability on statistical data at the expense of other sources. Yet, the increasing acceptance of the centrality of demographic factors in social history reflected the well-established approach of the 'Annales' school (Stoianovich, 1976). This influential group of French historians sought an interdisciplinary 'total' history which would 'explicitly employ theoretical concepts, imaginative interpretations and interdisciplinary approaches' (Baker, 1984, p. 2). The explicit inclusion of demographic issues was matched by the ready assimilation of varied and difficult sources and large amounts of quantitative data. The 'Annales' approach allowed the sort of all-embracing explanations and the search for long swings in social and economic development which historical demographers increasingly sought. Thus, Willigan and Lynch (1982) conclude their text on historical demography with a consideration of the 'Annales' approach, using as an example Le Roy Ladurie's (1978) study of Montaillou, 'a reconstruction of the "total history" of the demographic, biological, ecological and socio-economic levels of French peasant life ...' They further suggest that 'individual or community-level data permit the reconstruction of the concreteness of daily life and human interaction shaped by demographic events, belief systems, or form of local distribution and cultivation ...' (Willigan and Lynch, 1982, p. 440). Indeed, a marked feature of historical demography – or perhaps we should employ the term demographic history – is an expansion of the scope of interest well beyond the processes of fertility and mortality.

A central question concerns what Wrigley (1981a, p. 216) has referred to as the 'logical status of population history', that is whether population change is relegated to the role of a dependent variable or whether population trends themselves may be considered to influence wider social and economic change. The Annalistes would certainly argue for demography as an integral, defining feature of different systems of material life and would attach less importance to the primacy of the economic as suggested by their Marxist critics. Thus, it may be argued that while the discovery of sources and of techniques for their analysis is crucial, equally important is 'a new appreciation of the significance of the interplay of the forces which govern the population characteristics of a community. Just as the rise of economic history as a subject for investigation testifies to the recognition of the central importance of production to a society, so the parallel development of population history reflects a similar recognition of the significance of reproduction' (Wrigley, 1981a, p. 221). An appreciation of the reciprocal relationship between production and reproduction has been made possible by recent research on marriage and fertility and has been reflected in the search for general theoretical statements.

The development of historical demography has not, of course, gone unchallenged. Whilst the voices raised against the computerisation and quantification have been partly stilled, Cobb's (1976, p. 8) eloquent plea that 'the dark mechanised forces of the Social Sciences' should not rob history of its attention to individuals did not fall entirely on stony ground. Many authors (see, for example, Anderson, 1980, p. 30) have tackled the perennial problem of the significance of statistical averages which may gloss over crucial differences of geography, of social class or occupational group and others have challenged the whole approach, to family history for example, of simple measurement of size and structure. Hill's (1978, p. 452) view that 'wherever parish records can be checked against other sources, they turn out to be hopelessly inaccurate' has not received wide support, being dismissed by Palliser (1982, p. 341) for example as venturing 'beyond proper scholarly caution into statistical Luddism'. On the wider methodological front, some English historians have reacted against the volume of work in population history. The implicit borrowing from the Annales school has also been attacked. For Stedman Jones (1972, p. 110) 'even some of "the new ways of history" seem already to have been warped by the English historical climate. The influence of the Annales has finally been transmitted to England through the medium of the Cambridge school of historical demography - in the form of clumsy parody'. His view may well have been moderated by the extraordinary productivity of the small group of scholars gathered around the Cambridge group, although his point that the Annales approach merely forms the basis for another specialisation in English history rather than a total historical interpretation remains valid. Much controversy was generated over works such as Laslett (1965) or Laslett and Wall (1972), but gradually the recognition awarded to the major works by Wrigley and Schofield (1981) or Wall et al. (1983) and to the disciples of the Cambridge Group, for example, Smith (1985), Snell (1985), Macfarlane (1986), Levine (1977), Wrightson and Levine (1979) or Kussmaul (1981) has shown that historical demography has rapidly come of age.

It is an interesting aspect of the process of coming of age that geographical perspectives have recently come to the forefront of concern amongst historical demographers. A concern with, on the one hand, parish-level studies and, on the other, national totals has left a gap at the regional level.

HISTORICAL DEMOGRAPHY

Thus, a conference in Oxford in 1985 (Ogden, 1986) specifically addressed this problem which poses difficulties both for
the successful use of sources and for the appropriate adaptation of theory. Two aspects of the links between historical
demography and geography are worth comment. First,
Perrenoud (1985) points out that some 20% of all publications
in the International Bibliography of Historical Demography
were concerned with spatial aspects. Principal amongst these
was the study of migration, a point also made by the present
author (Ogden, 1984, p. 63) in an analysis of the Journal of
Historical Geography for the period 1975-1982. But geographers have also contributed to historical demography directly,
without too fussy a regard for disciplinary boundaries. In
Britain, several members of the Cambridge Group have a
geographical training; in France, the traditional closeness of
teaching and research in history and geography has influenced the method and content of many major works.
Secondly, historical demographers aided or not by a specifically geographical training, have seen the importance of
recognising geographical variability in demographic phenomena.
    The great expansion of interest in historical demography
has become very evident in the quantity of publication both
within, and increasingly beyond, the west European
countries. The two decades following the publication of, for
example, Glass and Eversley's (1965) collection of essays have
seen a number of introductory texts (Wrigley, 1966; 1969;
Hollingsworth, 1969; Guillaume and Poussou, 1970; Marcilio
and Charbonneau, 1979; Willigan and Lynch, 1982; or
Dupâquier, 1984); a new journal, the Annales de Demographie
Historique; and a mass of material in books and scattered
historical periodicals. Perrenoud (1985) noted that historical
demography occupies an important place in demography generally: some 13% of titles listed in Population Index between
1977 and 1984 were in historical subjects. His survey of the
International Bibliography of Historical Demography for the
same period records that most titles concerned periods from
the sixteenth century onwards and that the regional dominance of Europe is still overwhelming. Over 80% of studies
were on the developed world generally, although the most
recent years saw an acceleration of research on Africa, Asia,
Central America and the Caribbean.

THE SEARCH FOR THE GRAND DESIGN

As the empirical results of detailed research in historical
demography have appeared, it has become both desirable and
necessary to see them in broader contexts. There are many
themes which historical demography has generated, but we
may take four as examples, each of which is supported by

vigorous debate and publication both in Britain and in many other countries. All of these aspects are of importance to our understanding of regional population dynamics in the past, and several of them have implications for contemporary demography.

Malthus, Fertility and Marriage

Whatever the subsequent disputes over Malthus's stated or presumed beliefs, he posed key questions about the relationship in traditional societies between population and resources, between man and environment. In particular, much debate has focused on the operation at different periods and in different places of the 'positive' checks to population growth, principally disease and famine, and the 'preventive' checks of later marriage and therefore lower fertility. Recent work on England has indicated that understanding the links between demography and economy not only helps us to understand short-term population fluctuations but may also provide the key to understanding 'one of the most fundamental of all changes in the history of society' (Wrigley and Schofield, 1981, p. 458), that is the industrial revolution of the late eighteenth and early nineteenth centuries. Their work spanning the period 1541-1871 allows an analysis of demographic trends before and after industrial transformation and their findings are of fundamental significance.

Four recent works have, in addition to Wrigley and Schofield, brought Malthus's ideas under fresh and dispassionate scrutiny. James's (1979) biography gave us much of interest in setting the man's views against his background and times. Petersen (1979), on the other hand, provided a punchy review of Malthus's main ideas and separated very usefully what he actually said from what people think, or hope, he said. The scope of the Malthusian legacy, however, was shown in a conference held in Paris in 1980 whose abbreviated proceedings were published by Dupâquier et al. (1983), selected from 164 papers presented at the meeting. Malthus's direct, and indirect, legacy was shown to be immense, for his attempt at a general theory of population change has been taken up in a host of ways, in economics, sociology, history, politics, biology and even theology. Some of these aspects were further reviewed at a conference in Cambridge in 1984 (Coleman and Schofield, 1986).

The real contribution of Wrigley and Schofield's (1981) masterly work is that it has led to a reassessment of the relationship between fertility and mortality in English population growth. Their book was a technical tour de force, the details of which are discussed below, and it succeeded in altering the whole basis on which the debate on population history is now conducted. It was received with great acclaim, although the authors admit that it raised as many questions as it answered. The main book has been supplemented by many contributions by the authors elsewhere, for example, Wrigley and Schofield (1983), Wrigley (1981b; 1983a,b; 1985a), Schofield (1983, 1985a), the 1985 essays being part of the special issue of the <u>Journal of Interdisciplinary History</u> (Spring 1985); and the findings have already begun to be disputed and reinterpreted (Lindert, 1983; Weir, 1984; Goldstone, 1986).

One of the key findings of the major work has been to elucidate the workings of Malthus's 'preventive' check. Schofield (1985a, p. 573) succinctly summarised the Malthusian argument thus:

> On the one hand, where a preventive check prevailed, nuptiality and fertility would vary, keeping the tension between population and resources within reasonable bounds while living standards oscillated at a relatively high level. On the other hand, societies in which marriage was unresponsive to economic circumstances and fertility was constant would experience low standards of living and, at the limit, be wracked by spasms of high mortality.

Wrigley and Schofield's work has shown that mortality was not the crucial determinant, an idea held with fervour by some (for example, McKeown and Brown, 1955; McKeown, 1976; and see Schofield's refutation, 1983) and, in particular, their work has drawn attention away from the long-standing speculation about the causes of mortality decline in the eighteenth century. Rather, Wrigley and Schofield focus on fertility change, and the primacy of nuptiality in determining that change. Figure 8.1 indicates the sort of relationships that may exist in the Malthusian scheme, with current emphasis being on the preventive check. As Wrigley (1983a, p. 149) observed:

> marriage was the hinge on which the demographic system turned, and, given the crucial importance of the tension between production and reproduction which affected all pre-industrial societies, its significance was far wider than the purely demographic.

Thus, in contrast to the mortality-dominated high-pressure equilibrium sometimes seen as the pre-industrial norm, England experienced a fertility-dominated low-pressure system. For the debate on population growth in the later eighteenth century, their findings have proved momentous; for this they ascribe not to a decline in mortality, but to a rise in fertility, itself explained by a quite substantial reduction in the age of women at first marriage. The prime determinant of fertility changes was marriage itself rather

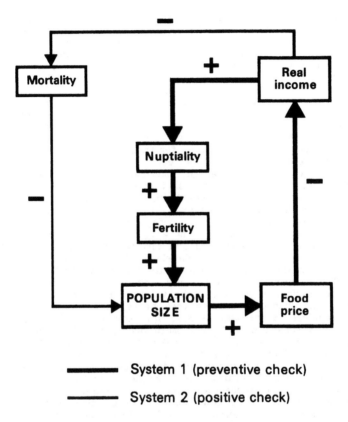

Figure 8.1: The Malthusian positive and preventive checks (Source: Wrigley and Schofield, 1981, 458).

than birth-control within marriage. This is not to say that mortality and mortality crises were not significant, but they are less significant than previously assumed. Wrigley (1983b, p. 122) considers, however, that Malthus identified correctly the major elements in the system; his judgement was 'remarkably sound, especially as he expressed it in the later editions of the Essay. As a historian he emerges largely unscathed from the tests made possible by the subsequent accumulation of greater knowledge of early modern England ...' although, of course, he was less good at foreseeing the effects of changes which were already in train in his day.

Although Weir (1985, p. 343) has noted that the 'logical flaws in Wrigley's evidence for a strong preventive check are a topic of current debate', there is a powerful argument for much more research on the history of marriage patterns (Dupâquier et al., 1981), particularly on whether and when men and women married and how far this was a function of an economic calculus. We need to know much more about how different were social groups, regions, age-groups and so on. Macfarlane (1986) has both summarised the current state of knowledge on England and illustrated the ways in which recent findings impose new interpretations of the English experience. Of course, research in marriage behaviour was already well-established before Wrigley and Schofield and indeed was part of the stimulus for their work.

An excellent example of an early, influential attempt at a general descriptive model of marriage behaviour, with a strong geographical flavour, was that of Hajnal (1953; 1965). In the latter paper he argued that the European pattern, west of a line from Leningrad to Trieste had long been characterised by high age at marriage and a high proportion of people who never married at all. This had a fundamental effect on household formation and fertility, many ramifications of which were persuasively developed by Chambers (1972), the uniqueness of the European pattern being implicitly linked to its exceptional economic and social development. Subsequent work has reinforced the validity of Hajnal's principal ideas: the Princeton studies (see below) have confirmed its geographical extent and others, for example, Smith (1979; 1981; 1983) have delved much more deeply into its historical origins, suggesting in particular that the European pattern is much older than first assumed. No-one maintains, of course, that Hajnal's pattern implies strict homogeneity through what was a very varied cultural realm: Wrigley (1981b) in a magisterial treatment of these issues has set the English experience against that of France and Sweden to show that there were significantly variant forms within the European pattern. For example, Kussmaul's (1981) treatment of servants in England - ploughmen, carters, dairymaids, apprentices and so forth, hired into the families of their employers, showed that they 'constituted around 60% of the population, aged fifteen to twenty-four' (p. 3) and this clearly helps explain the pattern of late and non-universal marriage: marriage could be late and, when service disappeared, marriage age dropped.

Although much recent emphasis has been given to studies of nuptiality and fertility, work on mortality crises and long-term mortality trends in pre-industrial communities has, of course, continued. Woods (1982, pp. 87-102) provides a neat summary and we may also refer to the recent work of Slack (1985) on the impact of plague in Tudor and Stuart England or some stimulating work on hunger in history from Rotberg and Rabb (1985) or, as a European example, Perrenoud (1979; 1980; 1981). Beier and Finlay (1986) have brought together some useful contributions on mortality in London between 1500 and 1700.

The Demographic Transition

Changing assumptions about the functioning of the pre-industrial demographic system and of the determinants of change in England in the later eighteenth century thus have profound implications for the notion of the demographic transition. This theory seeks to describe the way in which supposedly high, uncontrolled levels of mortality and fertility in traditional societies gave way to low, controlled levels in industrial societies, which have passed through the process most frequently, if imprecisely, labelled 'modernisation'. Rapid population growth was said to have occurred in the early stages of transition when mortality fell rapidly, but fertility remained high and was reduced only slowly. There is no doubt that a transition took place, that it was revolutionary in terms of the population increase it generated, largely irreversible and that it was affecting most parts of the presently developed world by the later decades of the nineteenth century. The theory of demographic transition was first expounded by Thompson (1929) and developed by Landry (1934; 1945) and in its fullest form by Notestein (1945; 1948; 1950; 1953). Whilst the notion itself remains an attractive one, the theory has not emerged unscathed, especially for England. It has remained a topic of debate, however, not only because it attempts to account for past experience but because it may or may not aid an understanding of contemporary processes in the third world.

Objections to demographic transition theory have come from many quarters, attacking its vision of pre-transitional demography, its concept of 'modernisation', the precise nature of the stages of transition it suggests and the way in which it tends to overemphasise the dependent nature of population structure. These objections have been usefully summarised by Woods (1982, pp. 158-73), drawing on some of the ideas raised by, for example, Coale (1973) and Wrigley (1972b) and further developed by Caldwell (1981; 1982a,b). Some new perspectives have been provided by Seccombe (1983); and the descriptive usefulness of the theory confirmed by Noin (1983). Schofield (1985a, p. 577) is particularly scathing in his attack on transition theory:

> In place of an understanding of historical change, it offers assertions about population dynamics based on schematic, and wholly unhistorical, assumptions about demographic behaviour. Careless of context, it lumps together all pre-transitional societies, past and present, disregarding the mediating influence of specific charac-

teristics of economic and family structure, institutions and value systems.

In other words, it was not one of the inspirations for Wrigley and Schofield (1981), because it significantly under-rated the variability and importance of nuptiality and fertility in pre-industrial societies, with mortality the only dynamic variable, leaving 'the positive check as the only mechanism by which accommodation might be reached between a population and its limited resources' (Schofield, 1985a, p. 577). The importance of rising fertility in explaining population growth in late eighteenth and early nineteenth-century England has, for that country at least, completely overturned the theory of demographic transition, although for other European countries mortality may well have played the critical role in population increase.

Significant progress has been made over the last twenty years in determining the rate and pace of declining fertility in the later stages of the demographic transition. The series of volumes from Princeton University on the European countries has applied standard demographic indices at the regional level. Under Coale's general guidance (e.g., 1973), volumes have appeared on Germany (Knodel, 1974), Belgium (Lesthaeghe, 1977), Italy (Livi-Bacci, 1977), Russia (Coale, Anderson and Härm, 1979), Britain (Teitelbaum, 1984) and the rather different case of France (van de Walle, 1974). Coale's view of fertility decline suggested (see Woods, 1982, pp. 102-30) three general conditions preceding a decline in marital fertility: that the decision-making on fertility should be a conscious choice; that reduced fertility should be seen as a social and economic advantage; and that the technical ability to reduce fertility must be present. Thus, interpreted at its simplest, Coale's view would suggest that only a population exhibiting these three characteristics will possess all the preconditions for a general reduction in fertility.

Some further examples of the regional implications of the Princeton and other works are given below, but Knodel and van de Walle (1979) have summarised the major findings. They make the important point that fertility decline in Europe took place under remarkably diverse socio-economic and demographic conditions and that there is no simple correlation between development and population transition:

> The striking factor that the countries of Europe had in common when fertility declined was time itself ... with the exception of the forerunner, France, and a few stragglers, such as Ireland and Albania, the dates of the decline were remarkably concentrated. The momentous revolution of family limitation began in two-thirds of the province-sized administrative area of Europe during a thirty-year period, from 1880-1910 (pp. 235-6).

Although they admit that there were regional clusters 'particularly resistant to the modernisation of reproduction' (p. 240), they thus give more weight to culture than to economic development and point out the importance of the geographical diffusion of attitudes to fertility and of contraceptive practice. From this they go on to draw controversial implications for current attitudes to the problem of high fertility in the third world. Thus (p. 240), they urge that 'current recommendations to shift the emphasis away from family planning programmes and toward development efforts as a means of reducing fertility should be viewed with considerable caution'. They recognise of course that the historical record for Western Europe does not help as a guide to the success of governmental efforts to control fertility, since most were staunchly pronatalist. We may also look to Smith's (1981, p. 612) long-term analysis which suggests parallels between Marx's model of nineteenth century England and the debate over the relative importance of internal and external influences in the contemporary third world.

Family and Household

Identification of marriage as a key to understanding fertility and population growth has been matched by vigorous investigation of family and household in the past. It is clear that some of the demographic patterns identified above gave rise to, and were the product of, distinctive patterns of family formation. In England, for example, while the researchers at the Cambridge Group divided into those primarily interested in the dynamics of fertility and mortality, and those interested in the history of the family, there has never been any doubt about the interdependence of the research findings. For example, the work of Laslett and Wall (1972) on household structure revealed that marriage in pre-industrial England involved the setting up of a new household, while Macfarlane (1978) indicates that one crucial aspect of English 'individualism' is the nuclear family, the independent single family household. The distinctiveness of the English, as with the European, pattern of marriage and family formation is both deep-rooted and has profound implications for the interpretation of industrialisation: nuclear families may have been more adaptable to change, more geographically and socially mobile.

The study of family and household has also gone well beyond the demographic into the wider study of social behaviour, for example, the broad portraits of the family (Shorter, 1975; Houlbrooke, 1984) of sex and marriage in England by Stone (1977) or Macfarlane (1985) or in France by, for example, Flandrin (1984) and Segalen (1980). Few aspects of the history of sexuality have remained untouched: for example, bastardy (Laslett, Oosterveen and Smith, 1980);

age at sexual maturity, pre-marital sex, orphans (all in Laslett, 1977); spinsters, feminism and sexuality (Jeffreys, 1985); or the book by Shorter (1982) on the history of women's bodies. Anderson (1980) in his brief but informative review of the history of the western family from 1500-1914 is able to identify three approaches which recent scholarship has embraced: the demographic approach, concerned with marriage, childbearing and household size; the 'sentiments' approach, concerned less with statistics and measurement than with the emotional meaning of marriage and the family; and, finally, the 'household economics' approach which seeks to 'interpret households and families above all in the context of the economic behaviour of their members' (p. 65).

Whilst these various approaches have to some extent proceeded from a critique of the demographic approach, the latter has yielded important results, very well illustrated in the works of Laslett and Wall (1972) and Wall, Robin and Laslett (1983). The general influence of Laslett in these matters is shown in the festschrift edited by Bonfield et al. (1986). The crucial question posed had two facets: how had household size and structure changed over time and to what extent could broad geographical patterns be discerned? The crucial finding was to explode the myth of the large, extended, pre-industrial family which had supposedly given way to the nuclear family at the hands of industrialisation. In Britain and north-west Europe in particular, 'the typical domestic group has been small and simple in structure at least since medieval times; where this group was larger and more complex (which was usually only among the more affluent) it was servants and not kin who were responsible'. The conjugal pair was the centre of family life and 'at no time in the past one thousand years ... has kinship been the dominant basis for social organisation, certainly in Britain and probably anywhere in western Europe'. Thus, it is grossly oversimplified to suggest 'any simple model of industrialisation causing family change' (all quotes from Anderson, 1979, pp. 50-51, italics in original).

There were certainly strong geographical variations. At the grandest scale, north west Europe emerged as highly distinctive, Hajnal (1983), for example, tying in his earlier (1965) views on age at marriage and celibacy to a wider interpretation of the uniqueness of the European pattern. England seems to have been fairly uniform, although the variation was much greater in continental Europe. Within western Europe, the small conjugal family household was very common but far from universal:

> Married brothers and their families often shared the same house in parts of France and Italy ... while stem-family systems were quite widely found in Austria, France and Germany. In parts of eastern Europe, huge and complex

households predominated with extensions both laterally and vertically (Wrigley, 1977, p. 78).

Mediterranean regions also had distinctive family forms (for example, Smith, 1981b). Whilst Laslett and Wall (1972) had set out clear methodological and theoretical procedures, not everyone was convinced. Berkner (1972; 1975) for example, argued that the evidence used for the pre-industrial nuclear family was not conclusive. It may be misleading to take a simple, cross-sectional view of household at a particular moment, since many 'nuclear' families may go through an 'extended' phase. More attention, therefore, needed to be given to the life-cycle approach to trace individual families over long periods (Mitterauer and Sieder, 1979; 1982). Wrigley (1977, p. 72) has explained that because a substantial proportion of young people of both sexes spent some years in service or apprenticeship between leaving home and eventually marrying, 'the life-style of many people therefore included membership of at least three families and frequently of many more', and indeed of more than one type. Others have criticised the complex questions of definition of family and household, while Wall et al. (1983), in recognising these and other criticisms (p. 1) attempt in particular to give more attention to the broad socio-economic context in which the household was set and the precise influence of occupation on its structure. This new volume has attracted praise, but also renewed criticism for its constrained geographical scope: while it covers northern, central and eastern Europe, it excludes the Mediterranean, or worse still, implies that it is a part of the western European pattern (Kertzer, 1985, p. 99).

Migration and Mobility in the Past

Progress in the study of migration in the past has been considerable, with geographers playing a leading role in both theoretical and empirical work. It is probably fair, however, to say that these studies have been rather less remarkable in the breadth of issues raised than studies in fertility, mortality or the household. Seminal works have appeared on the nature and volume of migration in traditional rural society, on the contribution of migrants to urban growth, on international migration and slavery. Yet a whole range of questions remains unanswered and indeed many are as yet unasked. One example, which seems likely to be particularly fruitful for research, is the relationship between migration and the wider demographic system. Migration may create distorted age and sex structures at both origin and destination and so influence levels of nuptiality, fertility and mortality; migrants themselves may take on very distinctive demographic characteristics.

Sources may, of course, be problematic. Unlike birth, death or marriage, migration is less easily defined, happens with much less predictability and frequently escapes recording at all. We do not have the equivalent of the parish or civil registers to record a change of residence. It is not until the censuses of the nineteenth century (Baines, 1972; 1985) that systematic recording of birth places gives some impression for whole communities of the impact of migration. Yet, we know of the pitfalls in relying on birth-place data and of the need to qualify conclusions with respect to complementary sources. For example, several authors have used marriage registers, which may give occupation, place of birth and residence of spouses, as well as complementary data on occupation of parents and details of witnesses, to reconstruct the social geography of the city (Sewell, 1985 on Marseille) or of the countryside (Ogden, 1980). Migration studies have not lacked for ingenuity. Withers (1985) in his study of Scottish highland migration to Dundee, Perth and Stirling in the nineteenth century was able to rely on census material, while two other studies of Scotland (Lovett, Whyte and Whyte, 1985; Houston, 1985) used, respectively, apprenticeship records and testimonials, the movement certificates issued by the Scottish church, for the seventeenth and eighteenth centuries. Wareing (1980; 1981) and Kitch (1986) have also shown the usefulness of apprenticeship records for England in the seventeenth and eighteenth centuries. Laslett and Harrison (1963 ; and see Laslett, 1977) used population listings for 1676 and 1688 for a Nottinghamshire village to illustrate the extent of population turnover. Others, for example, Clark (1979) on England between 1660-1730 have used depositions in ecclesiastical courts, documents in which witnesses notified their place of birth as well as their present residence while Butcher (1974) used the freemen rolls for New Romsey in Kent in the fifteenth and early sixteenth centuries, indicating a large influx of immigrants. Most sources have the disadvantage of recording only one, perhaps minor, migration movement in an individual's life-time. Some, though, allow a reconstruction of migration stages, for example Kussmaul's (1981) use of the diary of Mayett, a servant in husbandry in Buckinghamshire at the turn of the eighteenth century. For contemporary migration in the twentieth century this has proved soluble by intensive questionnaires amongst a large sample asking them to reconstruct their exact migration histories (Courgeau, 1987) but this is scarcely possible for the past. However, recent work has shown that it may be possible to gauge the structure of migration over time by using family reconstitution methods (Souden, 1984), turning one of the oft-criticised limitations of the method - the absence of individuals from the parish registers - into an opportunity. Souden tentatively suggests, therefore, for sixteen widely dispersed cases in England in the seventeenth and eighteenth centuries, long-term trends in the proportions of 'movers' and 'stayers' in different communities.

The weight of empirical research on pre-industrial Western Europe has buried for good the notion of an immobile village population wedded to the soil. The attachment of families to holdings and of serfs to estates certainly existed. There was continuity of residence in villages from generation to generation. Yet stability has not proved a satisfactory characterisation of pre-industrial rural society as a whole. Even in communities where permanent residential migration was limited largely to moves for marriage or to neighbouring villages, seasonal mobility may have been very significant in connecting different agricultural economies and the country-side with the town, as Châtelain (1977) and Poitrineau (1983) have recently shown to great effect for France. For England before the Civil War, Clark (1979, p. 39) goes so far as to suggest that 'migration was an almost universal phenomenon affecting the great mass of the national population'. Further, while much mobility was within well-defined bounds, where 'servants, apprentices, would-be spouses and others' were 'out to better themselves, travelling fairly limited distances, to a neighbouring town or village' there was an increasing flow of long-distance movers, often the poor, pushed towards the towns. Thus did Hufton (1974) characterise migration as an industry of the poor, and as an acid corroding social links within communities.

Migration, of course, created new links and new com-munities within cities, which naturally come to dominate the literature on the eighteenth and nineteenth centuries. Thus, Poussou (1983) shows the impact of Bordeaux on its sur-rounding 'bassin démographique', Pooley (1977) gives much detail on the origins of Liverpool's population, as does Jackson (1982) for Duisberg or Meckel (1985) for Boston. Swift and Gilley (1985) provide useful discussions of the Irish in Victorian cities while Sewell (1985), in his painstaking study of Marseille, challenges Chevalier's earlier view of migration to Paris as 'an onslaught of impoverished and disorientated hordes that sank the city into the depths of social pathology'. Instead, he finds no permanent 'dangerous class':

> the mass of plebian immigrants ... were an extraordinary lot - competitive, ambitious, able and flexible; not the scum, but the salt, of the earth (p. 267).

As for the countryside, so in the city, the remarkable characteristic is the gross turnover of population. So often we have to turn to net migration rates which may conceal a great deal more than they show, and the total numbers in-volved may have much impact on social and demographic behaviour. For example, in Boston during the decade of the

233

1880s, total immigration into the city was more than twelve times greater than net immigration:

> Some three times as many families lived in Boston at some time during the 1880s as lived there at any single time in the decade (Sewell, 1985, p. 150, quoting Thernstrom and Knights, 1970).

In a wide range of nineteenth-century cities the proportion of the population living in a city that continued to live there ten years later was only 30 to 50 per cent.

Finally, we must draw attention to a rather different, but equally fertile, strand of research into the history of migration. International migration, and particularly the history of slavery, has posed fascinating questions both on the availability and interpretation of sources, on the demographic impact of migration at origin and destination and must of course form a part of our wider understanding of the processes of economic growth in Europe in the seventeenth and eighteenth centuries. The history of forced migrations, of which slavery is the most significant, has been treated from a wide variety of perspectives. It is the transatlantic trade in slaves that has attracted most scholarly interest: Curtin's (1969) general view has been supplemented by works appearing during the 1970s, for example, Engerman and Genovese (1975) or Marsh (1974). Much interest has focused on the Caribbean. For example, Higman's (1976) study of Jamaica was extended by him to cover the British Caribbean generally (Higman, 1984) between 1807 and 1834, using the slave registration and compensation records that exist for British colonies for the interval between the abolition of the Atlantic slave trade and slave emancipation. Higman's (1984) study is indeed, in part, an exercise in historical demography: model life-tables are applied to estimate and allow for the extent of under-registration and then a full analysis is attempted of age, sex, fertility, mortality and patterns of natural increase or decrease. He points to the significance of mortality, particularly in areas of sugar cultivation where 'the optimum size of sugar plantations promoted a form of labour organisation - gang labour and driving - that not only maximised labour productivity but maximised mortality as well' (see Gemery, 1985, p. 664). Sheridan (1985) has also recently treated slavery in the British West Indies, but for the period 1680-1834, and provides much detail on medical as well as demographic history. Thus, he looks at the effects on slave health of culture shock, diet, work loads, punishment, housing, clothing and sanitation, as well as at theories of disease causation and the education of doctors, midwives and nurses. Again he indicts the sugar plantations:

Here was found the vicious circle of disease and racism, whereby the debility of blacks was often misconstrued by planters and doctors as racial characteristics of laziness and the shamming of illness (Sheridan, 1985, p. 342).

Clearly the study of slavery has both posed questions of the first importance in understanding the relationships between migration, demography and economy and sparked controversy. One example of the latter is the debate over the influence of slavery on the role and nature of the black family in the modern United States, as discussed for example by Gutman (1976) for the period 1750-1925. Another is the refutation of the view that whatever the horrible details of the trade, slavery did not greatly disrupt African economy and society. Inikori (1982) and his contributors have argued strongly that the demographic effects were dramatic and that the under-population of sub-Saharan Africa, together with the political chaos produced by the slave trade, stopped economic advance. Much hinges around numbers: Inikori suggests that the total of slaves exported was much higher than previous estimates and, given that up to half the slaves were women, he suggests that sub-Saharan Africa's population in 1880 would have been 112 million more than it actually was. Thus, the cumulative effect of the slave trade may have been to halve Africa's potential population (see Havinden, 1983), a provocative view which will doubtless stimulate much further research.

TOWARDS A HISTORICAL POPULATION GEOGRAPHY

Our final section draws out and reflects upon a pervasive point of considerable relevance for the future of research in historical demography: the question of geographical variation and the extent to which this may be explored within existing sources, methodology and theory. The majority of research in historical demography has been devoted to north-west Europe and, whilst there has been considerable extension of interests in recent years, the historical demography of whole countries and, indeed, continents in the present Third World is scarcely known beyond basic trends. While the distinctiveness of the west European pattern of demographic behaviour in the past has been established, there is still much to be learnt about intra-European variation, and in particular the extent to which well-documented trends in England and France are typical. Of equal importance is the degree of variation within individual countries, that is by geographical region, by social class, and in countryside or town.

As indicated above, recent research has indeed left a notable gap in our knowledge of processes operating at the regional scale. Whilst censuses and other aggregate data are

HISTORICAL DEMOGRAPHY

available for the nineteenth century onwards in many
countries, the nature of parish registers and of the technique
of family reconstitution has meant that pre-1800 historical
demography has been largely at the parish scale or at the
realm of national reconstructions and speculations. Flinn
(1981) was able to review some 600 parish-level analyses
covering periods from 1500 to 1820 in Europe. These have the
advantage of a common technique and, therefore, compar-
ability of findings but they represent 'only a tiny fraction of
the European population ... a few hundred parishes out of
several hundred thousand in early modern Europe. Nor are
they evenly distributed by country, geographically within
countries, or by type of society - urban, rural, industrial,
agricultural, peasant, estate, servile or free' (Flinn, 1981, p.
5). Not less than half of the reconstitutions available to
Flinn, indeed, came from France alone. The poor quality of
the old parish registers in Ireland or Scotland has proved a
hindrance, although there has been a gradual expansion of
work in Western, though less in Eastern, Europe.
   The approach used by the Cambridge Group in planning
their work on England is instructive. The nature of the
techniques and the importance of the results has been referred to above, but the way they coped with the question of
geographical variability is also of importance and has attracted criticism. The national reconstruction contained in
Wrigley and Schofield (1981) is based on 404 parishes widely
distributed over England whose registers were aggregated by
a mass of local historians. These aggregations were complemented by a much smaller number of family reconstitutions,
some of which were used in the main volume and thirteen of
which have been discussed in Wrigley and Schofield (1983).
The files of the Cambridge Group are described in Laslett
(1983, pp. 287-91), who records that returns from some 750
parishes had been gathered by that date and some 30 reconstitutions undertaken. The general view from the Cambridge
Group has been that regional divergence was remarkably small
in England compared to other European countries: for
example, Laslett (1985, p. 537) remarked that the 'English
behaved in these matters like the red coats on parade in
front of Buckingham Palace, every unit in step with every
other, and all changing direction at the same time'. Nevertheless, Anderson (1985, p. 600) has drawn attention to the
fact that the 404 parishes exclude London, small communities
and those with major discontinuities in registration and
include a regional bias: '12.5 per cent of the parishes are
from Bedfordshire and none is from Cornwall', although
Wrigley and Schofield did adopt rigorous procedures for
correcting imbalances. Anderson has also (1985, pp. 602, 605)
drawn attention to the very important problem of inferring
individual and family behaviour, or we might add regional
patterns, from national aggregates. Thus, he suggests that

the family reconstitution data indicate that the national average figures are an aggregate of different underlying tendencies and that there are differences among parishes in for example nuptiality, and infant and child mortality:

> Yet the causal model developed by Wrigley and Schofield does not address differences of time, place, and economic structure, apparently assuming a homogeneous underlying experience (Anderson, 1985, p. 605).

Goldstone (1986, p. 31) too has urged caution in 'relating national trends in fertility and nuptiality to national aggregate data on wages and industrialisation ... A complete theory of England's demographic revolution is most likely to emerge from a regional mapping of real wages, industrialisation, agricultural change, and shifts in nuptiality and fertility ...' Schofield (1985a, p. 591), however, whilst recognising the necessity, and difficulty, of further regional work, suggests that:

> it does not follow that the relationships involved can only be properly understood at a regional or local level even if the experience of the national aggregate proves not to have been typical of any one part of the country. For 'nation', 'region', and 'locality' are abstractions and each level of abstraction has its own meaning. To move from a national investigation to a regional one is, therefore, to add a new level of understanding. The latter need neither negate nor supercede the earlier, rather as the conceptual relationships of chemistry are neither negated nor superseded (sic) because they are reducible to those of physics.

It is worth noting here that across the Channel, the larger number of available reconstitutions and the ready appreciation of regional differences has encouraged speculation on regional demographic trends. Thus, the special number of Population (1975) revealed the progress of research in France which included family reconstitutions for a sample of forty rural parishes for the 17th and 18th centuries, analysed in detail for the south-west (Henry, 1972), south-east (Henry, 1978), north-west (Henry and Houdaille, 1973) and north-east (Houdaille, 1976). Biraben (1985) has recently shown the potential for historical demography for the period before 1670, until recently considered rather impenetrable because of the poor quality of the records. Dupâquier (1981), on the other hand, has initiated a new approach to the nineteenth century. He suggests that a simple extension of Henry's 40 villages is not sufficient to embrace the urban changes that characterised France at this period and proposes a more genealogical approach, taking 3000 couples formed

HISTORICAL DEMOGRAPHY

under the First Empire and tracing the masculine line to the present. In all, this will involve 20,000 couples, using the civil registers and census enumerators' books, and will provide evidence on a vast range of socio-demographic measures by social group, geographical location and occupation.

The reconstruction of the historical population geography of the nineteenth and twentieth centuries has, because of the availability of the sources of civil registration and the censuses, proceeded. Even for Europe, however, this work is still in its infancy and Adams (1979, p. 118) has indicated that the technical virtuosity of family reconstitution has led to the relative neglect of the nineteenth century until recently. He has also reminded us that the regional level of analysis is still a very coarse mesh, since the units include great variability of culture, education, occupation and so forth and he approves of the attempts by Spagnoli (1977a; 1977b) further to refine this analysis. For the Lille arrondissements of 129 communes in the period 1859-63 Spagnoli (1977b) finds that there is as much variation as in France as a whole in demographic behaviour.

This sort of consideration must inevitably underlie our interpretation of the Princeton project on the decline of European fertility which used 500 sub-areas. As Seccombe (1983, p. 35) has indicated, the problem with such national and regional divisions is that they are used:

> without any sustained attempt to generate regional and class breakdowns on the basis of relevant socio-economic categories. The multi-class and mixed-region totals which are compiled, statistically manipulated and interpreted, inevitably mask structural variation ...

The authors in the Princeton series are not, of course, unaware of these problems of interpretation. Thus, Knodel and van de Walle (1979, p. 236) have indicated the importance of cultural factors such as a common dialect and common customs in determining the different pace of fertility decline, rather than socio-economic variations:

> there is greater similarity in fertility trends among provinces within the same region but with different socio-economic characteristics than is true among provinces with similar socio-economic characteristics but located in different regions.

Thus, in Russia, the persistence of customs and attitudes unfavourable to family limitation helps to explain why eastern minorities in European Russia and rural populations in Central Asia were slow to reduce their fertility. For France, Wrigley (1985b) has recently produced an intriguing discussion of regional data to elucidate the peculiar, early, decline of fertility in France, attempting to see how far regional variation may be masked in national totals.

A final, and particularly instructive, example of the problems of analysing geographical variations may be drawn from recent work on fertility and mortality in England and Wales in the nineteenth century. Woods (1984, p. 40) quotes persuasive evidence for regional mortality:

> a male baby born in an inner area of Liverpool in 1861 could be expected to live 26 years whilst a female might be expected to live an additional year. In Okehampton, Devon, comparable life expectations would have been 57 years and 55 years for males and females respectively.

He suggests, therefore, that the range of mortality in England and Wales in the 1860s was as great as that between England and Wales in the 1840s and in the 1960s; and that the experience of mortality decline was highly place-specific. Woods uses data from the Registrars' General <u>Annual Reports</u> together with that of the population Censuses of England and Wales to build up a picture of mortality in the 631 registration districts, a feat made possible by computerisation. This analysis draws attention to the low life-expectancies of the cities including parts of London and industrial areas and the rather better placed rural districts of, for example, the south, West Midlands and South West, showing the strong environmental influence on mortality in the nineteenth century. Those living in many rural districts in 1861 thus had life expectancies that would be equalled nationally only in the 1920s:

> even in 1931, 20 of the 84 county boroughs in England and Wales had male life expectancies at birth in the low fifties, a figure which the inhabitants of several rural areas had attained some seventy years previously (Woods, 1984, p. 64).

An analysis of fertility on a similar geographical basis (Woods, 1982; Woods and Smith, 1983) reveals further evidence for the 'compositional effect' of local variations underlying national aggregates. The decline of marital fertility showed distinct regional patterns. The calculations reveal how complex the geographical variation and its explanation are: it is 'not merely a matter of urban-rural differences or North versus South. Similar fertility levels have been reached by populations living in very diverse areas' (Woods and Smith, 1983, p. 212). The authors go on to try to solve at least some of the social-class specific variations in marital fertility by using data from the census enumerators' books for 1851 to 1871 for three English towns, Sheffield, Sunderland and Cheltenham.

HISTORICAL DEMOGRAPHY

CONCLUSION

This survey of recent developments in historical demography
has attempted to illustrate, albeit with an extremely selective
choice of examples, some of the breadth and depth of recent
scholarship. The volume of literature which has lately
appeared is ample evidence of the view that historical demography, far from being a minor historical specialisation, has
assumed the character of a separate discipline. Three aspects
of its present status are worthy of emphasis. First is its
degree of technical virtuosity which has allowed growing
mastery over a range of source materials. Secondly, the
breadth of interpretation of results of research has led to a
reassessment of the role of demography in past economies and
societies and to a flood of publications where the statistical
threads of demography are woven into a much broader canvas
of social and economic relations. Finally, perhaps the greatest
research challenge lies in the discovery and interpretation of
geographical variations, that is in the creation of a historical
population geography.

REFERENCES

Adams, P.V. (1979) 'Towards a Geography of French Histor-
    ical Demography: Problems and Sources', French Histor-
    ical Studies, XI, 1, 108-30
Anderson, M. (1980) Approaches to the History of the
    Western Family 1500-1914, Macmillan, London
Anderson, M. (1985) 'Historical Demography after "The Popul-
    ation History of England"', Journal of Interdisciplinary
    History, XV, 4, 595-608
Baines, D.E. (1972) 'The Use of Published Census Data in
    Migration Studies' in E.A. Wrigley (ed.), Nineteenth-
    Century Society. Essays in the Use of Quantitative
    Methods for the Study of Social Data, Cambridge
    University Press, Cambridge
Baines, D.E. (1986) Migration in a Mature Economy: Emi-
    gration and Internal Migration in England and Wales
    1861-1900, Cambridge University Press, Cambridge
Baker, A.R.H. (1984) 'Reflections on the Relations of Histor-
    ical Geography and the Annales School of History' in
    A.R.H. Baker and D. Gregory (eds.), Explorations in
    Historical Geography, Cambridge University Press,
    Cambridge, pp. 1-27
Beier, A.L., and Finlay, R. (eds.) (1986) London 1500-1700.
    The Making of the Metropolis, Longman, London
Berkner, L. (1972) 'The Stem Family and the Developmental
    Cycle of the Peasant Household - An Eighteenth-Century
    Austrian Example', American Historical Review, 77,
    398-418

Berkner, L. (1975) 'The Use and Misuse of Census Data for the Historical Analysis of Family Structure', Journal of Interdisciplinary History, 5, 721-738

Biraben, J-N. (1985) 'Le Point de l'Enquête sur le Mouvement de la Population en France avant 1670', Population, 40, 1, 47-70

Bonfield, L., Smith, R., and Wrightson, K. (eds.) (1986) The World We Have Gained: Histories of Population and Social Structure, Blackwell, Oxford

Butcher, A.F. (1974) 'The Origins of Romney Freemen 1433-1523', Economic History Review, 2nd Series, XXVII, 16-27

Caldwell, J.C. (1981) 'Mechanisms of Demographic Change in Historical Perspective', Population Studies, 35, 5-27

Caldwell, J.C. (1982a) Theory of Fertility Decline, Academic Press, London

Caldwell, J.C. (1982b) 'The Failure of Theories of Social and Demographic Change to Explain Demographic Change: Puzzles of Modernisation or Westernisation', Research in Population Economics, 4, 297-332

Chambers, J.D. (1972) Population, Economy and Society in Pre-Industrial England, Oxford University Press, Oxford

Châtelain, A. (1976) Les Migrants Temporaires en France de 1800 à 1914: Histoire Économique et Sociale des Migrants Temporaires des Campagnes Françaises au XIXe siècle au Début du XXe siècle, Publications de l'Université de Lille III, Lille

Clark, P. (1979) 'Migration in England During the Late Seventeenth and Early Eighteenth Centuries', Past and Present, 83, 57-90

Coale, A.J. (1973) 'The Demographic Transition Reconsidered' in International Population Conference Liège, IUSSP Liège, Volume 1, 53-72

Coale, A.J., Anderson, B.A., and Härm, E. (1979) Human Fertility in Russia Since the Nineteenth Century, Princeton University Press, Princeton, N.J.

Cobb, R. (1976) Tour de France, Duckworth, London

Coleman, D., and Schofield, R. (1986) The State of Population Theory: Forward from Malthus, Blackwell, Oxford

Courgeau, D. (1987) 'Recent Developments in French Migration Research' in P.E. Ogden and P.E. White (eds.), Migrants in Modern France, George Allen and Unwin, London (forthcoming)

Curtin, P.D. (1969) The Atlantic Slave Trade. A Census, The University of Wisconsin Press, Madison

Dupâquier, J. (1981) 'Une Grande Enquete sur la Mobilité Géographique et Sociale aux XIXe et XXe Siècles', Population, 36, 6, 1164-1167

Dupâquier, J. (1984) Pour la Démographie Historique, PUF, Paris

HISTORICAL DEMOGRAPHY

Dupâquier, J., and Dupâquier, M. (1985) L'Histoire de la
Démographie, Librairie Académique Perrin, Paris
Dupâquier, J., Fauve-Chamoux, A., and Grebenik, E. (1983)
(eds.), Malthus Past and Present, Academic Press,
London
Dupâquier, J., Hélin, E., Laslett, P., Livi-Bacci, M., and
Sogner, S. (1981) (eds.), Marriage and Remarriage in
Populations of the Past, Academic Press, London
Elton, G.R. (1984) The History of England. Inaugural Lecture
Delivered 26 January 1984, Cambridge University Press,
Cambridge
Engerman, S.L., and Genovese, E. (1975) (eds.), Race and
Slavery in the W. Hemisphere, Princeton University
Press, Princeton
Flandrin, J-L. (1984) Familles, Parenté, Maison, Sexualité
dans l'Ancienne Société, Seuil, 2nd edition, Paris
Fleury, M., and Henry, L. (1958) 'Pour Connaître la Popul-
ation de la France Depuis Louis XIV', Population, 13,4,
663-86
Fleury, M., and Henry, L. (1965) Nouveau Manuel de Dé-
pouillement et d'Exploitation de l'Etat Civil Ancien,
INED, Paris
Flinn, M.W. (1981) The European Demographic System 1500-
1820, John Hopkins University Press, Baltimore
Gemery, H.A. (1985) 'Review of B.W. Higman (1984)', Econ-
omic History Review, 38,4, 663-4
Glass, D.V., and Eversley, D.E.C. (1965) (eds.), Population
in History, Essays in Historical Demography, Arnold,
London
Goldstone, J.A. (1986) 'The Demographic Revolution in
England: A Re-Examination', Population Studies, 40, 1,
5-33
Guillaume, P., and Poussou, J-P. (1970) Démographie His-
torique, Armand Colin, Paris
Gutman, H.G. (1976) The Black Family in Slavery and Free-
dom 1750-1925, Basil Blackwell, Oxford
Hajnal, J. (1953) 'Age at Marriage and Proportions Marrying',
Population Studies, 7, 2, 111-36
Hajnal, J. (1965) 'European Marriage Patterns in Perspective'
in D.V. Glass and D.E.C. Eversley (eds.), Population in
History, Arnold, London, pp. 101-143
Hajnal, J. (1982) 'Two Kinds of Preindustrial Household
Formation System', Population and Development Review,
8, 3, 449-94
Havinden, M. (1983) 'Review of J.E. Inikori (1982)', Economic
History Review, 36, 2, 319
Henry, L. (1972) 'Fécondité des Mariages dans le Quart
Sud-Ouest de la France de 1720 à 1829', Annales ESC.,
3, 612-39; 4, 977-1023
Henry, L. (1978) 'Fécondité des Mariages dans le Quart

Sud-Est de la France de 1670 à 1829', Population, 25, 4-5, 856-83

Henry, L., and Houdaille, J. (1973) 'Fécondité des mariages dans le Quart Nord-Ouest de la France de 1670 à 1829', Population, 28, 4-5, 873-922

Higman, B.W. (1976) Slave Population and Economy in Jamaica 1807-1834, Cambridge University Press, Cambridge

Higman, B.W. (1984) Slave Populations of the British Caribbean 1807-1834, Johns Hopkins University Press, Baltimore

Hill, C. (1978) 'Sex, Marriage and the Family in England', Economic History Review, XXI, 3, 450-63

Hollingsworth, T.H. (1969) Historical Demography, Cornell University Press, Ithaca, New York

Houdaille, J. (1976) 'La Fécondité des Mariages dans le Quart Nord-Est de la France de 1670 à 1829', Annales de Démographie Historique, 341-92

Houlbrooke, R.A. (1984) The English Family 1450-1700, Longman, London

Houston, R. (1985) 'Geographical Mobility in Scotland 1652-1822: The Evidence of Testimonials', Journal of Historical Geography, 11, 4, 179-94

Hufton, O. (1974) The Poor in Eighteenth-Century France 1750-1798, Oxford University Press, Oxford

Inikori, J.E. (ed.) (1982) Forced Migration: The Impact of the Export Slave Trade on African Societies, Hutchinson, London

Jackson, J.H. (1982) 'The Occupational and Familial Context of Migration in Duisberg 1867-1890', Journal of Urban History, 8, 235-70

James, P. (1979) Population Malthus: His Life and Times, Routledge and Kegan Paul, London

Jeffreys, S. (1985) The Spinster and Her Enemies: Feminism and Sexuality 1880-1930, Pandora, London

Kertzer, D. (1985) 'Future Directions in Historical Household Studies', Journal of Family History, 10, 1, 99-107

Kitch, M.J. (1986) 'Capital and Kingdom: Migration to Later Stuart London' in A.L. Beier and R. Finlay (eds.), London 1500-1700. The Making of the Metropolis, Longman, London, 224-251

Knodel, J.E. (1974) The Decline of Fertility in Germany 1871-1939, Princeton University Press, Princeton, N.J.

Knodel, J., and van de Walle, E. (1979) 'Lessons from the Past: Policy Implications of Historical Fertility Studies', Population and Development Review, 5, 2, 217-45

Kussmaul, A. (1981) Servants in Husbandry in Early Modern England, Cambridge University Press, Cambridge

Landry, A. (1934) La Révolution Démographique: Etudes et Essais sur les Problèmes de la Population, Sirez, Paris

Landry, A. (1945) Traité de Démographie, Payot, Paris

Laslett, P. (1965) The World We Have Lost, Methuen, London

Laslett, P. (1976) Family Life and Illicit Love in Earlier Generations, Cambridge University Press, Cambridge

Laslett, P. (1983) The World We Have Lost - Further Explored, Methuen, London

Laslett, P. (1985) 'Review of Teitelbaum (1984)', Population and Development Review, 11, 3, 534-37

Laslett, P., and Harrison, J. (1963) 'Clayworth and Cogenhoe' in H.E. Bell and R.L. Ollard, Historical Essays 1600-1750 Presented to David Ogg, pp. 157-184

Laslett, P., Oosterveen, K., and Smith, R.M. (eds.) (1980) Bastardy and its Comparative History, Arnold, London

Laslett, P., and Wall, R. (eds.) (1972) Household and Family in Past Time, Cambridge University Press, Cambridge

Le Roy Ladurie, E. (1978) Montaillou: The Promised Land of Error, Brazillier, New York

Lesthaeghe, R.J. (1977) The Decline of Belgian Fertility 1800-1970, Princeton University Press, Princeton, N.J.

Levine, D. (1977) Family Formation in An Age of Nascent Capitalism, Academic Press, New York

Lindert, P.H. (1983) 'English Living Standards, Population Growth and Wrigley-Schofield', Explorations in Economic History, 20, 131-55

Livi-Bacci, M. (1977) A History of Italian Fertility During the Last Two Centuries, Princeton University Press, Princeton, N.J.

Lovett, A.A., Whyte, I.D., and Whyte, K.A. (1985) 'Poisson Regression Analysis and Migration Fields: The Example of the Apprenticeship Records of Edinburgh in the Seventeenth and Eighteenth Centuries', Transactions of the Institute of British Geographers N.S., 10, 3, 317-32

Macfarlane, A. (1978) The Origins of English Individualism: The Family, Property and Social Transition, Blackwell, Oxford

Macfarlane, A. (1986) Marriage and Love in England. Modes of Reproduction 1300-1840, Blackwell, Oxford

Marcilio, M.L., and Charbonneau, H. (1979) Démographie Historique, Presses Universitaires de France, Paris

Marsh, H. (1974) Slavery and Race: The Story of Slavery and Its Legacy, David and Charles, Newton Abbot

McKeown, T. (1976) The Modern Rise of Population, Arnold, London

McKeown, T., and Brown, R.G. (1955) 'Medical Evidence Related to English Population Changes in the Eighteenth Century', Population Studies, IX, 119-41

Meckel, R.A. (1985) 'Immigration, Mobility and Population Growth in Britain 1840-1880', Journal of Interdisciplinary History, XV, 3, 393-417

Mitterauer, M., and Sieder, R. (1979) 'The Development Process of Domestic Groups: Problems of Reconstruction and Possibilities of Interpretation', Journal of Family History, 4, 257-84

Mitterauer, M., and Sieder, R. (1982) The European Family, Blackwell, Oxford (orig. ed. 1977, Munich, revised 1982)

Noin, D. (1983) La Transition Démographique dans le Monde, Presses Universitaires de France, Paris

Notestein, F. (1945) 'Population: The Long View' in T.W. Schultz (ed.), Food for the World, Chicago University Press, Chicago

Notestein, F.W. (1948) 'Summary of the Demographic Background of Problems of Underdeveloped Areas', Millbank Memorial Fund Quarterly, 26, 249-55

Notestein, F. (1950) 'The Population of the World in the Year 2000', Journal of the American Statistical Association, 45, 335-45

Notestein, F. (1953) 'Economic Problems of Population Change' in Proceedings of the Eighth International Conference of Agricultural Economists 1953, Oxford University Press, London, pp. 13-31

Ogden, P.E. (1980) 'Marriage, Migration and the Collapse of Traditional Peasant Society in France' in P. White and R. Woods (eds.), The Geographical Impact of Migration, Longman, London, pp. 152-179

Ogden, P.E. (1984) 'Historical Population Geography' in J.I. Clarke (ed.), Geography and Population. Approaches and Applications, Pergamon, Oxford, pp. 61-7

Ogden, P.E. (1986) 'Regional Demographic Patterns in the Past', Journal of Historical Geography, 12, 2 (forthcoming)

Palliser, D.M. (1982) 'Tawney's Century: Brave New World or Malthusian Trap?', Economic History Review, XXXV, 3, 339-53

Perrenoud, A. (1979) La Population de Gèneve du Seizième au Début du Dix-Neuvième Siècle. Etude Démographique. Tome I: Structures et Mouvements, Editions de la Société d'Histoire et d'Archéologie

Perrenoud, A. (1980) 'Contribution à l'Histoire Cyclique des Maladies. Deux Cents Ans de Variole à Genève (1580-1810)' in A.E. Imhof (ed.), Mensch und Gesundheit in der Geschichte, Mattheisen Verlag, Husum

Perrenoud, A. (1981) 'Surmortalité Féminine et Condition de la Femme XVIIe-XIXe Siècles. Une Vérification Empirique', Annales de Démographie Historique, 89-104

Perrenoud, A. (1985) 'Où va la Démographie Historique? Analyse de Contenu de la Bibliographie Internationale', unpublished paper to the Table Ronde on Historical Demography at the General Conference of the International Union for the Scientific Study of Population, Florence, 5-12 June 1985

Petersen, W. (1979) Malthus, Heinemann, London

Poitrineau, A. (1983) Remues d'Hommes. Les Migrations

Montagnardes en France 17e-18e Siècles, Aubier Montaigne, Paris

Pooley, C.G. (1977) 'The Residential Segregation of Migrant Communities in Mid-Victorian Liverpool', Transactions of the Institute of British Geographers N.S., 2, 364-82

Poussou, J-P. (1983) Bordeaux et le Sud-Ouest au XVIIIe Siècle. Croissance Économique et Attraction Urbaine, Editions de l'Ecole des Hautes Etudes en Sciences Sociales, Paris

Rotberg, R., and Rabb, T. (eds.) (1985) Hunger and History. The Impact of Changing Food Production and Consumption Patterns of Society, Cambridge University Press, Cambridge

Schofield, R.S. (1983) 'The Impact of Scarcity and Plenty on Population Change in England 1541-1871', Journal of Interdisciplinary History, XIV, 265-291

Schofield, R.S. (1985a) 'Through a Glass Darkly: The Population History of England as an Experiment in History', Journal of Interdisciplinary History, XV, 4, 571-94

Schofield, R.S. (1985b) 'English Marriage Patterns Revisited, Journal of Family History, 10, 1, 2-20

Seccombe, W. (1983) 'Marxism and Demography', New Left Review, 137, 22-47

Segalen, M. (1983) Love and Power in the Peasant Family, Blackwell, Oxford (First ed. 1980, Flammarion, Paris)

Sewell, W.H. (1985) Structure and Mobility. The Men and Women of Marseille 1820-1870, Cambridge University Press and Maison des Sciences de l'Homme, Paris

Sheridan, R.B. (1985) Doctors and Slaves. A Medical and Demographic History of Slavery in the British West Indies 1680-1834, Cambridge University Press, Cambridge

Shorter, E. (1975) The Making of the Modern Family, Basic Books, New York

Shorter, E. (1982) A History of Women's Bodies, Basic Books, New York

Slack, P. (1985) The Impact of Plague in Tudor and Stuart England, Routledge and Kegan Paul, London

Smith, R.M. (1978) 'Population and its Geography in England 1500-1730' in R.A. Dodgshon and R.A. Butlin (eds.), An Historical Geography of England and Wales, Academic Press, London, pp. 199-237

Smith, R.M. (1979) 'Some Reflections on the Evidence for the Origins of the "European Marriage Pattern" in England' in C. Harris (ed.), The Sociology of the Family, Keele, pp. 74-112

Smith, R.M. (1981a) 'Three Centuries of Fertility, Economy and Household Formation in England', Population and Development Review, 7, 595-622

Smith, R.M. (1981b) 'The People of Tuscany and Their

Families in the Fifteenth Century: Medieval or Mediterranean?', Journal of Family History, 6, 1, 107-28

Smith, R.M. (1983) 'Hypothèses sur la Nuptialité en
Angleterre aux XIIIe-XIVe Siècles, Annales: ESC., 38,
107-36

Smith, R.M. (eds.) (1985) Land, Kinship and Life-Cycle,
Cambridge University Press, Cambridge

Snell, K.D.M. (1985) Annales of the Labouring Poor: Social
Change and Agrarian England 1660-1900, Cambridge
University Press, Cambridge

Souden, D. (1984) 'Movers and Stayers in Family Reconstitution Populations', Local Population Studies, 33, 11-28

Spagnoli, P. (1977a) 'Population History from Parish Monographs. The Problem of Local Demographic Variation',
Journal of Interdisciplinary History, VII, 427-52

Spagnoli, P. (1977b) 'High Fertility in Mid-Nineteenth
Century France: A Multivariate Analysis of Fertility
Patterns in the Arrondissement of Lille', Research in
Economic History, 2, 281-336

Stedman Jones, G. (1972) 'History: The Poverty of
Empiricism' in R. Blackburn (ed.), Ideology in Social
Science, Fontana, London, pp. 96-115

Stoianovitch, T. (1976) French Historical Method: The
Annales Paradigm, Cornell University Press, Ithaca, New
York

Stone, L. (1977) The Family, Sex and Marriage in England
1500-1800, Weidenfeld and Nicolson, London

Swift, R., and Gilley, S. (eds.) (1985) The Irish in the
Victorian City, Croom Helm, London

Teitelbaum, M.S. (1984) The British Fertility Decline.
Demographic Transition in the Crucible of the Industrial
Revolution, Princeton University Press, Princeton, N.J.

Thernstrom, S., and Knights, P. (1970) 'Men in Motion: Some
Data and Speculation about Urban Population Mobility in
Nineteenth-Century America', Journal of Interdisciplinary
History, 1, 7-36

Thompson, W.S. (1929) 'Population', American Journal of
Sociology, 34, 959-75

Tranter, N.L. (1985) Population and Society 1750-1940.
Contrasts in Population Growth, Longman, London

Van de Walle, E. (1974) The Female Population of France in
the Nineteenth Century: A Reconstruction of 82 Departments, Princeton University Press, Princeton, N.J.

Wall, R., Robin, J., and Laslett, P. (eds.) (1983) Family
Forms in Historic Europe, Cambridge University Press,
Cambridge

Wareing, J. (1980) 'Changes in the Geographical Distribution
of Apprentices to the London Coampanies 1486-1750',
Journal of Historical Geography, 6, 241-9

Wareing, J. (1981) 'Migration to London and Transatlantic

Emigration of Indentured Servants 1683-1775', Journal of Historical Geography, 7, 356-78

Weir, D. (1984) 'Rather Never Than Late: Celibacy and Age at Marriage in English Cohort Fertility 1541-1871', Journal of Family History, 9, 341-55

Weir, D. (1985) 'Review of J. Dupâquier (et al.) (1983)', Population and Development Review, 11, 2, 341-3

Willigan, J.D., and Lynch, K.A. (1982) Sources and Methods of Historical Demography, Academic Press, New York

Withers, C.W.J. (1985) 'Highland Migration to Dundee, Perth and Stirling 1753-1891', Journal of Historical Geography, 11, 4, 395-418

Woods, R.I. (1982) Theoretical Population Geography, Longman, London

Woods, R.I. (1984) 'Mortality Patterns in the Nineteenth Century' in R. Woods and J. Woodward (eds.), Urban Disease and Mortality in Nineteenth Century England, Batsford, London, pp. 37-64

Woods, R.I., and Smith, C.W. (1983) 'The Decline of Marital Fertility in the Late Nineteenth Century: The Case of England and Wales', Population Studies, 37, 207-25

Wrightson, K., and Levine, D. (1979) Poverty and Piety in an English Village, Terling 1525-1700, Academic Press, New York

Wrigley, E.A. (1966a) 'Family Limitation in Pre-Industrial England', Economic History Review, 2nd Series, 19, 1, 82-109

Wrigley, E.A. (1966b) 'Family Reconstitution' in E.A. Wrigley (ed.), An Introduction to English Historical Demography, Weidenfeld and Nicolson, London, pp. 96-159

Wrigley, E.A. (ed.) (1966c) An Introduction to English Historical Demography, Weidenfeld and Nicolson, London

Wrigley, E.A. (1968) 'Mortality in Pre-Industrial England: The Example of Colyton, Devon over Three Centuries', Daedalus, 97, 546-80

Wrigley, E.A. (1969) Population and History, Weidenfeld and Nicolson, London

Wrigley, E.A. (ed.) (1972a) Nineteenth-Century Society: Essays in the Use of Quantitative Methods for the Study of Social Data, Cambridge University Press, Cambridge

Wrigley, E.A. (1972b) 'The Process of Modernisation and the Industrial Revolution in England', Journal of Inter-disciplinary History, 3, 225-59

Wrigley, E.A. (1977) 'Reflections on the History of the Family', Daedalus, 106, 71-85

Wrigley, E.A. (1981a) 'Population History in the 1980s', Journal of Interdisciplinary History, 12, 2, 207-26

Wrigley, E.A. (1981b) 'Marriage, Fertility and Population Growth in Eighteenth-Century England' in R.B. Outhwaite (ed.), Marriage and Society: Studies in the

Social History of Marriage, Europa Publications, London, pp. 137-85

Wrigley, E.A. (1983a) The Growth of Population in the Eighteenth-Century: A Conundrum Resolved', Past and Present, 98, 121-50

Wrigley, E.A. (1983b) 'Malthus's Model of a Pre-Industrial Economy' in J. Dupâquier et al. (eds.), Malthus Past and Present, Academic Press, London, pp. 111-124

Wrigley, E.A. (1985a) 'Urban Growth and Agricultural Change: England and the Continent in the Early Modern Period', Journal of Interdisciplinary History, 15, 4, 683-728

Wrigley, E.A. (1985b) 'The Fall of Marital Fertility in Nineteenth Century France: Exemplar or Exception?', European Journal of Population, Part I: 1, 1, 31-60; Part II: 1, 2, 141-177

Wrigley, E.A., and Schofield, R.S. (1981) The Population History of England 1541-1871: A Reconstruction, Arnold, London

Wrigley, E.A., and Schofield, R.S. (1983) 'English Population History from Family Reconstitution: Summary Results', Population Studies, 37, 157-84

Chapter Nine

URBAN MORPHOLOGY

J.W.R. Whitehand

Of recent geographical papers on the internal structure of the city, about 12 per cent are on morphology (Whitehand, 1986). For a subject that includes the town plan, building form, and the pattern of land and building utilisation, this is a modest share of the research literature. It should be seen in the light of the fact that in the English-speaking world urban morphology is poorly integrated into urban geography as a whole, finding a home in a rather different set of academic journals than mainstream social and economic aspects of urban geography (Whitehand, 1986). It belongs as much to historical geography as to urban geography; a fact that reflects the longevity of the urban landscape that is the urban morphologist's object of study.

In the German-speaking countries the subject is much closer to the mainstream of urban geography, yet still within historical geography. This reflects a weaker distinction in central Europe between the study of present-day towns and the study of their historical aspects. Like urban geography as a whole, urban morphology has major intellectual roots in the German-speaking countries, and this is important in understanding progress in it, not least in recent times.

Traditions in Urban Morphology

The majority of research in urban morphology has come from three geographical areas - central Europe (the home of the German morphogenetic tradition), Great Britain and North America. The literature from outside these areas, though by no means negligible in absolute terms, is comparatively small. Despite the importance of urban geography in France, urban morphology is a relatively minor interest in that country at the present time (Pinchemel, 1983, pp. 305, 308) and has seldom attracted many researchers, in spite of the early work of Blanchard (1912) and the long tradition in France of research on rural settlement forms.

Two of the three main 'national schools' have considerable internal diversity: indeed only the oldest - the central
European one - has sufficient unity to justify treating it as
an entity (Figure 9.1). It has an essentially unbroken lineage
that may be traced back to the end of the nineteenth
century, notably to the early work of Schlüter (1899), who
postulated a morphology of the cultural landscape (Kulturlandschaft), as the counterpart in cultural geography of the
already rapidly developing field of geomorphology. The
Schlüter school gathered momentum after the First World War,
most notably following Geisler's (1924) Die deutsche Stadt,
which included comprehensive classifications of the sites, town
plans and building types of German towns. At much the same
time, Hassinger (1927) was broadening the scope of his
morphological studies, which in the 1910s had been primarily
concerned with the historic architectural styles of Vienna, to
give greater attention to land and building utilisation and
residential densities in Basel. Geisler's work proved particularly influential, but it also generated controversy. This was
fuelled by the study of the plan formation of the settlements
of Germany by Martiny (1928). In their attempt to produce
comprehensive surveys of town plans over the whole of
Germany both Geisler and Martiny allowed themselves to be
pushed by the enormous scope of their projects into essentially morphographic classification without due consideration of
relevant advances in urban history and the history of town
planning (Whitehand, 1981a, p. 4).

This reliance on the classification of present-day forms
without tracing their historical development, combined with a
failure to use large-scale topographical plans, led to the
overlooking of the period compositeness of town plans. It
precluded the identification of genetic plan units and
seriously impeded the recognition of functional differences
between streets. The fact that until the middle inter-war
years research by architects on the history of town planning
remained, like the work of urban historians, largely unknown
to German urban geographers had a stultifying effect on the
study of town plans. There was the additional problem that
urban geographers themselves had not yet produced a sufficiently coherent body of knowledge about the social and
economic organisation of towns to make possible a more pene-
trating urban morphology. Major advances came in the middle
inter-war years as a result of two developments. The first
was the recognition by urban geographers of the work of
urban historians such as Meier and Rörig, and of architects,
such as Klaiber and Siedler, who were interested in the
history of town planning (Whitehand, 1981a, pp. 5-6). The
second was the wider recognition, partly as a result of the
work of Bobek (1927), of the illogicality of concentrating
attention on the forms in the urban landscape at the expense
of the forces creating them. This second development ulti-

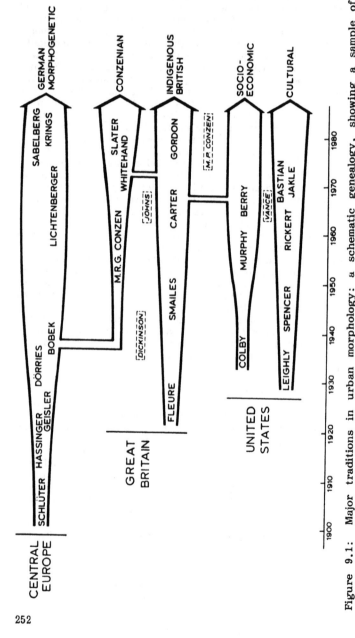

Figure 9.1: Major traditions in urban morphology: a schematic genealogy, showing a sample of authors.

mately had the effect, following Christaller's (1933) sharply focused view of one aspect of it, of shifting the focus of attention in German urban geography from form to function.

Post-war urban morphology in central Europe is a direct descendant of the German morphogenetic tradition of the inter-war years. Its key features are the integrated treatment of individual cities - less frequently several cities or parts of cities - including the interweaving of non-morphological aspects, and the extensive use of graphical, including cartographical, representations of facts and concepts. It is seen at its best in two major studies of Vienna, one of Bobek and Lichtenberger (1966) and the other by Lichtenberger (1977). A major recent example is the study of the inner areas (including the city centres) of Brussels, Bruges, Gent, Antwerp, Louvain and Malines by Krings (1984), which deals with all aspects of urban form. Of comparable importance, but concentrating primarily on historic buildings, is the study of four Italian cities by Sabelberg (1984), which is a step towards the development of a typology of Italian cities. These scholarly monographs, essentially cultural-genetic in perspective, are lavishly illustrated, much of the evidence being presented cartographically, and represent a tradition that can be traced back without major break to the work of Schlüter.

The lineage of British urban morphology is more heterogeneous. Two distinctive strands can be recognised; one imported, the other indigenous (Figure 9.1). The imported one is a direct descendant of the German morphogenetic tradition, brought to Britain in 1933 by M.R.G. Conzen, a pupil in Berlin of, among others, Louis and Bobek.

Conzen's writing on urban morphology was slow to have an impact in either his homeland or his country of adoption. Much of it was in English and not readily accessible to German-speaking geographers. Furthermore, it represented an intellectual tradition that was not readily understood by English-speaking geographers, despite the attempt by Dickinson (1942a; 1942b; 1945; 1948) to present in English the findings of much of the German work of the inter-war years from which it sprang. Arguably the three contributions of Conzen that were to prove the most influential were published during the period of five years 1958-62. His study of Whitby (Conzen, 1958) was not only a unique record of the building types and land and building utilisation of a whole town but a demonstration of how a detailed elucidation of a town's physical development can form the basis for conservation of the urban landscape. It was followed by Alnwick, Northumberland: a study in town-plan analysis (Conzen, 1960), a monograph that has proved itself to be the major contribution to urban morphology in the English language.

The achievements of the Alnwick study, many of them built upon two years later in a study of central Newcastle upon Tyne (Conzen, 1962), can be summarised under five heads (Whitehand, 1981a, pp. 12-13): first, the establishment of a basic framework of principles for urban morphology; secondly, the adoption for the first time in the geographical literature in the English language of a thorough-going evolutionary approach; thirdly, the recognition of the individual plot as being the fundamental unit of analysis; fourthly, the use of detailed cartographic analysis, employing large-scale plans in conjunction with field survey and documentary evidence; and fifthly, the conceptualisation of developments in the townscape. The tripartite division of the townscape into town plan, building forms, and land use that was recognised has since become widely accepted. The sub-division of the town plan for analytical purposes into streets and their arrangement in a street system, plots and their aggregation in street-blocks, and buildings, or more precisely their block-plans, has become a further standard way of reducing the complexity of reality to manageable proportions.

Conzen's conceptualisations of aspects of the development of the town plans of Alnwick and central Newcastle have a variety of derivations (Whitehand, 1981a, pp. 14-15). The fringe-belt concept is a major development of the Stadtrandzone of Louis (1936). Not only is it arguably the most important construct in Conzen's concept of a developing urban area but it has been the subject of considerable attention in recent years. Put simply, urban fringe belts are the physical manifestations of periods of slow movement or actual standstill in the outward extension of the built-up area and in the initial stages of their development are made up of a variety of extensive uses of land, such as by various kinds of institution, public utilities and country houses, usually having below average accessibility requirements to the main part of the built-up area. Developments of Louis's original notion include the recognition of a distinction in the case of a fringe belt associated with a town wall, or similar sharply-defined limitation to growth, between the restricted intramural zone, consisting mainly of secondary development within the generally close-grained morphological frame of a traditional plot pattern, and the extramural zone, consisting of more open, sometimes dispersed, development associated with the greater topographical freedom afforded by the relatively large-grained rural field pattern. Developments taking place following the creation of a fringe belt, notably the way in which the belt is perpetuated in spite of being encompassed by the built-up area but also the absorption of certain of its plots by advancing residential accretions, are an integral part of the concept.

Some of Conzen's concepts, though consistent with the German research tradition in which he grew up, have no direct precursors. The burgage cycle is a case in point. It consists of the progressive filling in with buildings of the backland of burgages and terminates in the clearing of build-

ings and a period of 'urban fallow', a term derived from
Hartke's (1953) Sozialbrache or 'social fallow' and applied by
him to rural field or vineyard land lying partly or wholly
waste while the owner works elsewhere. Completion of the
burgage cycle is followed by a redevelopment cycle. Thus the
burgage cycle is a particular variant of a more general
phenomenon of building repletion where plots are subject to
increasing pressure, often associated with changed functional
requirements, in a growing urban area.

This conceptualisation of processes is essentially foreign
to the second, indigenous, strand in British urban morphology. This owes its initial impetus to Fleure (1920), who
provided valuable descriptive generalisations about the layout
of European towns. Most of the work, however, has been
undertaken since the Second World War. It has been influenced to a large degree by the writing of Smailes (1955).
He emphasised the characterisation of present townscapes in
broad terms, recommending rapid reconnaissance surveys
rather than building-by-building surveys. History was
important in so far as it left a tangible residue. The perspective that he eloquently put forward co-existed with the
detailed, developmental analyses of Conzen during much of
the 1950s and 1960s. It was succeeded briefly in the late
1960s and early 1970s by quantitative, predominantly morphographic, analyses (Davies, 1968; Johnston, 1969) in which
historical geography remained subordinate or absent. Carter
(1965) was almost alone among British-born urban geographers at this time in giving a major place to historical
geography in its own right.

Both imported and indigenous strands within British
urban morphology were subjected to considerable influence
from America during the 1960s and 1970s. American urban
morphology had itself been characterised by two distinctive
strands (Figure 9.1), at least from the middle inter-war
years; a cultural geography strand having much in common
with rural settlement geography and the Berkeley School
(Leighly, 1928; Spencer, 1939), and a somewhat eclectic
socio-economic perspective, emphasising land-use studies
(Colby, 1933; Murphy, 1935). It was the second of these that
had by far the greater impact on British urban morphology.
Having by the 1950s adopted the concentric zone model of
sociologist Burgess (1925) and the sector model of land
economist Hoyt (1939) and created their own multiple-nuclei
model (Harris and Ullman, 1945), the followers of this branch
of American urban geography exported their ideas widely to
other English-speaking countries in the 1960s. Their perspective was morphological only in its concern with land-use
patterns: town plan and building form were generally treated
as land-use containers, if considered at all. Perhaps its most
important effect on British urban morphology was to generate

interest in the theoretical explanations of neo-classical economics.

The cultural geography strand in America has always been comparatively weak within urban geography. It has arguably provided the only true urban morphology within America. Starting with Leighly (1928), a line can be traced through the work of Spencer (1939; 1947) to that of Rickert (1967), Bastian (1980) and Jakle (1983). Although the main concern of this research has been with architectural styles, links with the long, continuing line of central European research on this aspect have been minimal.

This antecedence of separate, but sometimes interacting, schools is important in understanding recent and current research in urban morphology. It is this research that will now be considered under three main headings: town-plan analysis, the cyclical approach, and agents of change. While it is possible to include under one or other of these headings most of the main developments in geographical urban morphology in the past 20 years, the treatment does not purport to be exhaustive. Two major studies (Johns, 1965; Vance, 1977) do not fit satisfactorily under these headings, nor do they belong to the major traditions depicted in Figure 9.1. The particular attention given to work in the Conzenian tradition, or having antecedents in it, reflects both the author's direct involvement in this work and the pace of research stemming from this tradition in recent years. No attempt is made to consider studies of urban landscape perception, and reference to urban morphological research in other disciplines is made only where it clearly overlaps with that in geography.

Town-Plan Analysis

After the conceptual richness and analytical depth of Conzen's contributions to town-plan analysis in the early 1960s, this aspect of urban morphology has developed only slowly in the past 20 years. In the light of the central European ante-cendence of Conzen's work it is remarkable that most of the scholars taking up his ideas have been British. When his publications are viewed in terms of their accessibility to scholars in various countries, however, this is more under-standable. For little of his work was published in German, and at the time that the studies of Alnwick and central Newcastle were published German geographers were apparently referring hardly at all to the English-language literature (Whitehand and Edmondson, 1977). Within English-speaking countries the weight of scholarship that Conzen's studies demonstrated and their unfamiliar intellectual perspective from the standpoints of Britain and America may well have deterred potential emulators. In the 1960s geography in English-speaking countries was entering a phase of its development in which there was generally less interest in maps and landscapes. Moreover, when there was a revival of interest in the landscape in the second half of the 1970s, the concern was more with the perception of landscape than with its historical development (Whitehand, 1981b, p. 138).

Although Straw (1967) wrote a thesis on the lines of the Alnwick study, the investigation of the Scottish town of St Andrews by Brooks and Whittington (1977) was until recently the nearest approach in the English-language literature to a major published analysis of the plan of an individual town since Conzen's study of central Newcastle. To draw parallels with either the Alnwick study or that of central Newcastle, however, would be misleading, since Brooks and Whittington tend to eschew the conceptualisation of physical developments. Furthermore, they consider the medieval period only, and are concerned to a large extent with evaluating a 'Bird's Eye View' of St Andrews. More representative of the Conzenian tradition of town-plan analysis is the work of Slater (1985; 1986). Using techniques of analysis and reconstruction similar to Conzen's, he has elucidated the stages in the medieval plan development of two English towns, Hedon in east Yorkshire and Lichfield in the west Midlands. In teasing out the early topographical development of these towns particular use is made of Conzen's concept of 'plan-units', namely individual-ised combinations of streets, plots and buildings distinct from their neighbours. Both the longevity of key elements in plan development and the complexity and compositeness of super-ficially unitary parts of town plans confirm Conzen's findings. Nearly all the recent town-plan analyses of comparable depth have come from central Europe. German historians have been particularly active in using the topographical and cartographic record, often as an essential complement to archaeological and documentary evidence, in uncovering the subtleties of town-plan development (Keyser, 1958; Stoob, 1973, 1979; Fahlbusch, 1984), although the long-established interest of German geographers in town-plan analysis continues (Lafrenz, 1977; Meynen, 1984).

An important basis for the comparative study of town plans is being provided by the international project for the publication of historical town-plan atlases (Hall and Borgwik, 1978). The most sophisticated contributions so far have come from West Germany, where there is a long tradition of com-parative research. The first attempt at a comparative study in Britain using Conzenian terminology was by Whitehand and Alauddin (1969). In attempting a preliminary survey of at least some aspects of all the major town-plan elements (street systems, plot patterns and building block-plans) for the whole of Scotland, this study drew attention to both the potentialities of such studies and the dangers of succumbing to a naive morphography without the combination in each town of ground survey, cartographic analysis and documentary evidence.

URBAN MORPHOLOGY

An attempt has recently been made by Slater (1982) to mitigate this problem. Concentrating on just two English counties, Warwickshire and Worcestershire, he restricted attention to two key-elements in the town plan - market places and burgage series. Limiting the field of investigation in this way still does not make it practicable to undertake a proper morphogenetic study of all towns in the study area, but Slater's descriptive categorisations of market-place shapes are at least informed by selective information on origins and evolution. Variations in the breadth to depth ratio for burgages would seem to be related to whether plots were laid out as part of a planned new town or developed from former village tofts. Such studies are at least beginning to fill the gap in our knowledge of British town plans, but they are still largely descriptive accounts informed by a detailed knowledge of Conzenian concepts (Whitehead, 1981b, p. 129).

Recognising this, Slater has undertaken detailed analyses of burgages, the basic 'cells' of the medieval town plan. Conzen's analyses of large-scale plans, suggesting an original standard burgage width in Alnwick of 28-32 feet, revealed in survivals of this width and in such fractions and multiples of it as 0.5, 0.75, 1.25 and 1.5, have been followed up by Slater (1981) in his analysis of modern plot frontages especially in the English midlands and south-west England. A convincing case is made for plot measurement and recon-struction as a method of establishing an outline pattern of development that could only be established otherwise from comprehensive documentation or by extensive archaeological excavation. In addition the method enables Conzen's hypoth-esis of a 'standard' burgage width to be tested in towns whose archives lack the very large-scale early maps with which he worked.

Despite the growing interest in burgage patterns, the burgage cycle has received comparatively little attention. The only major exception in the English language is the study of central Nottingham by Straw (1967). The most detailed attempt in recent years to apply the same idea is the study in Polish of central Łódź (Koter and Wiktorowska, 1976). In this instance the modern central business district developed not in the medieval core but in an area that had been laid out as an industrial-residential area in the early nineteenth century. None the less, developments comparable to the institutive, repletive, climax and recessive phases that had been recog-nised by Conzen in his burgage cycle are demonstrated in an analysis based on cartographic evidence for seven cross-sections in time between 1824/27 and 1973.

Outside the Conzenian tradition, there have been a variety of studies of individual town-plan elements, notably street plans (for example, Pillsbury, 1970), but each study has tended to stand on its own, rather than contribute to a wider framework of knowledge. An integrated body of concepts is absent (Conzen, 1978, pp. 147-52). A rare attempt to summarise the varied work on the United States and view it in the context of Conzenian concepts is that of M.P. Conzen (1980). Confining attention to the nineteenth century, he examines colonial antecedents, new town foundations, and mature town accretions and modifications both on urban fringes and within densely built-up urban cores. He also suggests a preliminary division of the century into three morphogenetic periods. This consists of an early phase, continuing colonial traditions of low density and weakly differentiated land-use structure, which gave way in the 1820s and 1830s to a transitional phase of major central density increases, centrifugal and centripetal land-use sorting, and major experiments in new building forms. This middle phase was superseded in the late 1870s by a third phase in which transport took on a particularly important articulating role and urban form was characterised by high density, CBD domination, and residential segregation.

The Cyclical Approach

The continuing vitality of town-plan analysis has depended on the efforts of a small number of historical geographers and historians. However, the significance of this research, at least that part of it in the Conzenian tradition, is out of proportion to the numbers pursuing it, for it has provided one of the seed-beds for studies of wider aspects of urban form, especially the cyclical character of land utilisation and building form.

The most important link between Conzenian town-plan analysis and the cyclical approach to the development of land-use patterns is the fringe-belt concept. The potentialities of this link had already been recognised by Conzen (1969, p. 125) some years before they were subjected to detailed investigation. Indeed Louis's (1936) Stradtrandzonen had been conceived primarily in terms of land use.

Instead of treating fringe belts as an integral part of a wider town study, Whitehand (1967) examined them in their own right. The basic idea, approach and terminology, however, were similar to those employed in Conzen's studies of Alnwick and central Newcastle. The alternating phases of residential accretion and fringe-belt creation and the subsequent evolution of the zones thereby created were a fundamentally different conception from the models of Burgess and Hoyt, which were so popular among English-speaking, especially American, urban geographers at the time. It is not surprising, therefore, that the first attempt to apply the fringe-belt concept to a North American city (M.P. Conzen, 1968) seemed foreign to its American audience. This application to Madison, Wisconsin was important in providing empirical support for the hypothesis that fringe belts become consolidated over time by attracting compatible land uses. It also drew attention to the need to integrate the fringe-belt concept with general theories of urban and economic growth.

The fringe-belt concept was taken up more widely in the 1970s. For the first time research on a French town (Clermont Ferrand) was undertaken (Whitehand, 1974), and revealed the pronounced differences in pattern on the ground, though not in formative processes, that occurred where the onset of the Industrial Revolution was markedly retarded. Setting the fringe-belt concept in a wider context, Carter (1978) treated it as a basis for schematising the internal structure of urban areas, concentrating on the industrial town in England and Wales between 1730 and 1900 and incorporating into his schema the development of the city centre and residential areas. In German-speaking Europe, however, there were no direct successors to Louis's original study of Berlin, despite the translation and development, albeit belatedly, of his idea in Britain.

The 1970s were characterised by attempts to relate the development of the form of urban areas to economic fluctuations generally and building cycles in particular. In his study of changes in the form of working-class housing, Forster (1972) recognised bye-law cycles which he related to building cycles. The links between building cycles and the fringe-belt concept were soon explored (Whitehand, 1974). The incorporation of rent theory into the discussion, however, introduced an element foreign to the morphogenetic tradition (Whitehand, 1972a). In particular the incorporation within morphogenetics of the deductive element that plays such an important role in the constructs of urban-rent theorists was a new departure.

The initial attempt to develop links between rent theory and cycles of urban growth was only indirectly concerned with fringe belts. Concentrating on the spatial pattern of land uses locating at the urban fringe, Whitehand (1972b) focused attention on the broad, heterogeneous category of institutional land uses that had been largely ignored by land economists. Noting their tendency to have linkages with the urban area that were weaker than those of housing areas and their comparatively large space requirements relative to their investment in buildings and other site improvements, he postulated that institutions would tend to have shallower bid-rent curves than residential developments and, ceteris paribus, less accessible locations. However, owing to the tendency for the development of land for institutional purposes to be less subject than housebuilding to large-scale, long-term fluctuations, the bid-rent curve of institutions could be envisaged as sliding less far down the rent axis during housebuilding slumps. Thus sites that in times of a buoyant land market and high land values would have been utilised for housebuilding became more susceptible to occupation by institutions. The long-term outcome viewed in terms of a transect from city centre to rural fringe was for the varying admixture of new housing and new institutional developments to present a landscape resembling alternating zones of residential accretion and fringe-belt development. The pattern of development in north-west Glasgow between 1840 and 1923 was found to be largely consistent with such a schema and the idea of alternating periods of housebuilding boom and slump being associated more generally with variations in the intensity of new development was found to be consistent with the growth of a largely residential part of the fringe of west London between 1826 and 1869 (Whitehand, 1975). Parkes and Thrift (1980, p. 427) subsequently presented a schematisation of the relations between building cycles, land values, fringe-belt formation, and variations in the density of residential accretion.

An extension of research on the urban fringe was the investigation of the changes that fringe areas underwent when they became encompassed within the built-up area. Noting the similarity between fringe belts and zones with a high admixture of institutions, Whitehand (1972a, 1974) examined the paradox that, despite the fact that it had been well established that fringe belts not only tended to remain in existence for lengthy periods but, under certain circumstances, expanded into adjoining areas, the bid-rent mechanism as conceived up to that time implied that institutions would be displaced by housing in the next, or at least a subsequent, housebuilding boom. He offered a theoretical explanation in terms of the increases that tended to take place in the bid rents of institutions for their own and adjacent sites relative to those of housebuilders. These increases were related to the tendency for more capital to be invested over time in institutional sites than in housing sites and for the strength of the linkages between institutions and the urban area to increase as institutions became well established on their sites. At the same time, as institutions grew they tended to relocate those parts of their activities having weaker linkages to the urban area, for example sports fields, on less accessible, cheaper land at the urban fringe. In this way locational decisions concerning different types of land use were effectively embodied within the decision-making of a single large organisation, much as they were in the case of local government.

Evidence supporting this reasoning is by no means complete, but certain aspects of the changes undergone by fringe belts after their initial establishment have been the subject of empirical studies. These have shown that it is not only institutions that are long-established on their sites that are able to compete successfully for internal sites. There may be a succession of similar uses on the same site, and, since there is a secular trend of rising land values, each tends to be more intensive than its predecessor (Whitehand 1974 pp. 48-9; Barke, 1976). Explanations that have been advanced for this include the fact that some of the assets accruing to a particular land use on a particular site, for example buildings and compatible environment, can be passed on to a related use (Whitehand, 1974, p. 48). This has been related to the displacement of some institutions to sites progressively farther from the city centre (Whitehand, 1974, p. 49). In his study of the sequence of land uses in fringe belts in the Scottish town of Falkirk Barke (1976) found considerable variations between both uses and locations relative to the town centre in their susceptibility to change. He noted, for example, how certain sequences of change (for instance, the change from public open space to community buildings) were associated with continuity of ownership, especially where the owner was the local authority.

Simultaneously with these empirical studies, Openshaw was attempting to construct a theoretical framework on a more ambitious scale, in the belief that fringe-belt theory could provide 'a unitary explanatory framework for the morphological and functional study of the townscape in a time extended framework' (Openshaw, 1974, p. 20). He was concerned with both the formation phases of fringe-belt development and those of subsequent change as fringe belts became enveloped within a built-up area, but he was especially interested in the functional linkages of fringe-belt land uses with other parts of the urban area. Broadening the scope from what he believed to have been an overconcern with the 'medieval fringe-belt model' (Openshaw, 1974, p. 6) in which developments are viewed in relation to a single-centre urban area with a long history, he felt the need to develop a more general model of the fringe-belt process that did not assume a specialised historical context and to show that the model was consistent with existing theories of urban economic structure. In particular, Openshaw drew attention to the significance of 'fringe-belt cores', which might not only be central business cores, around which fringe belts develop. He speculated on the process of relocation associated with the decline in the functional significance of a fringe-belt core, which could involve the 'capture' of a fringe belt of a declining core by a growing one, the transfer of core and fringe-belt functions to a growing core and fringe belts, or the development of a declining inner fringe belt into an incipient middle fringe belt for a growing core. Emphasis was placed on the various functional linkages that land uses within fringe belts might have - for example, distinguishing in the case of middle and outer fringe-belt uses between those linked to the urban area as a whole and those linked to local residential accretions. The variety of functional links that a fringe belt might have in a large urban area was regarded as crucial, especially in a conurbation containing a number of centres, and it was concluded that the location of fringe belts might not be related at all to the proximity of a town centre or to the economics of peripheral location. The empirical challenge that this conception presents has yet to be taken up.

Openshaw's schema was an attempt to construct a general theory of the development of urban form based on the fringe-belt concept after a period during which fringe belts had been studied largely as separate entities from other processes. This return to a more integrated view was confirmed in the studies by Slater (1978) and Carter and Wheatley (1979). Slater augmented the fringe-belt concept as it had become accepted by the late 1970s, especially its conception in primarily economic terms, by examining the relationship between the timing of the creation, modification and redevelopment of a single category of fringe-belt land use - the nineteenth century ornamental villa - and significant stages in the life cycle of the occupants. He concluded that the stage reached in the family life cycle was a significant factor in determining the timing of the adjustment of the landscape to changing intra-urban location and more general economic climate. Carter and Wheatley were also concerned with social considerations, emphasising the relationship between fringe-belt development and the development of the distribution pattern of social classes.

This return to a more integrated view is also apparent in Whitehand's (1977) attempt to provide the basis for a historico-geographical theory of urban form. Though not couched explicitly in terms of fringe-belt development, it bears a family resemblance to the studies just considered. It departs from them in assigning an explicit role to the adoption and diffusion of innovations. Indeed, it is difficult to envisage a realistic theory of the long-term development of industrial cities that does not take some account of the course of innovation.

The roles of innovation and diffusion complement the roles of building cycles and bid rents already discussed. It is possible to recognise periods of rapid adoption of innovations with a direct influence on the urban landscape, separated by periods of relative quiescence. The relationship between certain innovations appears to be fairly direct (for instance between certain town-planning innovations and innovations in building) whereas in other cases it is likely to be indirect (for instance between innovations in retailing and those in constructional materials for housing). The explanation most probably lies in the connecting links between a great variety of types of innovation and economic cycles. Indeed it seems plausible that innovation stimulates constructional activity, although the fact that innovations have sometimes lagged behind, rather than led, constructional activity suggests that other factors are at work. It should also be noted that the time spans over which individual innovations are adopted are considerably more variable in length than building cycles. Although some innovations are associated only with the annulus of building created by one particular boom in construction activity (Adams, 1970), others are associated with more than one boom (Whitehand, 1977, p. 406). The tendency for residential zones of similar architectural styles to be separated by fringe belts created during housebuilding slumps is further reduced by the fact that not all outward growth is contiguous.

Innovations in transport play a special role in that they change the nature of the accessibility surface which itself performs such an important role in the process whereby activities acquire sites. It might be expected that the adoption of such innovations as suburban railways and motor buses would have been associated with proportionately higher increases in peripheral land values, whereas during periods of little change in the nature of transport increases in land values would have been proportionately higher in the city centre (Hoyt, 1933, pp. 336-7). However, the fact that transport innovations tend to be associated with increased building activity, which is itself associated with increases in land values, makes it difficult to assess the relative strengths of the underlying influences.

This schema of urban development has implications for parts of the urban area to which little reference has been made, such as city centres (Whitehand, 1978; 1983a), and to inter-urban aspects of form (Whitehand, 1977, pp. 409-12). It has been reviewed by Gordon (1981; 1984), who sees the need for either the explicit incorporation of the behaviour of those whose decisions influence urban form or, preferably, an alternative framework in which they, rather than innovation, accessibility, and constructional cycles, are the primary elements. This alternative approach has been favoured in a number of studies and is an important complement to the perspectives already considered.

The Agents of Change
Within urban geography the individuals and organisations responsible for urban development have until recent years remained a largely unknown quantity. A number of studies by historians, however, have helped to draw the attention of geographers to their significance (for example, Dyos, 1968; Hobhouse, 1971; Thompson, 1974). Moreover, quantitative analyses of residential development, such as that by Craven (1969), and of the building industry by Aspinall (1982), though not directly concerned with the urban landscape, have made apparent the need for more rigorous analysis of the relationship between forms in the landscape and the organisations and individuals collectively responsible for the anonymous forces and cyclical processes that have attracted

so much attention. Forerunners of the increased interest among geographers in the agents of change were the study by Johns (1971) of the Devon towns of Dawlish and Chelston and Carter's (1970) 'decision-making' approach to town-plan analysis, although the latter was more concerned with decision making than with decision makers. The recent attempts by Gordon (1981; 1984) to produce an organisational framework for urban morphology in which decision-makers and decision-making are primary elements have generated further interest, and the study of 150 years of office development in Toronto by Gad and Holdsworth (1984) has shown the insights that can be obtained when building form is related to both the activities of individual organisations and the changing social and economic environment in which they exist. So far the main emphasis in the recent expansion of this work has been on the relationship between the various agents of change and building form.

Whereas the cyclical approach in urban morphology has been used predominantly in studies of the outward growth of urban areas, analyses of the interrelations between agents of change and the building forms created at the urban fringe are comparatively rare. Bastian (1980) is one of the few to have examined the connection between house style and the architect-client relationship, although recently Trowell (1985) has taken a major step forward by his detailed analysis of the roles of landowners, developers and architects in the development of a Leeds suburb between 1838 and 1914. Most analytical studies of the roles of the various agents of change in shaping the physical form of urban areas have been concerned with town- and city-centres, mostly within Britain, many of them having been undertaken by the Urban Morphology Research Group at Birmingham University. A brief review of the nature and some of the more important conclusions of this work to date will suffice to indicate both links with longer-established approaches and essentially new findings.

One respect in which these studies are consistent with research in the morphogenetic tradition is in the importance they attach to legacies from the past (Whitehand, 1984; Freeman, 1986a). Studies of the development since the First World War of examples of county towns and suburban towns in south-eastern and midland England suggest that changes to commercial cores have been heavily influenced by existing morphological frames. These acted as moderating influences on attempts to enlarge the scale of redevelopment. Changes were over-whelmingly adaptive rather than augmentative, to use Conzen's terms, although the sizes of redeveloped sites increased. The increasing scale of redevelopment produced a reaction in the 1970s, reflected in attempts to recreate traditional scale frontages by designing long-fronted buildings so that they had the appearance of more than one building.

Legacies from the past also exercised an influence through the antecedence of firms and organisations. As a major provincial central place well away from London, Northampton entered the twentieth century well endowed with locally-based firms and organisations - professional, commercial and cultural. These played a major part in determining the type and architectural style of redevelopments in the town centre in the late 1920s and 1930s. In contrast, Watford, a London suburb, though growing much more rapidly than Northampton in the 1920s, was comparatively poorly endowed with local firms and was more reliant for redevelopments in its town centre on speculative ventures by private individuals (Whitehand, 1984, pp. 10-28).

The reciprocal relations between form and function, which constitute a major underlying theme in the morphogenetic approach, are seen in a different light when attention is focused on the agents of change. Instead of thinking of function largely in terms of the use a building performs - for instance as a shop or an office - a prime consideration becomes the role that buildings perform in providing a financial return for those involved in creating, managing and using them. Some owners view buildings primarily as investments, while for others buildings are primarily containers for their businesses. The principal interests at work are often those of developers and investors rather than building users. Though the tradition of erecting new structures on speculation goes far back in history, its incidence in British commercial cores has tended to increase in recent decades, at least by comparison with the first half of this century. In his study of nine town- and city-centres in west Yorkshire between 1945 and 1968, Bateman (1971, p. 26) revealed that the median proportion of private-sector redevelopment that was speculative was two-thirds. In the case of office development in the city of London, Barras (1979, pp. 50-53) found that a similar proportion of schemes with over 10,000 m² of gross floorspace were speculative during the 1970s: this compared with about one-half in the 1950s and 1960s, although by the end of the 1970s users were again tending to increase their share of development. Over the last 50 years increasing proportions of redevelopments in the town centres of Northampton, Aylesbury and Wembley have been speculative, but not of those in central Watford (Whitehand, 1984; Freeman, 1986a). This trend has implications both for the timing of development, since in building in general it would seem that speculative building has held up less well in economic slumps than building for owner occupation (Whitehand, 1981c), and for its style, since it has been suggested, again in building generally, that 'fashions in speculative markets tended to follow those in the bespoke market at a distance' (Bowley, 1966, p. 390). This last view

is borne out by Whitehand (1984) but not by Freeman
(1986a).
Two secular changes emerge as being of major importance
in the development of commercial cores since the First World
War (Whitehand, 1984; Freeman, 1986a). First, there was the
large-scale entry of retail chainstores, many of them public
companies, into property development in the 1930s. Many
chainstores already operated nationally and had distinctive
house styles reproduced by their own architects' departments
or by architects with whom they had standing relationships.
It thus became increasingly common for buildings of similar
appearance to be erected in town centres over the whole of
Britain. The population catchment considered by a particular
chainstore to be necessary for it to function profitably became
an increasingly important factor underlying the admixture of
building styles present in a particular town centre. Secondly,
the involvement of property companies and insurance companies in the speculative development and ownership of
property grew slowly in the 1930s and then rapidly from the
late 1950s onward. Most of these companies were operating on
a national scale. Together with the retail and service chains
they largely replaced the local property developers and
owners. By the 1930s in the London suburban centres of
Watford and Wembley and at least by the mid 1970s in the
county towns of Northampton and Aylesbury, these national-
scale developers were mainly commissioning architects from
outside the towns in which the developments took place,
particularly London-based firms in the case of the suburban
centres. This, together with the tendency for the retail and
service chains to employ their own architects, meant that new
buildings designed by local architects were comparatively few.
There is at least an inference that this may be significant for
the sense of place of firms having a key role in townscape
change (Whitehand and Whitehand, 1984 p. 245).
These major secular changes are bound up with the
growing concentration of development activity nationally in the
hands of major firms operating countrywide. It is a reasonable
inference that six development companies accounted for well
over one-half of the major shopping schemes undertaken in
Britain between 1965 and 1978 (Hillier Parker Research, 1979,
p. 33), and Barras (1979, p. 55, App. 2) has shown that six
development companies accounted for about one-half of the
major office redevelopments in the City of London between
1959 and 1979. Though systematic research has yet to be
undertaken, a similar concentration is assumed to have taken
place in architectural design (Marriott, 1967, pp. 27-9).
However, this has not necessarily led to the concentration of
activity in the hands of fewer firms within individual town
centres. In Northampton and Watford, though not in
Aylesbury and Wembley, there has actually been a greater
dispersion of activity among architects in recent decades

(Whitehand, 1983b; Freeman, 1986b, p. 178). This should be
viewed in the context of the wider geographical spread in the
post-war period of the activities of all types of firms involved
in the development process (Whitehand and Whitehand, 1984).
Whereas in the inter-war period a few local firms were responsible for a sizeable proportion of the redevelopments
undertaken, the national firms of recent decades have seldom
undertaken more than two or three redevelopments in a
particular town centre. This is true of building owners,
architects and builders. Thus the notion that the supplanting
of local firms by national ones has necessarily created greater
homogeneity among additions to the townscape must be viewed
with caution. In fact since the 1960s there has been an
increasing diversity of architectural styles within the town
centres so far studied, although this probably reflects a
phase of increased pluralism in architecture rather than any
dispersion in the number of firms accepting commissions in
particular town centres (Whitehand, 1983b).

Urban Landscape Management
A danger with research that approaches urban morphology
through the agents of change is that the agents themselves
become the object of study and sight is lost of the very
places that are the object of the geographer's concern. The
morphogenetic tradition, rooted as it is in the evolving form
of places, provides a safeguard against this. It is appropriate
in conclusion to return to this tradition to consider the
suggestion that it can form a basis from which historical
analysis can be linked to the future management of the urban
landscape (Conzen, 1966; 1975; Whitehand, 1981b).
Although thorough-going morphogenetic analyses of
urban areas have been few in number in the last decade,
especially outside the German-speaking world, Conzen himself
has expressed the view that this inherently historico-
geographical perspective is capable of yielding a conceptual
basis for planning practice (Conzen, 1975). This view is
grounded in the notion that the key to informed management
of the urban landscape is understanding how the landscape
has evolved. The Conzenian townscape is a stage on which
successive societies work out their lives, each society
learning from, and working to some extent within the framework provided by, the experiments of its predecessors.
Viewed in this way urban landscapes represent accumulated
experience, old-established landscapes especially so, and are
thus a precious asset. This asset, according to Conzen, is
threefold. First, it has practical utility in providing
orientation: our mental map and therefore the efficiency with
which we function spatially is dependent on our recognition of
the identity of localities. Secondly, it has intellectual value
by helping both individual and society to orientate in time:

through its high density of forms a well-established urban landscape provides a particularly strong visual experience of the history of an area, helping the individual to place himself within a wider evolving society, stimulating historical comparison and thus providing a more informed basis for reasoning. Thirdly, and more contentiously, the combination of forms created by the piecemeal adaptation, modification and replacement of elements in old-established urban landscapes has aesthetic value: for example, in the maintenance of human scale, in the visual impact of and orientation provided by dominant features, such as churches and castles, and in the stimulus to the imagination provided by variations in street width and orientation. Clearly all three assets are interrelated, and emotional and aesthetic experiences are tightly intertwined with appreciation of historical and geographical significance.

If a key concept for future research is to be singled out from Conzen's papers on the management of the urban landscape, then prime consideration should be given to the townscape as the 'objectivation of the spirit of a society', viewed not at a moment in time but as a historical phenomenon. As change after change takes place on the ground, consisting of additions, modifications and subtractions, the urban landscape encapsulates the history of a society in a particular locale. An important aspect of this from the geographical standpoint is the variation between urban landscapes in the nature and intensity of their historical expressiveness. This can provide a basis for the identification of urban landscape units. These in turn provide a framework within which conservation priorities may be determined.

The practical aspects of this are as yet poorly worked out. An integrated study of town plan, building form, and land and building utilisation is a prerequisite, but in practice these three elements have tended to be the subject of separate investigations in the English-speaking world. One possible avenue of research would be to investigate the theoretical and practical problems of effecting for a sample of towns a comparison of theoretical developments based on Conzenian notions of urban landscape management with actual developments. The historical overtones of the current conservation movement and the continuing reaction against modern architecture and large-scale redevelopment provide a social climate favourable for such research. The opportunity exists to provide the sense of direction and conceptual basis that is largely lacking in recent attempts to conserve the urban landscape. What is particularly needed is a conception of how some parts of the urban landscape have a character distinctive from others that relates to their history and that of the community that created them, and of how individual developments from different historical periods fit together. Since Conzen's contribution to the management of the urban

URBAN MORPHOLOGY

landscape is concerned primarily with these matters, it can provide a theoretical underpinning where it is most needed.

As Conzen himself has suggested, the development and harnessing of his ideas for the purposes of managing the urban landscape is a natural extension of town-plan analysis and of some aspects of the cyclical approach. However, the connecting links between such ideas and the findings of recent analyses of the roles of agents of change are still in need of articulation. This task is near the top of the agenda for urban morphology in the next decade. Its satisfactory completion, however, awaits a thorough explication of the relationship between the changing form of the urban landscape and the agents of change. On this score the work that has been reviewed here reveals significant progress, but it also suggests that a sizeable task remains.

REFERENCES

Adams, J.S. (1970) 'Residential Structure of Midwestern Cities', Annals of the Association of American Geographers, 60, 37-62

Aspinall, P.J. (1982) 'The Internal Structure of the House-building Industry in Nineteenth-Century Cities' in J.H. Johnson and C.G. Pooley (eds.), The Structure of Nineteenth Century Cities, Croom Helm, London, pp. 75-105

Barke, M. (1976) 'Land Use Succession: A Factor in Fringe-Belt Modification', Area, 8, 303-6

Barras, R. (1979) The Development Cycle in the City of London, Centre for Environmental Studies, London

Bastian, R.W. (1980) 'The Prairie Style House: Spatial Diffusion of a Minor Design', Journal of Cultural Geography, 1, 50-65

Bateman, M. (1971) Some Aspects of Change in the Central Areas of Towns in West Yorkshire since 1945, Portsmouth Polytechnic Department of Geography, Portsmouth

Blanchard, R. (1912) Grenoble: Etude de Géographie Urbaine, Armand Colin, Paris

Bobek, H. (1927) 'Grundfragen der Stadtgeographie', Geographischer Anzeiger, 28, 213-24

Bobek, H., and Lichtenberger, E. (1966) Wien: Bauliche Gestalt und Entwicklung seit der Mitte des 19. Jahrhunderts, Böhlau, Graz

Bowley, M. (1966) The British Building Industry: Four Studies in Response and Resistance to Change, Cambridge University Press, Cambridge

Brooks, N.P., and Whittington, G. (1977) 'Planning and Growth in the Medieval Scottish Burgh: The Example of St Andrews', Transactions of the Institute of British Geographers N.S., 2, 278-95

Burgess, E.W. (1925) 'The Growth of the City' in R.E. Park,
E.W. Burgess and R.D. McKenzie (eds.), The City,
University of Chicago Press, Chicago, pp. 47-62
Carter, H. (1965) The Towns of Wales, University of Wales
Press, Cardiff
Carter, H. (1970) 'A Decision-Making Approach to Town Plan
Analysis: A Case Study of Llandudno' in H. Carter and
W.K.D. Davies (eds.), Urban Essays: Studies in the
Geography of Wales, Longman, London, pp. 66-78
Carter, H. (1978) 'Towns and Urban Systems 1730-1900' in
R.A. Dodgshon and R.A. Butlin (eds.), An Historical
Geography of England and Wales, Academic Press,
London, pp. 367-400
Carter, H., and Wheatley, S. (1979) 'Fixation Lines and
Fringe Belts, Land Uses and Social Areas: Nineteenth-
Century Change in the Small Town', Transactions of the
Institute of British Geographers N.S., 4, 214-38
Christaller, W. (1933) Die Zentralen Orte in Süddeutschland,
Fischer, Jena
Colby, C.C. (1933) 'Centrifugal and Centripetal Forces in
Urban Geography', Annals of the Association of American
Geographers, 23, 1-20
Conzen, M.P. (1968) 'Fringe Location Land Uses: Relict
Patterns in Madison, Wisconsin', unpublished paper
presented to the 19th Annual Meeting of the Association
of American Geographers West Lakes Division, Madison,
Wisconsin
Conzen, M.P. (1978) 'Analytical Approaches to the Urban
Landscape' in K.W. Butzer (ed.), Dimensions in Human
Geography: Essays on Some Familiar and Neglected
Themes, University of Chicago Department of Geography,
Chicago, pp. 128-65
Conzen, M.P. (1980) 'The Morphology of Nineteenth-Century
Cities in the United States' in W. Borah, J. Hardoy and
G. Stelter (eds.), Urbanization in the Americas: The
Background in Comparative Perspective, National Museum
of Man, Ottawa, pp. 119-41
Conzen, M.R.G. (1958) 'The Growth and Character of Whitby'
in G.H.J. Daysh (ed.), A Survey of Whitby and the
Surrounding Area, Shakespeare Head Press, Eton, pp.
49-89
Conzen, M.R.G. (1960) Alnwick, Northumberland: A Study in
Town-Plan Analysis, George Philip, London
Conzen, M.R.G. (1962) 'The Plan Analysis of an English City
Centre' in K. Norborg (ed.), Proceedings of the IGU
Symposium in Urban Geography, Lund 1960, Gleerup,
Lund, pp. 383-414
Conzen, M.R.G. (1966) 'Historical Townscapes in Britain: A
Problem in Applied Geography' in J.W. House (ed.),
Northern Geographical Essays in Honour of G.H.J.

Daysh, University of Newcastle upon Tyne, Newcastle upon Tyne, pp. 56-78

Conzen, M.R.G. (1969) Alnwick, Northumberland: A Study in Town-Plan Analysis, 2nd edition, Institute of British Geographers, London

Conzen, M.R.G. (1975) 'Geography and Townscape Conservation' in H. Uhlig and C. Lienau (eds.), Anglo-German Symposium in Applied Geography, Giessen-Würzburg-München, 1973, Lenz, Giessen, pp. 95-102

Davies, W.K.D. (1968) 'The Morphology of Central Places: A Case Study', Annals of the Association of American Geographers, 58, 91-110

Dickinson, R.E. (1942a) 'The Development and Distribution of the Medieval German Town. I: The West German Lands', Geography, 27, 9-21

Dickinson, R.E. (1942b) 'The Development and Distribution of the Medieval German Town. II: The Eastern Lands of German Colonisation', Geography, 27, 47-53

Dickinson, R.E. (1945) 'The Morphology of the Medieval German Town', Geographical Review, 35, 74-97

Dickinson, R.E. (1948) 'The Scope and Status of Urban Geography: An Assessment', Land Economics, 24, 221-38

Dyos, H.J. (1968) 'The Speculative Builders and Developers of Victorian London', Victorian Studies, 11, 641-90

Fahlbusch, F.B. (1984) 'Die Wachstumsphasen von Duderstadt bis zum Übergang an Mainz 1334-66' in H. Jäger, F. Petri and H. Quirin (eds.), Civitatum Communitas: Studien zum Europäischen Städtewesen, Bohlau, Koln, pp. 194-212

Fleure, H.J. (1920) 'Some Types of Cities in Temperate Europe', Geographical Review, 10, 357-74

Forster, C.A. (1972) Court Housing in Kingston upon Hull: An Example of Cyclic Processes in the Morphological Development of Nineteenth Century Bye-Law Housing, University of Hull Department of Geography, Hull

Freeman, M. (1986a) Town-Centre Redevelopment: The Roles of Developers and Architects, University of Birmingham Department of Geography, Birmingham

Freeman, M. (1986b) 'The Nature and Agents of Central-Area Change: A Case Study of Aylesbury and Wembley Town Centres 1935-1983', unpublished Ph.D. thesis, University of Birmingham

Gad, G., and Holdsworth, D. (1984) 'Building for City, Region and Nation' in V.L. Russell (ed.), Forging a Consensus: Historical Essays on Toronto, University of Toronto Press, Toronto, pp. 272-319

Geisler, W. (1924) Die Deutsche Stadt: Ein Beitrag zur Morphologie der Kulturlandschaft, Engelhorn, Stuttgart

Gordon, G. (1981) 'The Historico-Geographic Explanation of Urban Morphology: A Discussion of Some Scottish Evidence', Scottish Geographical Magazine, 97, 16-26

URBAN MORPHOLOGY

Gordon, G. (1984) 'The Shaping of Urban Morphology', Urban History Yearbook, 1-10

Hall, T., and BorgWik, L. (1978) 'Urban-History Atlases: A Survey of Recent Publications', Särtryck ur Historisk Tidskrift 1978, Norstedts Tryckeri, Stockholm, pp. 305-19

Harris, C.D., and Ullman, E.L. (1945) 'The Nature of Cities', Annals of the American Academy of Political and Social Science, 242, 7-17

Hartke, W. (1953) 'Die Soziale Differenzierung der Agrarlandschaft im Rhein-Main-Gebeit', Erdkunde, 7, 13-22

Hassinger, H. (1927) 'Basel: Ein Geographisches Städtebild' in F. Metz (ed.), Beitrage zur Oberrheinischen Landeskunde, (Karlsruher Geographentag), Breslau, pp. 103-30

Hillier Parker Research (1979) British Shopping Developments, Hillier Parker May and Rowden, London

Hobhouse, H. (1971) Thomas Cubitt: Master Builder, Macmillan, London

Hoyt, H. (1933) One Hundred Years of Land Values in Chicago, University of Chicago Press, Chicago

Hoyt, H. (1939) 'The Pattern of Movement of Residential Rental Neighborhoods' in The Structure and Growth of Residential Neighborhoods in American Cities, Federal Housing Administration, Washington D.C., pp. 112-22; reprinted in H.M. Mayer and C.F. Kohn (eds.), Readings in Urban Geography, University of Chicago Press, Chicago, pp. 499-510

Jakle, J.A. (1983) 'Twentieth Century Revival Architecture and the Gentry', Journal of Cultural Geography, 4, 28-43

Johns, E. (1965) British Townscapes, Edward Arnold, London

Johns, E. (1971) 'Urban Design in Dawlish and Chelston' in K.J. Gregory and W. Ravenhill (eds.), Exeter Essays in Geography in Honour of Arthur Davies, University of Exeter, Exeter, pp. 201-8

Johnston, R.J. (1969) 'Towards an Analytical Study of the Townscape: The Residential Building Fabric', Geografiska Annaler, Series B, 51, 20-32

Keyser, E. (1958) Städtegrundüngen und Städtebau in Nordwestdeutschland im Mittelalter, Bundesanstalt fur Landeskunde, Remagen

Koter, M., and Wiktorowska, D. (1976) 'Proces Przemian Morfologicznych Śródmieścia Lodzi (w Granicach Bytej Kolonii Tkackiej) Pod Wptyem Ksztattowania się Ogólnomiesjskiego Centrum Ustugowego', Acta Universitatis Lodziensis, Series II, 7, 41-88

Krings, W. (1984) Innenstädte in Belgien: Gestalt, Veränderung, Erhaltung (1860-1978), Ferd, Dümmlers, Bonn

URBAN MORPHOLOGY

Lafrenz, J. (1977) Die Stellung der Innenstadt im Flächennutzungsgefüge des Agglomerationsraumes Lubeck, Hirt, Hamburg

Leighly, J.B. (1928) 'The Towns of Mälardalen in Sweden: A Study in Urban Morphology', University of California Publications in Geography, 3, 1-134

Lichtenberger, E. (1977) Die Winer Altstadt: Von der Mittelalterlichen Bürgerstadt zur City, Franz Deuticke, Wien

Louis, H. (1936) 'Die Geographische Gliederung von Gross-Berlin' in H. Louis and W. Panzer (eds.), Länderkundliche Forschung: Krebs-Festschrift, Engelhorn, Stuttgart, pp. 146-71

Marriott, O. (1967) The Property Boom, Hamish Hamilton, London

Martiny, R. (1928) Die Grundrissgestaltung der Deutschen Siedlungen, Justus Perthes, Gotha

Meynen, E. (1984) 'Der Grundriss der Stadt Köln als Geschichtliches Erbe' in H. Jäger, F. Petri and H. Quirin (eds.), Civitatum Communitas: Studien zum Europäischen Städtewesen, Böhlau, Koln, pp. 281-95

Murphy, R.E. (1935) 'Johnstown and York: A Comparative Study of Two Industrial Cities', Annals of the Association of American Geographers, 25, 175-196

Openshaw, S. (1974) A Theory of the Morphological and Functional Development of the Townscape in a Historical Context, University of Newcastle upon Tyne Department of Geography, Newcastle upon Tyne

Parkes, D.N., and Thrift, N.J. (1980) Times, Spaces and Places: A Chronogeographic Perspective, Wiley, Chichester

Pillsbury, R. (1970) 'The Urban Street Pattern as a Cultural Indicator: Pennsylvania 1682-1815', Annals of the Association of American Geographers, 60, 428-46

Pinchemel, P. (1983) 'Geographers and the City: A Contribution to the History of Urban Geography in France' in J. Patten (ed.), The Expanding City: Essays in Honour of Professor Jean Gottman, Academic Press, London, pp. 295-318

Rickert, J.E. (1967) 'House Facades of the Northeastern United States: A Tool of Geographic Analysis', Annals of the Association of American Geographers, 57, 211-38

Sabelberg, E. (1984) Regionale Stadttypen in Italien, Franz Steiner, Wiesbaden

Schlüter, O. (1899) 'Bemerkungen zur Siedlungsgeographie', Geographische Zeitschrift, 5, 65-84

Slater, T.R. (1978) 'Family, Society and the Ornamental Villa on the Fringes of English Country Towns', Journal of Historical Geography, 4, 129-44

Slater, T.R. (1981) 'The Analysis of Burgage Patterns in Medieval Towns', Area, 13, 211-16

Slater, T. R. (1982) 'Urban Genesis and Medieval Town Plans in Warwickshire and Worcestershire' in T.R. Slater and P.J. Jarvis (eds.), Field and Forest: A Historical Geography of Warwickshire and Worcestershire, Geo Books, Norwich, pp. 173-202

Slater, T.R. (1985) 'Medieval New Town and Port: A Plan-Analysis of Hedon, East Yorkshire', Yorkshire Archaeological Journal, 57, 23-41

Slater, T. R. (1986) 'The Topography and Planning of Medieval Lichfield: A Critique', Transactions of the South Staffordshire Archaeological and Historical Society, 26

Smailes, A.E. (1955) 'Some Reflections on the Geographical Description and Analysis of Townscapes', Transactions of the Institute of British Geographers, 21, 99-115

Spencer, J.E. (1939) 'Changing Chungking: The Rebuilding of an Old Chinese City', Geographical Review, 29, 46-60

Spencer, J.E. (1947) 'The Houses of the Chinese', Geographical Review, 37, 254-73

Stoob, H. (1973) (ed.), Deutscher Städteatlas, Vol. 1, Grösschen, Dortmund

Stoob, H. (1979) (ed.), Deutscher Städteatlas, Vol. 2, Grösschen, Dortmund

Straw, F.I. (1967) 'An Analysis of the Town Plan of Nottingham: A Study in Historical Geography', unpublished M.A. thesis, University of Nottingham

Thompson, F.M.L. (1974) Hampstead: Building a Borough, 1650-1964, Routledge and Kegan Paul, London

Trowell, F. (1985) 'Speculative Housing Development in the Suburb of Headingley, Leeds, 1838-1914', Publications of the Thoresby Society, 59, 50-118

Vance, J.E. (1977) This Scene of Man: The Role and Structure of the City in the Geography of Western Civilization, Harper and Row, New York

Whitehand, J.W.R. (1967) 'Fringe Belts: A Neglected Aspect of Urban Geography', Transactions of the Institute of British Geographers, 41, 223-33

Whitehand, J.W.R. (1972a) 'Urban-Rent Theory, Time Series and Morphogenesis: An Example of Eclecticism in Geographical Research', Area, 4, 215-22

Whitehand, J.W.R. (1972b) 'Building Cycles and the Spatial Pattern of Urban Growth', Transactions of the Institute of British Geographers, 56, 39-55

Whitehand, J.W.R. (1974) 'The Changing Nature of the Urban Fringe: A Time Perspective' in J.H. Johnson (ed.), Suburban Growth: Geographical Processes at the Edge of the Western City, Wiley, London, pp. 31-52

Whitehand, J.W.R. (1975) 'Building Activity and Intensity of Development at the Urban Fringe: The Case of a London Suburb in the Nineteenth Century', Journal of Historical Geography, 1, 211-24

Whitehand, J.W.R. (1977) 'The Basis for a Historico-Geographical Theory of Urban Form', Transactions of the Institute of British Geographers N.S., 2, 400-16

Whitehand, J.W.R. (1978) 'Long-Term Changes in the Form of the City Centre: The Case of Redevelopment', Geografiska Annaler, Series B, 60, 79-96

Whitehand, J.W.R. (1981a) 'Background to the Urban Morphogenetic Tradition' in J.W.R. Whitehand (ed.), The Urban Landscape: Historical Development and Management, Academic Press, London, pp. 1-24

Whitehand, J.W.R. (1981b) 'Conzenian Ideas: Extension and Development' in J.W.R. Whitehand (ed.), The Urban Landscape: Historical Development and Management, Academic Press, London, pp. 127-52

Whitehand, J.W.R. (1981c) 'Fluctuations in the Land-Use Composition of Urban Development During the Industrial Era', Erdkunde, 35, 129-40

Whitehand, J.W.R. (1983a) 'Land-Use Structure, Built-Form and Agents of Change' in R.L. Davies and A.G. Champion (eds.), The Future for the City Centre, Academic Press, London, pp. 41-59

Whitehand, J.W.R. (1983b) 'Renewing the Local CBD: More Hands at Work than You Thought?', Area, 15, 323-26

Whitehand, J.W.R. (1984) Rebuilding Town Centres: Developers, Architects and Styles, University of Birmingham Department of Geography, Birmingham

Whitehand, J.W.R. (1986) 'Taking Stock of Urban Geography', Area, 18

Whitehand, J.W.R., and Alauddin, K. (1969) 'The Town Plans of Scotland: Some Preliminary Considerations', Scottish Geographical Magazine, 85, 109-21

Whitehand, J.W.R., and Edmondson, P.M. (1977) 'Europe and America: The Reorientation in Geographical Communication in the Post-War Period', Professional Geographer, 29, 278-82

Whitehand, J.W.R., and Whitehand, S.M. (1984) 'The Physical Fabric of Town Centres: The Agents of Change', Transactions of the Institute of British Geographers N.S., 9, 231-47

Chapter Ten

RURAL SETTLEMENT

B.K. Roberts

A survey of the historical geography of rural settlement has no direct predecessor, although progress in historical geography (Prince, in Cooke and Johnson, 1969, pp. 110-22; Baker, 1972 et seq; Prince, 1971; Prince, in Brown, 1980, pp. 229-250), rural settlement (Baker, in Cooke and Johnson, 1969, pp. 123-32; Cloke, 1979) and rural geography (Pacione, 1984) have been well charted. To sever the theme from the broader matrices of geography and historical geography is difficult, while the fact that historical geographers normally work closely with other disciplines (Prince, in Brown, 1980), pp. 235-38) adds further difficulties. Nevertheless, recognising that progress assessments are essentially an interpretation of events, a prime objective is to review some of the work in the field. The date of 1980 has been used as a primary threshold, but this has not been slavishly adhered to, and some substantially earlier key references are incorporated. Current research varies greatly in character: variations in scale and time are to be expected, but there are also important variations in objectives and methods. Within the published materials examined three distinct approaches can be identified; first, many studies are empirical, concerned with the use of often intractable sources to reconstruct some aspects of settlement conditions in the past (Baker et al., 1970). This is a legitimate exercise, indeed constitutes foundation work, for as the historian Elton pointed out (1969, p. 34), knowledge and judgement only come via 'the painful mapping of the advancing frontier'. Second, placing particular cases within a broader framework of scholarship always demands study of the processes affecting settlement. Third, as experience, understanding, and perception of possible explanatory concepts expands, theoretical components are increasingly included.

In practice, empirical investigations of the historical geography of rural settlement are essentially concerned with four areas of enquiry; first, the character of the physical structures, their shapes and dimensions, boundaries and

materials and the extent to which these represent contemporary elements or relict features, remembering that the word
'contemporary' can apply to 1086 as well as 1986!; second,
functional aspects and lateral relationships require consideration, whether the economy is subsistence or market orientated, the degree of integration within an estate or other
administrative or marketing framework, and the location
relative to routes, together with the demographic and social
characteristics of the inhabitants; third, settlements of a
given type will have a territorial extent, and exploring this
vital geographical dimension leads directly to classification,
for to create distributions similarities and differences must be
observed and defined; fourth, the historical geographer is
particularly concerned with the trajectory of rural settlements
through time, introducing to all the elements noted above
questions of stability and change, continuity and discontin-
uity. While this list affords no definition of rural settlement
geography as practised by historical geographers, it does
delimit common themes of empirical investigation. Empirical
studies lead inexorably towards the second approach, concentrating on processes, and in 1972 Baker, building on
questions devised by Berkhofer, framed lines of inquiry for
examining these (1972, pp. 16-17).

Process studies lay emphasis on three goals: the
development of an understanding of the general processes at
work in generating geographical change; second, an understanding of the way these operate in specific situations by
means of behavioural stance, leading to what has been termed
'the dynamics of historical change and of social groups in
conflict' (Baker and Billinge, 1982, p. 238); and thirdly, the
possibility of creating some conceptual and theoretical frameworks. None of these, of course, are exclusive categories. In
practice, while particular processes acting upon and affecting
individual settlements are singularly diverse in character,
they do, nonetheless, engender a surprisingly limited number
of changes. These may be listed as:

1. Conditions of stability or instability, which range in
   degree of intensity from slow almost imperceptible
   change to rapid, even cataclysmic, developments
   (Renfrew, in Green et al., 1978; Sporrong, 1985,
   pp. 27-31);

2. Conditions of expansion, involving both accretion to
   existing entities as well as diffusion and active
   colonisation, often associated with increasing individuality (Jones, M., 1977; McIntosh, 1981; Jamieson
   et al., 1983; Wonders, 1983);

3. Conditions of interaction and competition, involving
   concentration, reduction or congregation, in situ

reorganisation and ex situ replacement, and often increasing communality (Sutton, 1981; Gant, 1982; Gade and Escobar, 1982; Lovell, 1983; Dilsaver, 1985);

4. Conditions of active contraction, involving depopulation, shrinkage and destruction (Beresford and Hurst, 1971; Loffler, 1979; Gissel et al., 1981; Sveinbjarnardottir et al., 1982; Pozorski et al., 1983);

5. The appearance of cyclic activity, either short term, involving movement and periodic site reoccupation, or longer term, trends determined by secular changes of a physical or economic character (Noble, W.A., 1977; Able, 1966; 1980; Parry, 1978).

'What is happening here?' and 'What has happened here?' remain legitimate primary questions, even though the answers may always be partial, and ultimate understanding limited. The value of this approach has been noted by Guelke, who significantly draws upon archaeological exemplars (1982, pp. 48-9; in Baker and Billinge, 1982, pp. 189-96; see also Barker, 1977, pp. 40-1). Empirical information, garnered via the areas of study described above and an awareness, derived from experience, that one, some, or, in the fullness of time, all of these processes may have been involved in creating a given settlement situation can then be used in explanatory contexts, some of which can contain theoretical components. Once again, the list is surprisingly brief, for many ideas can be subsumed under (1) core periphery models; (2) diffusion models; (3) hierarchical models. This is not because rural settlement is simple, indeed the reverse, for following the ideas of Weaver, cited by Wilson (in Brown, 1980, pp. 203-4), elements of three types of system can be detected: simple systems, based upon a few variables, systems of disorganised complexity, in which many variables interact only weakly, and systems of organised complexity, in which large numbers of components and variables are present, and some of which interact strongly.

Here, then, are several possible frameworks for discussion, which allow the identification of varied categories of content within published work. However, all of these approaches are applied within varied geographical contexts. No review of this length could be truly world wide, but neither should it be wholly Eurocentric. Nevertheless, the rise of Christian Europe and the 'diaspora of some 60 million Europeans over the past 150 years' (Lemon and Pollock, 1980, preface) associated with the many facets of worldwide colonial expansion and with the widespread implanting of the idea that willed change was possible' (Roberts, J.M., 1985, p. 431) has

gradually superimposed aspects of occidental urban civilisation, and the characteristics of rural settlement which are
its corollaries, over diverse earlier indigenous patterns.
Nevertheless, the writing of historical geography depends
upon the nature and quality of the sources available, and
these are themselves a product of the circumstances in which
they were created. Inquiries using the rich documentary and
archaeological resources of Western Europe will, inevitably,
differ substantially from those in regions where westernisation
is only represented by a thin colonial veneer and where
prehistory lies close to the surface. Between these two extremes there are many possibilities. This image affords one
pragmatic context within which regional schools of inquiry
must be viewed. Thus, at any period of time the settlement of
any region may be thought of as possessing a unique sub-
surface profile, reaching backwards in time, and reflecting
varied and complex historical and prehistoric situations.
These remarks have profound practical implications: in the
formative stages of this essay the author was influenced by a
remark by William Norton: 'With only a few exceptions the
North American theoretical and conceptual contributions have
received little application in older settled areas' (Norton,
1985, p. 128) and this helpful irritant lies at the core of
much of the argument presented here. One may in fact contend that settlement evolution within new, well-documented,
landscapes is substantially easier to deal with than that in
old, long settled regions, where the operating rules for
models need to be more sophisticated, and where a 'colonial'
datum or clear cut 'frontier' is absent. It is clearly no
accident that many of the models used by settlement geographers which are cited most often are derived either from
North American contexts (Hudson, 1969) or from wholly
peripheral, marginal environments of the old world (Bylund,
1960; 1972) where a relative simplicity prevails. Work in a
variety of contexts is revealing the inherent limitations of
such models (Grossman, 1971 et seq.; Haining, 1982;
Grossman, 1983). It is no criticism of the great North
American republic to hint that its advanced urban settlement
systems may be seen as merely a vigorous and successful
form of colonial activity, where the indigenous systems were
sufficiently submerged to be unable to reassert themselves!
What such areas have to tell the old world may be inherently
limited by the relatively shallow soil in which they are rooted.

Figure 10.1 depicts some aspects of settlement realities
in an old settled region. What appears is a compromise, for it
would be more informative in plan and even better in colour,
but the presentation adopted is guaranteed to survive the
rigours of black and white publication. The shading is
designed to suggest what would lie beneath the feet of a
person walking a strip of terrain cutting at right angles
across the chalk escarpment of the Yorkshire Wolds in the

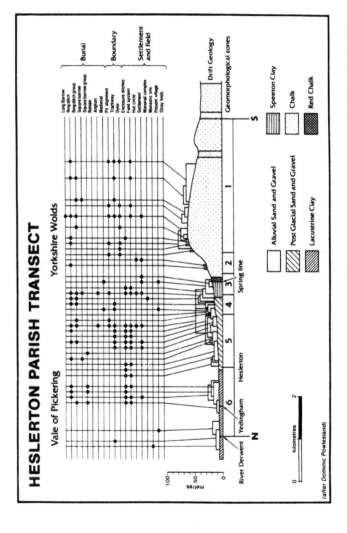

Figure 10.1: Settlement Features along a Transect in Heslerton Parish, Yorkshire.

North East of England (Powlesland, 1985). The grid gives
some impression of the stacked antiquity present, although,
of course, this is inevitably over-emphasised by the small
scale of the drawing. Nevertheless, to take one element,
settlement: what the diagram cannot show clearly is that there
is a northern zone, comprising a ribbon or chain of super-
imposed successive occupations, extending spatially for nearly
ten kilometres along the scarp foot (this limit merely representing the edges of the 10 x 10 km sample square). Archaeological evidence, the excavation of a sample area of nearly
two hectares, shows that occupation extends in time from the
Mesolithic to the Anglo-Saxon period. Quite remarkably, many
of the features in this key zone are stratified, each being
separated from the other by thin layers of blown sand, which
has drifted in from the basin of the former Lake Pickering to
the north. There are signs that the later Anglo-Saxon and
medieval occupation may lie in a second zone, somewhat
upslope of the earlier ribbon, where one Anglian site at least
30 acres (12 hectares) in extent is known and the present
villages are located.

Of course, the present landscape is largely composed of
features from the recent past, visible in the eighteenth and
nineteenth century dispersed farmsteads and hedged field
boundaries (Allison, 1976) and in the vernacular architecture
of the brick and chalk clunch villages (Harris, 1961), while
the fodder crops and grass of the fields attest successful and
continuing husbandry. Nevertheless, a deep well of time lies
beneath, sometimes present as antecedent structures, defining
frameworks for later arrangements (Hurst, 1984, Figure
10.1), but also as elements wholly at discord with all later
developments, seen as vivid crop marks when farmers plant
lucerne to feed their stock. The essential point is that this is
an old landscape, and apart from the basic dowry of geomorphology, climate and biological potential, it is a manu-
factured landscape - made by man during many millenia of
essentially continuous occupation. Here is no tabula rasa, no
isotropic surface, in either physical or cultural terms, but a
landscape in which at least six, and probably nearer seven
thousand years of near continuous usage have allowed man to
alter the soils, aspects of the geomorphology, the agricultural
potential and the elements of the visible scene. There have
been no 'new beginnings' (Darby, 1973, p. 1), and while the
word 'continuity' should never be applied without careful
consideration of the kaleidoscopic pattern of possible meanings
(Finberg, 1964, pp. 1-20; Widgren, 1983, p. 10), in many
parts of Britain comparable evidence for sustained usage
through millennia rather than centuries is now emerging. This
has both social and economic implications. In Heslerton, the
medieval and post medieval documentation has yet to be
examined, but while the place name is today applied to two
villages, the Domesday Book of 1086 lists no less than twelve

separate entries, implying the presence of a remarkable tenurial complexity which has yet to be related to the on ground discoveries. None of this complex data has yet been incorporated into any general model. If we acknowledge that the circumstances along the southern edge of the Vale of Pickering may offer exceptional conditions for the survival of stratified evidence there are, nevertheless, no grounds for believing that complexity of former settlement visible here is in any way exceptional. Such complexity is attested by numerous recent studies in diverse British environments (Miles and Benson, 1974; Bowen and Fowler, 1978; Miles, 1982; Spratt, 1982; Mercer, 1981; Fowler, 1981/3; Knight, 1984). In an essay entitled 'Wildscape to Landscape' (in Mercer, 1981, pp. 9-54), essentially a discussion of 'landtaking' in the prehistoric period, involving the translation of the wildwood into fields and farms, Peter Fowler has captured something of the excitement of current views and new perceptions. Where men farm they must, of course, live, but he touches a causal factor which always parallels landtaking: 'Well-farmed land, however, always gives more than income; it gives power. As always, the basic question of farming in prehistoric Britain comes down to "Who owned the land?"' (Fowler, in Mercer, 1981, p. 47). Of course, neither landownership nor tenure can be excavated, but archaeologists are paying increasing attention to land boundaries, from the scale of the individual house plot and field to larger territorial units, while a recent study by Bradley (1984) has stressed the positive advantage of a long time perspective when evaluating limited evidence. He argues that settlement plans may show their greatest variation during phases when burial rites are fairly uniform, while conversely when settlements are less complex cemeteries show a more elaborate structure. This rational stretching of the limits of inference inherent in the archaeological method, asking successive questions of empirical data, each basically simple, but becoming cumulatively complex, has important implications for the historical geographer (1984, pp. 158-9; Drury, in Delano-Smith and Parry 1981, pp. 40-53; Guelke, in Baker and Billinge, 1982).

This is not to deny that most other parts of the world possess an equally long history of interaction between man and the land, not least the ancient hearths of civilisation in the Middle East. Set against such time spans, the post Roman European cultural developments involved in the rise western Europe are themselves rooted in a relatively shallow soil. Nevertheless, the time depths represented in studies drawn from such old world regions (Clout, 1977) must be contrasted with those from peripheral (Mead, 1981) or colonial contexts (Davidson, 1981), where de facto, much early development and even initial European contacts (Dyson, 1985) can only be known from archaeological work. 'The Frontier' is an attractive concept (Norton, 1984, pp. 87-102). In many contexts a ruthless progression of 'sequent occupance' can indeed be seen, not least in the USA, where the work of Turner (1920), Brown (1948) and Billington (1960) still provide essential foundations. The process was conceptualised by Whittlesley as early as 1929, while Fontana (in Schuyler, 1978, pp. 23-24) and Meinig (in Baker and Billinge, 1982, pp. 71-78) have respectively summarised the archaeological and geographical elements of the cultural interactions which can ensue. Christopher (1984) adopts a version of this concept as a framework for reviewing the impact of European colonisation upon the whole of Africa south of the Sahara. However, and this is an important point, while this latter study concentrates upon the many and diverse European groups, the role of indigenous African communities is to be the subject of a further study. Superficially at least, much European activity in Africa seems to owe little to its pre-literate precursors, although where Islamic sources are available, in West Africa for instance (Levtzion, 1973), the complexity, indeed sophistication, of such antecendents is well attested. Christopher stresses that not only did individual colonial powers and peoples vary in their actions and policies, often with little regard for the contemporary ideas of legality in Europe or regard for the rights of their African compatriots, but similarly 'settler Rhodesians and indigenous Zimbabweans did things differently' (1984, p. 7) and these elements of continuity are reflected in the landscapes. Guelke's (1982) examination of South African settlement adds a level of subtlety in understanding the processes at work, pointing out that human actions are a reflection of changing ideas and assumptions as well as external forces.

SCALE IN SETTLEMENT STUDIES

In 1972 Baker intimated that historical geography might move towards comparative work, problem rather than source based (1972, p. 28). In this respect scale is crucial, irrespective of the source materials and the contexts in which these are applied. Here, a subjective threshold has been adopted: small scale studies are considered to fall within a single square kilometre, while large scale studies are in excess of this. This crude division derives from Figure 10.2, but reflects, in the pattern of shading on the lower band, the traditional form/pattern dichotomy of rural settlement studies. The figure derives from three sources: first, the ideas of scale in the writings of Haggett (Chorley and Haggett, 1964, pp. 364-68; Haggett, 1965, pp. 4-9; Haggett and Frey, 1977, pp. 6-10 and passim); second, the splendid visual presentations of settlements found in the work of the Greek architect Doxiadis (1968); third, the hierarchy of elements of settlement study found in the author's own work (in Hooke, 1985). The three horizontal bands define pragmatic frames within which empirical, process and theoretical studies are pursued. The internal divisions of each category are shown as broken lines to emphasise the crudity of the classification involved and the extent to which complex transitions exist. The lower scale is logarithmic, and the overlap along the spine of the book should be noted. The lower band concerns the varied physical entities, buildings, farmsteads, hamlets, villages, towns, conurbations and even 'dynamegalopolis' itself (according to Doxiadis, the eastern seaboard of the USA) which are subjected to many methods of study, from archaeological dissection at the lower end of the scale (for dirt archaeology is inherently limited to the study of at most a few hundred square metres, even when the work represents a sampling strategy conducted over a considerable area), to mathematical dissection at the higher.

PLACE, FORM AND FUNCTION: SMALL SCALE STUDIES

All settlements comprise scatters or concentrations of individual dwellings. Their architecture reflects diverse environment and cultural factors (Oliver, 1971; Singh, 1972, part IV; Fenton, 1978), but they are also important for what they represent in the lives of individuals and families. As Hughey noted 'Together, house and yard form a nucleus within which culture expresses itself, is perpetuated and reiterated' (1978, p. 131). It is at this scale of inquiry that rural settlement studies most closely touch social anthropology, and there are sound reasons for believing that this was so in former centuries (le Roi Lauderie, 1978, pp. 24-52, p. 295). Interior arrangements, living spaces, and perhaps even the bedspaces, may be seen as representing important controlling mechanisms - touching aspects of inheritance systems, the numbers of children, the farm labour budget, although in geography the social interactions at this level are normally treated in general rather than in detail. Architectural studies, are important at this lower threshold of settlement study, and often contain a spatial element. Building types and styles, the character and dimensions of ancillary buildings and farmstead layouts normally reflect status variations and levels of prosperity as well as the local physical environment and cultural traits (Brunskill, 1971 et al.; Fenton, 1978; Pollock in Lemon and Pollock, 1980, pp. 16-23; Barley and Smith in Thirsk, 1985). In England, foundation work by Barley (1961) is now encouraging the physical and documentary study of rural buildings in their local contexts (Somerset etc., 1984), adding an essential dimension to studies of village morphology, while archaeology is giving a view of the historical evolution of the peasant house (Hurst,

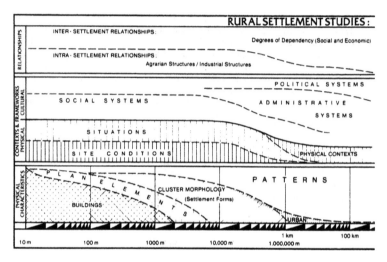

Figure 10.2: Rural Settlement Studies:

1984). Recent work in Holland on the distribution of settlement types has also included earlier work on farmstead layouts (Renes, 1982) while old and new worlds have been linked
by Jordan's comparison of log cabin construction techniques
in mid-America and Fenno-Scandinavia, the geography of
architectural styles being associated with documented patterns
of long distance migration (1983), as is Noble and Seymour's
study of barn types in the north eastern United States
(1982). Usually the proponents of such studies are concerned
with the recording of survivals before they disappear or with
the preservation of relict structures as elements within the
contemporary landscapes (Newcombe, 1979; Schmaal, 1982).
    Morphological studies are now as much part of landscape
archaeology as historical geography (Sheppard, 1966 et seq.;
Beresford and St. Joseph, 1979; Harvey, 1980 et seq.;
Taylor, 1983; Roberts and Glasscock, 1983; Roberts, in
Hooke, 1985; Austin, 1985; Chapelet and Fossier, 1985;
Roberts, 1986 forthcoming). One important presentation,
forming an exciting link between traditional architectural
styles, settlement morphologies, and the value of such traditional features in the landscapes of the present is to be
seen in the Brinkenboek, a study of village greens in the
province of Drenthe, Holland (Houting, 1981). In their
crowded country the Dutch are of necessity aware of the
practical relevance of historical geography to contemporary
problems (Haartsen and Renes, 1982). Morphological work in
Britain is now demonstrating both an important dichotomy

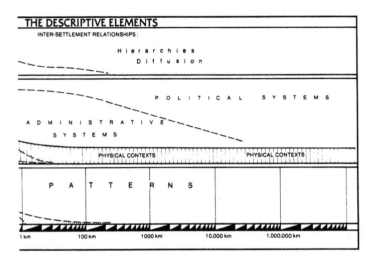

THE DESCRIPTIVE ELEMENTS

INTER-SETTLEMENT RELATIONSHIPS:

Hierarchies
Diffusion

POLITICAL   SYSTEMS

ADMINISTRATIVE
SYSTEMS

PHYSICAL CONTEXTS   PHYSICAL CONTEXTS

PATTERNS

1 km   100 km   1000 km   10,000 km   1,000,000 km

the Descriptive Elements.

between planned and unplanned types, and the presence of
villages made up of several nuclei of growth - composite or
polyfocal plans (Taylor, 1983). Village origins in farmstead
groupings, which are subjected to varying degrees of reorganisation and regulation as population increases and as
field systems become larger and need more complex operating
rules, have long been understood in Sweden (Hannerberg,
1984; Sporrong, 1981; 1985) where increasing attention is now
being applied to the more primitive antecedent arrangements
(Widgren, 1983; Kristiansen, 1984). In Denmark Jeppersen
(1981), drawing upon materials from Holland and northern
Germany, and using the techniques of landscape archaeology,
has demonstrated the through time trajectories of a group of
villages in northern Fyn. In Holland, Steegh's atlas of
settlement forms (1985) is a warning of the diversity which
remains to be discovered and fully examined.

The dichotomy between forms and patterns has tended to
condition scales of rural settlement study, but in England
recent developments are serving to open a door to the possibility of studies bridging this threshold. Thus, work by
archaeologists, generally at the scale of the parish, are
revealing that while historical settlement may be dominated by
a single village, middle and late Anglo Saxon pottery scatters
are probably indicative of an antecedent pattern which was
made up of dispersed farmsteads and hamlets (Rowley, 1974;
Taylor, 1983). A process of aggregation into larger clusters
has clearly taken place, raising questions of the forces which

generated such changes, which must have been closely linked with developments in field systems as well as society and tenure (Rowley, 1981; Roberts, 1986 forthcoming). In English, no graceful term exists to describe this phenomena, but the Spanish 'congregación' has undoubted attractions (Lovell, 1983). Roberts, in a summary paper (1983, p. 45) touching on critical controls in the balance between nucleation and dispersion has talked of 'communality of assent', 'communality to economise', and 'communality of enforcement', which derive from social rather than environmental circumstances.

Studies of the broad relationships between rural settlements and the physical environment are now less fashionable amongst British historical geographers, although place-name scholars have continued the tradition (Fellows Jensen, 1972; 1978). However, the recent discoveries concerning the late appearance of villages is stimulating re-appraisals (Fellows Jensen, 1984), and bringing extreme caution concerning topographical interpretations drawn from the site of the historic village. The perception of the terrain beneath settlements as an archaeological site, to which materials were brought from a adjacent territory draws from both archaeological and geographical roots (Chisholm, 1962; Welinder in Kristiansen, 1984, pp. 1-25) and remains a powerful tool of investigation even when excavation is not involved, for soil phosphate levels will reflect site usage (Kiefmann, 1979; Widgren, 1983; Edwards and Simms, 1983; Eidt, 1984). Whatsoever their disposition, farmsteads are supported by the land adjacent. Where communality is not strong the land of the individual farmstead is the smallest unit of exploitation, but when communality is present, be the actual entities of settlement dispersed or nucleated, then the group territory will form the smallest entity in any functional hierarchy and often the lowest element in an administrative hierarchy. Here is one answer to the place name problem; in many old settled regions the name refers to this territory rather than to what may be relatively ephemeral physical entities of settlement within this. To cut through many niceties of definition, the village, its fields, meadows and other lands over which it (or its owner) possess usage rights, constitute an indivisible functional entity: there is always an intimate and mutual relationship between a village, its field system, and its inhabitants (Yates, 1982).

Where men live, they must support themselves, and with few exceptions in rural contexts this means operating farms, engaging in craftwork, or in trade. While fishing villages undoubtedly occupy a peculiar position between economies based upon hunting and gathering and communities bound to the soil, most rural settlements have been based upon land resources. The list of empirical investigations relevant to the study of rural settlement forms and patterns (above) are

RURAL SETTLEMENT

equally applicable to field systems (Steensberg, 1983), but
recent work has concentrated upon several aspects: first,
investigations in Scandinavia and Britain (both owing much to
earlier German work - Uhlig and Lienau, 1972; Mayhew, 1973)
have begun to demonstrate the close bonds which link settlement plans, the distribution of field strips (Hall in Rowley,
1981, pp. 22-38; Hall in Roberts and Glasscock, 1983, pp.
115-131; Rowley in Cantor, 1982, pp. 25-55; Nash, 1982) and
social structures. Many types of rural settlements are involved, from loose farmstead clusters with a relatively low
level of communality to highly organised planned villages
(Demidowicz, 1985). More importantly, perhaps, recent work
is suggesting that the historical village can indeed be studied
in something like the intimacy possible within anthropological
or sociological investigations of a contemporary settlement.
Some historians have included plan analysis in their examin-
ations of particular villages (Harvey, 1965 and Ravensdale,
1974), although the important work of Toronto school lacks a
strong geographical feeling (Dewindt 1971). This stands in
contrast to a recent study where Howell (1983), also an
historian, has focussed upon a single English village,
Kibworth Harcourt, and its population. Her investigations
have included field systems, but also extended to settlement
morphology. She has attempted to map the tenurial units
present within a reconstructed plan of 1086. Linked with
family reconstitution this provides a remarkable and challenging picture (Howell, in Goody et al., 1976, pp. 112-155),
demonstrating the fundamental linkages between morphology
and social structures, expressed both through the demesne
land/tenant land division and the presence of separate
manors.

A general picture of many aspects of British field
systems was assembled in 1973 (Baker and Butlin). In a
concluding discussion, which emphasised the basic properties
common to all systems, i.e. structure, function, equilibrium
and change, they formulated questions which moved beyond
simplistic analyses of morphology towards the processes
generating, sustaining and destroying such arrangements.
The genesis of communal field systems continues, and in a
short but important paper Campbell (in Rowley, 1981, pp.
112-129) in effect laid the foundations for a whole programme
of new geographical work. He pointed out that the models
which exist to describe what was, in reality, a plurality of
types are very limiting, and in order to achieve a full
assessment many functional attributes must be carefully
appraised. He lists fourteen, ranging from field layout,
holding layout, cropping regulations, and modes of regulation, to the rules affecting fallow grazing and the relationship between the improved land and the wastes. His aim was
to 'clarify definitions, remove preconceptions and pose
questions' and he stresses that because of defective knowl-

edge of what really is present the primitive distribution maps currently available have but a limited use. Dodgshon (1980), in an important if at times opaque study, has argued that the emergence of the physical structures of communally organised subdivided field systems may have preceded the tenurial arrangements with which they became associated, that super-imposed assessment systems played an important part, and that tenurial reorganisation associated with township splitting, itself a by-product of organised land division, was a further important formative factor. In separate yet parallel work the economic historian Dahlman (1980) argued for the importance of the 'property rights paradigm' in understanding the characteristics of field systems involving a balance between private and collective rights.

The 'property rights paradigm' also forms an important bond between individual places and larger scale patterns (Bond, 1979; Roberts, in Reed, 1984), for together with administrative frameworks these constitute the man-made infrastructure. Mills (1980) has shown the extent to which the varied relationships between land and proprietors affected the social, economic and ultimately the physical characteristics of rural settlements in nineteenth century Britain, while Clemenson (1982) in a discussion of landed estates has given a general view of the English scene between the late eighteenth century and the present day. Such studies touch important points: property rights extend over all settlements, even purely subsistence settlements, for in this way the peasant surplus was diverted to the use of the possessors of these rights. Seen from the viewpoint of the inhabitants of a particular locality, a situation of dependency normally existed, and while this was often not at first a market dependency, this eventually emerges, formalising, redefining and often adjusting bonds which at first were social and tenurial, and measurable in terms of rent flows (Gissel et al., 1981, pp. 29-77). These threads of dependency bound indi-vidual places into larger entities. Early estate patterns in Britain have been the subject of numerous studies by Glanville Jones and the problems of defining these have given rise to a recent vigorous but largely inconclusive debate (Gregson, 1985; Jones, 1985). This may be expected to continue. However, the historian Kapelle (1979), in a regional analysis of Norman settlement in the north of England, has shown that the evidence of the estate patterns of Domesday Book can be used to create explanatory frameworks extending over large areas even when the troubled questions of their origins are by-passed.

SETTLEMENT PATTERNS: LARGE SCALE ANALYSIS

As the scale of inquiry increases, and as more and more entities of settlement are drawn into consideration, the possibilities for generalisation and theoretical and conceptual analysis increase. One of the oldest settlement models, perceived, as Glacken (1967, pp. 116-19) has shown, because it was for millenia a part of everyday experience, involves the idea of core and periphery; tamed and domesticated land and the untamed apparently natural world, the settled husbanded heartland and the peripheral frontier zone; the focus of the state and its edges. This perception underlies many of the explanatory concepts and theories present in settlement geography: dependency relationships exist between settlements within cores, classically seen in the ideas of Christaller, and between cores and peripheries. Within both varied importance and varied success can be analysed in terms of rank and hierarchy (Barker, 1980), while diffusion models are associated with questions concerning all aspects of colonisation, be this peaceful internal activity, filling in the earth's untamed spaces, or vigorous military or economic adventurism, imposing the control and the values of a heartland state upon varied types of peripheral region.

Diffusion and hierarchy interlock in complex ways, and are associated with models created by Turner, Christaller and Losch, Bylund and Hudson. Unwin's work (1981) on 'rank size distribution in medieval English taxation hierarchies' reveals, in the clumsiness of its title, a key problem: documentation does not necessarily refer to settlement as such, but to taxation units, tenurial structures, and administrative systems. Unwin is compelled to battle manfully to relate the realities of settlement, of which he is well aware, to the theoretical perspective with which he wishes to experiment. Hudson's 1969 paper is still stimulating further work. Haining (1982) has compared Hudson's Iowan data with some from Sweden, but Grossman in studies involving both contemporary and historical contexts, is using empirical evidence as a powerful critical tool. He points out (1971) the abnormality of the level terrains of Iowa, the high degree of individualism present, the complex and highly diversified origins of the population making up the rural society, and in another context (1983), that no man-made infrastructure is taken into account. To these, even if we ignore the fact that the frontier had passed through the Iowa study area as recently as 1890, may be added another: Hudson's arguments were concerned with farmsteads, not villages. From much of Grossman's work there emerges a strong indication that general laws are not likely to emerge in rural settlement studies, because these latter are essentially culture bound. This again leads to an important conclusion, that a key settlement control depends upon the balance between individuality and the degree of communal control, and this is a historically generated social factor.

Norton suggested that the three variables most evident in frontier studies, time, location and population help frame

four useful sets of ideas: namely length of residence concepts, migration field concepts, land-use competition concepts
and innovation diffusion concepts (1984, p. 96). In fact, at a
more basic geographical level the frontiers of landtaking can
be internal or external to the body of any state, although it
may be unwise to stress this useful distinction too heavily,
for frontier occupation is an essential ingredient of successful
state building. Nevertheless, the presence of internal
frontiers were important in generating many of the small scale
regional differences found in England, mapped by Thirsk as
farming regions (1967, p. 4). Parry's meticulous empirical
work on the altitudinal limits of cultivation in the Lammemuir
Hills, southern Scotland, led him to a broader consideration
of the impact of climate on agriculture and settlement (Parry,
1978; see also Delano-Smith and Parry, 1981), while the
present author has argued for the identification of two
internal frontiers, one between regions of good and intermediate land quality, and the other between intermediate and
poor land qualities (Roberts, 1982). In a region where historical and cultural changes have generated complex patterns
of continuity and change, there is never a simple process of
intensification of settlement within heartlands (with spread
and competition following initial colonisation) from which waves
of peripheral colonisation then diffuse throughout the frontier
regions. In an important and large scale regional study of
desertion and land colonisation in the Nordic countries
between 1300 and 1600 (Gissel et al., 1981) the scholars
involved adopted a comparative approach and identify settlement (i.e. the physical evidence) and rent changes (i.e. the
most basic documentary evidence of settlement) as key
components in the overall complex of problems that constitute
the focus of their comparative study (Gissel et al., 1981, p.
27). They contend that they ought to be able to relate rural
farming settlement and its changes to factors that are components of four major groupings:

1. Population - i.e. its size, composition, migration
    tendency, household structure, etc.

2. Land - geographical conditions, the supply of arable
    land in relationship to the available technology and
    technical know-how, systems of land utilisation,
    effects on acreage due to changes in climate, fertil-
    isation, etc.

3. Political and economic factors - as for example,
    taxation structure and tax burdens, conditions
    related to a market economy and the incentives they
    may have entailed for settlement redistribution,
    colonisation and contraction.

4. Legal and administrative factors - as for example,
rules for land apportionment, inheritance principles,
ownership structure, and other relatively institutionalised power factors (Gissel et al., 1981, p. 57).

Within this framework, whose general relevance to the
historical geography of rural settlement can hardly be disputed, they recognise the need to define 'zone variables' and
'individual variables' as means of inter-regional comparison
and a way of focusing upon individual places (Gissel et al.,
1981, p. 71). Their conclusions are complex, perhaps even
diffuse, touching upon economic changes and climatic factors,
but significantly they point out that they may not have taken
sufficient account of the development of class conditions in
the agrarian societies (Brenner, 1976) as an explanatory
factor.

REFLECTIONS AND PROSPECTS

This limited essay is demonstrably incomplete, for the potential material content is vast, and is widely scattered amid a
host of studies on diverse themes. Varied degrees of bias in
the selection of studies included here are evident enough to
require no further comment. It is the author's simple contention that advance in each of the three fundamental approaches, empirical, process analysis and theoretical explanatory concepts, is essential for the health of the field. While
there is general agreement about the areas of inquiry, and a
consistent tendency to seek an explanation for observed
regional variations in the spatial impact of the diverse processes in human behaviour and judgement, no universally
agreed set of objectives exists. Any critique of a field always
tends to reflect personal predictions, and while finding no
particular salvation in Marx, the author would agree that
again and again the direction of current inquiry shows that
the explanations of settlement characteristics lie in the area of
interaction between human social behaviour and the potentialities of a given environment.

However, the particular range of literature reviewed
here reveals some problems: most disturbingly, there is an
indisputable credibility gap between empirical studies and
existing theory. This is substantial, and, in Britain at least,
not getting narrower: if, on one hand, there is an obsession
with empirical details, there is on the other an equal
obsession with defining theoretical methodological stances.
Few, if any, studies demonstrate with complete success how
the two can be brought together. As a corollary, there is
still a general absence of sound comparative work, and this
would form an essential foundation for sound theory. Here,
the earnest debates of Scandinavian interdisciplinary scholars

seem to offer one pragmatic solution, for the collective effort of creating truly comparative studies both posed questions and imposed practical constraints (Gissel et al., 1981). To put it simply, while the data used as a basis for conceptual and theoretical arguments must be drawn from soundly based empirical studies, the concepts and theories which emerge must be seen, not least by scholars involved in empirical investigations, to possess an acceptable level of credibility and be applicable in more than a limited range of real world contexts.

There is more than a suspicion here that this dichotomy is sustained and strengthened by modes of publication: historical geography lacks the type of publication service that is currently offered by British Archaeological Reports. Large format, so that drawings and maps can become a significant and interpretable element of the publication; highly specialist items, incorporating essential primary data, for while interpretations may fade, accessible data is likely to experience reinterpretation: relative cheapness, when balancing print-run against degree of specialism. Neither subject, however, has yet achieved even the moderate uniformity of presentation used by scholars concerned with such topics as pollen analysis or sea level changes: perhaps this is an unrealistic objective. Here the author must admit that in his own approach to historical geography he is indisputably influenced by procedures in natural science and archaeology.

Whatsoever the approaches adopted by individuals or schools of inquiry, historical geography is concerned with the identification, definition, analysis, and explanation of those spatial differences and similarities which are, and have always been, part of life on earth. In this context maps are important creative tools. Speaking only about one limited area of inquiry, the author can only deplore the fact that the only general map of rural settlement in Britain is still that created by Harry Thorpe in 1964, with all its obvious limitations. Even for the nineteenth century, when complete data is available, no map yet exists, yet it would be of inestimable use for both retrogressive and prospective studies. The same arguments can be applied to field systems. Thirsk, an economic historian, has seen the importance of such national scale perspectives, for two of her major studies contain important maps of farming regions (Thirsk, 1967, p. 4; 1985, pp. xx-xxi). Whatsoever their limitations, such maps serve to provide contexts which bind together a multitude of specific empirical investigations and form an essential foundation for the bridge between these and more theoretical work. There are strong indications here that geographers should, in addition to seeking inspiration within other fields, continue to explore the possibilities inherent within the techniques of their own subject.

RURAL SETTLEMENT

Ultimately, the main challenge in the field of the historical geography of rural settlement is to create models suited to the complex situations found within old settled regions. This is not a task that can be undertaken rapidly, for, as the empirical evidence from Heslerton and elsewhere clearly shows, it will necessitate discovering, evaluating and then using new data which will lead to radical reassessments of what we really do know, even about seemingly well-explored areas.

REFERENCES

Able, W. (1966/80) Agricultural Fluctuations in Europe, trans. O. Ordish, Methuen, London
Allison, K.J. (1976) The Making of the English Landscape: The East Riding of Yorkshire, Hodder and Stoughton, London and Sydney
Austin, D. (1985) 'Doubts About Morphogenesis', Journal of Historical Geography, 11(2), 201-9
Baker, A.R.H. (ed.) (1972) Progress in Historical Geography, David and Charles, Newton Abbot
Baker, A.R.H. (1977) 'Historical Geography Progress Report', Progress in Human Geography, 3, 465-74, Arnold, London
Baker, A.R.H. (1978) 'Historical Geography Progress Report', Progress in Human Geography, 3, 495-501, Arnold, London
Baker, A.R.H. (1979) 'Historical Geography: A New Beginning', Progress in Historical Geography, 3, 560-7, Arnold, London
Baker, A.R.H., and Billinge, M. (eds.) (1982) Period and Place, Cambridge University Press, Cambridge
Baker, A.R.H., and Butlin, R.A. (1973) Studies of Field Systems in the British Isles, Cambridge University Press, Cambridge
Baker, A.R.H., Hamshere, J.D., and Langton, J. (eds.) (1979) Geographical Interpretations of Historical Sources, David and Charles, Newton Abbot
Barker, D. (1980) 'Structural Change in Hierarchic Spatial Systems in South West England between 1861 and 1911', Geografiska Annaler, 62B (1), 1-9
Barker, P. (1977) Techniques of Archaeological Excavation, B.T. Batsford Ltd., London
Barley, M.W. (1961) The English Farmhouse and Cottage, Routledge and Kegan Paul, London
Beresford, M.W., and Hurst, J.G. (1971) Deserted Medieval Villages, Lutterworth Press, London
Beresford, M.W., and St. Joseph, J.K.S. (1979) Medieval England: An Aerial Survey, Cambridge University Press, Cambridge

RURAL SETTLEMENT

Billington, R.A. (1960) Westward Expansion, 2nd edition, Macmillan Company, New York
Bond, C.J. (1979) 'The Reconstruction of the Medieval Landscape: The Estates of Abingdon Abbey', Landscape History, 1, 59-75
Bowen, H.C., and Fowler, P.J. (eds.) (1978) Early Land Allotment, British Archaeological Reports, British Series 48
Bradley, R. (1984) The Social Foundations of Prehistoric Britain: Themes and Variations in the Archaeology of Power, Longman, London and New York
Brenner, R. (1976) 'Agrarian Class Structure and Economic Development in Pre-Industrial Europe', Past and Present, 70, 30-75
Brown, E.H. (1980) Geography Yesterday and Tomorrow, Oxford University Press, Oxford
Brown, R.H. (1948) Historical Geography of the United States, Harcourt Brace and World Inc., New York
Brunskill, R.W. (1971) Illustrated Handbook of Vernacular Architecture, Faber and Faber, London
Brunskill, R.W. (1974) Vernacular Architecture of the Lake Counties, Faber and Faber, London
Brunskill, R.W. (1982a) Houses, Collins, London
Brunskill, R.W. (1982b) Traditional Buildings of Britain, Gollancz, London
Brunskill, R.W. (1982c) Traditional Farm Buildings of Britain, Gollancz, London
Bylund, E. (1960) 'Theoretical Considerations Regarding the Distribution of Settlement in Inner North Sweden', Geografiska Annaler, 42B, 225-31
Bylund, E. (1972) 'Generation Waves and Spread of Settlement' in W.P. Adams and F.M. Helleiner (eds.), International Geography, Vol. 2 IGU, Montreal, pp. 1306-9
Cantor, L. (1982) The English Medieval Landscape, Croom Helm, London and Canberra
Chapelot, J., and Fossier, R. (1985) The Village and House in the Middle Ages, B.T. Batsford, London
Chisholm, M. (1962) Rural Settlement and Land Use, Hutchinson, London
Chorley, R.J., and Haggett, P. (eds.) (1964) Frontiers in Geographical Teaching, Methuen, London
Christopher, A.J. (1984) Colonial Africa, Croom Helm, London and Canberra
Clemensen, H.A. (1982) English Country Houses and Landed Estates, Croom Helm, London and Canberra
Cloke, P. (1979) Key Settlements in Rural England, Methuen, London
Clout, H.D. (1977) Themes in the Historical Geography of France, Academic Press, London, New York, San Francisco

Cooke, R.U., and Johnson, R.H. (1969) Trends in Geography, Pergamon Press, Oxford, London

Dahlman, C.J. (1980) The Open Field System and Beyond, Cambridge University Press, Cambridge

Darby, H.C. (ed.) (1973) A New Historical Geography of England, Cambridge University Press, Cambridge

Davidson, B.R. (1981) European Farming in Australia, Elsevier Scientific Publishing Company, Amsterdam

Delano-Smith, C., and Parry, M. (1981) Consequences of Climatic Change, Department of Geography, University of Nottingham

Demidowicz, G. (1985) 'Planned Landscapes in North-East Poland: The Suraz Estate, 1550-1760', Journal of Historical Geography, 11(1), 21-47

Dewindt, E.B. (1971) Land and People in Holywell-cum-Needingworth, Pontifical Institute of Medieval Studies, Toronto

Dilsaver, L.M. (1985) 'After the Gold Rush', Geographical Review, 75(1), 1-20

Dodgshon, R.A. (1980) British Field Systems, Academic Press, London, New York

Doxiadis, C.A. (1968) Ekistics, An Introduction to the Science of Human Settlements, Hutchinson, London

Dyson, S.L. (1985) Comparative Studies in the Archaeology of Colonialism, British Archaeological Reports, International Series 233, Oxford

Edwards, K.J., Hamond, F.W., and Simms, A. (1983) 'The Medieval Settlement of Newcastle Lyons, County Dublin: An Interdisciplinary Approach', Proceedings of the Royal Irish Academy, 83C, No. 14, 351-76

Eidt, R.C. (1984) Advances in Abandoned Settlement Analysis, The Center for Latin America, University of Wisconsin, Milwaukee

Elton, G.R. (1967) The Practice of History, 3rd impression, Collins, London and Glasgow

Fellows Jensen, G. (1972) Scandinavian Settlement Names in Yorkshire, Akademisk Forlag, Copenhagen

Fellows Jensen, G. (1978) Scandinavian Settlement Names in the East Midlands, Akademisk Forlag, Copenhagen

Fellows Jensen, G. (1984) 'Place-Names and Settlements: Some Problems of Dating as Exemplified by English Place-Names in -by', Nomina, 8, 29-39

Fenton, A. (1978) The Northern Isles: Orkney and Shetland, John Donald Ltd., Edinburgh

Finberg, H.P.R. (1964) Lucerna, Macmillan, London

Fowler, P. (1981/83) The Farming of Prehistoric Britain, Cambridge University Press, Cambridge

Gade D.W., and Escobar, M. (1982) 'Village Settlement and the Colonial Legacy in Southern Peru', The Geographical Review, 72, 430-49

Gant, R. (1982) 'Abandon et reconquete de l'habitat rural

dans le Black Mountains (Pays de Galles)', Norois, 29 (113), 2-34
Gissel, S., et al. (1981) Desertion and Land Colonisation in the Nordic Countries c. 1300-1600, Almqvist and Wiksell International, Stockholm
Glacken, C.J. (1967) Traces on the Rhodian Shore, University of California Press, Berkeley, Los Angeles, London
Goody, J., Thirsk, J., and Thompson, E.P. (eds.) (1976) Family and Inheritance: Rural Society in Western Europe 1200-1800, Cambridge University Press, Cambridge
Green, D., Haselgrove, C., and Spriggs, M. (eds.) (1978) Social Organisation and Settlement, British Archaeological Reports, International Series, Supplementary 47, 2 vols.
Gregson, N. (1985) 'The Multiple Estate Model: Some Critical Questions', Journal of Historical Geography, 11 (4), 339-51
Grossman, D. (1971) 'Do We Have a Theory for Settlement Geography: The Case of Iboland', The Professional Geographer, 23, 197-203
Grossman, D. (1981) 'Population Growth in Reference to Land Quality: The Case of Samaria, 1922-75', The Geographical Journal, 147 (2), 188-200
Grossman, D. (1982) 'Northern Samaria: A Process-Pattern Analysis of Rural Settlement', Canadian Geographer, 26 (2), 110-27
Grossman, D. (1983a) 'Settlement Patterns in Judea and Samaria', Geojournal, 7 (3), 299-312
Grossman, D. (1983b) 'Comment upon "Describing and Modelling Rural Settlement Maps" by Robert Haining', Annals of the Association of American Geographers, 73, 298-302
Grossman, D., and Safari, Z. (1980) 'Satellite Settlements in Western Samaria', Geographical Review, 70 (4), 447-61
Guelke, L. (1982) Historical Understanding in Geography, Cambridge University Press, Cambridge
Haartsen, A.J., and Renes, J. (1982) 'Naar een Historisch-Geografische Typologie van het Nederlandse Landschap', Geografisch Tijdschrift, 16, 456-75
Haggett, P. (1965) Locational Analysis in Human Geography, Arnold, London
Haggett, P., Cliff, A.D., and Frey, A. (1977) Locational Models, Arnold, London
Haining, R. (1982) 'Describing and Modelling Rural Settlement Maps', Annals of the Association of American Geographers, 72, 211-23
Hannerberg, D. (1984) Gardar, Bol och Vangar i Hageestad: Olding och Organisation under 2000 ar, Meddelanden Serie B 60, Kulturgeografiska Institutionen Stockholms Universitet
Harris, A. (1961) The Rural Landscape of the East Riding of

Yorkshire: A Study in Historical Geography, Oxford University Press, for the University of Hull
Harvey, M. (1980) 'Regular Fields and Tenurial Arrangements in Holderness, Yorkshire', Journal of Historical Geography, 6, 3-16
Harvey, M. (1982) 'Regular Open Field Systems on the Yorkshire Wolds', Landscape History, 4, 29-39
Harvey, M. (1983) 'Planned Field Systems in Eastern Yorkshire: Some Thoughts on Their Origins', Agricultural History Review, 31, 92-103
Harvy, P.D.A. (1965) A Medieval Oxfordshire Village: Cuxham 1240-1400, Oxford University Press, Oxford
Hooke, D. (1985) (ed.), Medieval Villages, Oxford University Committee for Archaeology, Monograph 5
Houting, E., de Poel, K.R., and Van der Vaart, J. (1981) Brinkenboek: Een Vrkenning van de Brinken in Drenthe, Van Gorcum, Assen
Howell, C. (1983) Land, Family and Inheritance in Transition, Cambridge University Press, Cambridge
Hudson, J.C. (1969) 'A Location Theory for Rural Settlement', Annals of the Association of American Geographers, 59, 461-95
Hughey, D.V. (1978) 'Caymanian House-and-Yard Patterns', Caribbean Studies, 18, 1 and 2, 131-5
Hurst, J. (1984) 'The Wharram Research Project: Results to 1983', Medieval Archeology, 28, 77-111
Jamieson, E., Jeans, D.N., and Forster, J.J.H. (1983) 'The Spatial Scale of European Colonisation and Land Alienation in Northern New South Wales', Australian Geographer, 15, 322-5
Jeppersen, T.G. (1981) Middelalder Landsbyens Opstaen, Fynske Studier XI, Odense Bys Museer, Odense
Jones, G.R.J. (1985) 'Multiple Estates Perceived', Journal of Historical Geography, 11 (4), 352-63
Jones, M. (1977) Finland, Daughter of the Sea, Archon Books, Dawson, Folkestone
Jordan, T.G. (1983) 'A Reappraisal of Fenno-Scandinavian Antecedents for Midland American Log Construction', The Geographical Review, 72, 58-94
Kapelle, W.E. (1979) The Norman Conquest of the North: The Region and its Transformation 1000-1135, Croom Helm, London
Kiefmann, H-M. (1979) 'Die Phosphatemethode-Ergebniss Neurer Untersuchen', Geografiska Annaler, 61B, 1-7
Knight, D. (1984) Late Bronze Age Settlement in the Nene and Great Ouse Basins, British Archaeological Reports, British Series, 130, 2 vols.
Kristiansen, K. (1984) Settlement and Economy in Later Scandinavian Prehistory, British Archaeological Reports, International Series, 211
Lemon, A., and Pollock, N. (eds.) (1980) Studies in Over-

seas Settlement and Population, Longman, London and New York

Le Roi Lauderie, E. (1978) Montaillou: Cathars and Catholics in a French Village 1294-1324 trans. Barbera Bray, Scholar Press, London

Levtzion, N. (1973) Ancient Ghana and Mali, Methuen & Co., London

Loffler, G. (1979) 'Quantitative Methoden der Wustungsforschung', Geografiska Annaler, 61B, 81-9

Lovell, W. G. (1983) 'Settlement Change in Spanish America: The Dynamics of Congregacion in the Cuchamatan Highlands of Guatemala, 1541-1821', Canadian Geographer, 28, 2, 161-74

Mayhew, A. (1973) Rural Settlement and Farming in Germany, B.T. Batsford, London

McIntosh, C. B. (1981) 'One Man's Sequential Land Alienation on the Great Plains', The Geographical Review, 71, 425-445

Mead, W.R. (1981) An Historical Geography of Scandinavia, Academic Press, London and New York

Mercer, R. (ed.) (1981) Farming Practice in British Prehistory, Edinburgh University Press

Miles, D. (ed.) (1982) The Roman-British Countryside: Studies in Rural Settlement and Economy, British Archaeological Reports, British Series 103

Miles, D., and Benson, D. (1974) The Upper Thames Valley: An Archaeologial Survey of the River Gravels, Oxfordshire Archaeological Unit, Survey no. 2

Mills, D. R. (1980) Lord and Peasant in Nineteenth Century Britain, Croom Helm, London

Nash, A. (1982) 'The Medieval Fields of Strettington, West Sussex, and the Evolution of Land Division', Geografiska Annaler, 64B (1), 41-9

Newcombe, R. M. (1979) Planning the Past, Dawson, Archon Books, Folkestone

Noble, A. G., and Seymour, G. A. (1982) 'Distribution of Barn Types in North Eastern United States', The Geographical Review, 72, 155-70

Noble, W. A. (1977) 'Settlement Patterns and Migrations Among Nilgiri Herders, South India', Journal of Tropical Geography, 44, 57-70

Norton, W. (1984) Historical Analysis in Geography, Longman, London and New York

Oliver, P. (ed.) (1971) Shelter in Africa, Barrie and Jenkins, London

Pacione, M. (1984) Rural Geography, Harper and Row, London

Parry, M. (1978) Climatic Change, Agriculture and Settlement, Dawson, Archon Books, Folkestone

Parry, M., and Slater, T. (1980) The Making of the Scottish Countryside, Croom Helm, London and Montreal

Powlesland, D. (1985) The material cited is derived from unpublished excavation reports. Supported by funds from both Britain and America this work is directed by Dominic Powlesland, The Old Abbey, Yedingham, Nr. Malton, Yorkshire YO17 8SW. It is used with kind permission

Pozorski, T., Pozorski, S., MacKey, C.J., and Klymshyn, A.U., 'Pre-Hispanic Ridged Fields of the Casma Valley, Peru', The Geographical Review, 72, 407-416

Prince, H. (1971) 'Real, Imagined and Abstract Worlds of the Past', Progress in Geography 3, Edward Arnold, London

Ravensdale, J.R. (1974) Liable to Floods, Cambridge University Press

Reed, M. (1984) Discovering Past Landscapes, Croom Helm, London

Renes, J. (1982) Typologiee'n van bewonings- en perceelsvormen, Centrum voor Landbouwpublokaties en Landbouwdocumentatie, Wagingen

Roberts, B.K. (1982) Rural Settlement: An Historical Perspective, Historical Geography Research Series, 9, Geo Books, Norwich

Roberts, B.K. (1983) 'Nucleation and Dispersion: Towards an Explanation', Medieval Village Research Group Report, 31, 44-5

Roberts, B.K. (1986) The Making of the English Village, Longman, London, forthcoming

Roberts, B.K., and Glasscock, R.E. (eds.) (1983) Villages, Fields and Frontiers: Studies in European Rural Settlement in the Medieval and Early Modern Periods, British Archaeological Reports, International Series 185

Roberts, J.M. (1985) The Triumph of the West, British Broadcasting Corporation, London

Rowley, T. (ed.) (1974) Anglo-Saxon Settlement and the Landscape, British Archaeological Reports, British Series 6

Rowley, T. (ed.) (1981) The Origins of Open Field Agriculture, Croom Helm, London

Schmaal, A.P. (ed.) (1979/82) Looking at Historic Buildings in Holland, Riksdienst voor de Monumentenzord, Zeist

Schuyler, R.L. (1978) Historical Archaeology: A Guide to Substantive and Theoretical Contributions, Baywood Publishing Co. Inc. Farmingdale, New York

Sheppard, J.A. (1966) 'Pre-enclosure Field and Settlement Patterns in an English Township - Wheldrake, near York, Geografiska Annaler, 48B, 59-77

Sheppard, J.A. (1974) 'Metrological Analysis of Regular Village Plans in Yorkshire', Agricultural History Review, 22, 3-20

Sheppard, J.A. (1976) 'Medieval Village Planning in Northern England: Some Evidence from Yorkshire', Journal of Historical Geography

Singh, R.L. (ed.) (1972) Rural Settlements in Monsoon Asia,
    National Geographical Society of India, Varanasi 5
Somerset and South Avon Vernacular Building Research
    Group, Somerset Villages: West and Middle Chinnock,
    Somerset Rural Life Museum
Sporrong, U. (1981) Kann Ditt Land: Jordbruksbygd,
    Svenska Turistforeningen, Wiking Tryckeri AB,
    Sodertalje, Sweden
Sporrong, U. (1985) Malarbygd, Meddalanden Serie B 61,
    Kulturgeogafiska Institutionen Stockholms Universitet
Spratt, D.A. (ed.) (1982) Prehistoric and Roman Archaeology
    of North-East Yorkshire, British Archaeological Reports,
    British Series 104
Steegh, A. (1985) Monumentan Atlas van Netherland, de
    Walberg Pers
Steensberg, A. (1983) Borup, A.D. 700-1400, The National
    Museum, Copenhagen
Sutton, K. (1981) 'The Influence of Military Policy on
    Algerian Rural Settlement', The Geographical Review, 71,
    379-394.
Sveinbjarnardottir, G., Buckland, P.C., and Gerrard, A.J.
    (1982) 'Landscape Change in Eyjafjallasveit, Southern
    Iceland', Norsk Geografisk Tidskrift, 36 (2), 75-88
Taylor, C. (1983) Village and Farmstead, George Philip,
    London
Thirsk, J. (ed.) (1967) The Agrarian History of England and
    Wales, IV 1500-1640, Cambridge University Press
Thirsk, J. (ed.) (1985) The Agrarian History of England and
    Wales, V 1640-1750 2 vols, Cambridge University Press
Turner, F.J. (1920) The Frontier in American History, New
    York
Uhlig, H., and Lienau, C. (1972) 'Rural Settlements' in Die
    Siedlungen des Landlichen Raumes, Materialen zur
    Terminologie der Agrarlandschaft, vol 2, Lenz-Verlag,
    Giessen
Unwin, P.T.H. 'The Rank-Size Distribution of Medieval
    English Taxation Hierarchies with Particular Reference to
    Nottinghamshire', Professional Geographer, 33 (3),
    350-60
Weller, J. (1982) History of the Farmstead, Faber and Faber,
    London
Whittesley, D. (1929) 'Sequent Occupance', Annals of the
    Association of American Geographers, 19, 162-5
Widgren, M. (1983) Settlement and Farming Systems in the
    Early Iron Age, Acta Universitatis Stockholmensis,
    Stockholm Studies in Human Geography 3, Stockholm
Wonders, W.C. (1983) 'Mot Kanadas Nordvast: Pioneer Settle-
    ment by Scandinavians in Central Alberta', Geografiska
    Annaler, 65B (2), 129-52
Yates, E.M. (1982) 'The Evolution of the English Village',
    Geographical Journal, 148 (2), 182-206

NOTES ON CONTRIBUTORS

Professor R.A. Butlin, Department of Geography, Loughborough University, England.

Dr R. Dennis, Department of Geography, University College London, England.

Dr J.D. Hamshere, School of Geography, University of Manchester, England.

Dr P.E. Ogden, Department of Geography, Queen Mary College, University of London, England.

Dr M. Pacione, Department of Geography, University of Strathclyde, Glasgow, Scotland.

Dr C.G. Pooley, Department of Geography, University of Lancaster, England.

Dr B.K. Roberts, Department of Geography, University of Durham, England.

Dr J.R. Walton, Department of Geography, University College of Wales, Aberystwyth, Wales.

Dr J.W.R. Whitehand, Department of Geography, University of Birmingham, England.

Dr G. Whittington, Department of Geography, University of St Andrews, Scotland.

Dr I.D. Whyte, Department of Geography, University of Lancaster, England.

INDEX

Milton Keynes UK
Ingram Content Group UK Ltd.
UKHW020000071024
449327UK00031B/2596